PROMPTING SCIENCE
AND ENGINEERING STUDENTS IN PRACTICAL
TRIGONOMETRY

PROMPTING SCIENCE
AND ENGINEERING STUDENTS IN PRACTICAL
TRIGONOMETRY

GEORGE NORMAN REED

Prompting Science and Engineering Students in Practical Trigonometry
by George Norman Reed

ISBN 978-1-952027-36-9 (Paperback)
ISBN 978-1-952027-37-6 (Hardback)

Printed in the United States of America.

New Leaf Media, LLC
175 S. 3rd Street, Suite 200
Columbus, OH 43215
www.thenewleafmedia.com

TABLE OF CONTENTS

18/1. The vital importanceof the student getting to know his lathe.

18/2. Work holding devices explained.

18/3. Removing and replacing the gap in the center lathe's bed.

18/4. Setting up the face-plate on the lathe.

18/5. Using the lathe's collet chuck.

18/6. Using the 4 jaw chuck.

18/7. Using the sticky pin to centralize the work-piece.

18/8. Securely holding irregular shaped components.

18/9. Preventing damage occurring to the 3 jaw chuck's jaws.

18/10. Using the fixed steady.

18/10a. Helpful notes on drilling and reaming on the lathe.

18/11. Using the alternative set of soft jaws on the lathe.

18/12. The correct way to remove and replace the jaws of the 3 jaw chuck.

18/13. Important safety checks that need to be carried out on the center-lathe and its equipment.

18/14. Repair work that needs to be undertaken by the machinist.

18/15. Re-checking the lathe's parallelism.

18/16. Rectifying bell-mouthing of the chuck's jaws.

18/17. Solving the problems of turning very small diameters.

18/18. The set-up required to produce tapered work on the lathe.

18/19. Turning a work-piece held between centers.

18/20. Checking the tail-stocks alignment for turning long work-pieces,and the authors selection of the most useful lathe tools.

18/21. Inspection, sharpening, and honing of lathe tools. How one obtains an accurate point radius on an external screw-cutting tool.

18/22. Using a test piece to obtain true parallelism of the tail-stock.

18/23. Selecting the correct lathe tools for the requirements of the work-piece.

18/24. Replacing a screw-cutting tool midway through a screw-cutting procedure.

For more details please refer to the HOW TO USE THIS BOOK, (positioned near the index) This section contains more comprehensive in depth information.

This book contains many practical answers to the calculation problems found during the design and manufacturing stages of research, development, and production for engineering components.

A thorough study of this book will allow not only the less-able students, but also the trainee teachers of engineering who are currently working in our schools and colleges to augment their trigonometry skills and to improve their practical engineering expertise in the workplace. The 49 informative and descriptive drawings will be found to assist in the calculation and subsequent manufacture of sheet metal components with their requirement to possess extremely accurate bends and precise dimensions, including components that are fabricated, precision milled, drilled, jig-bored, turned, or screw-cut.

The book's contents will also provide a full explanation and solve many of the engineering problems that are encountered normally in the engineering workshop. A study of the methods used will also provide the necessary practical approach to the problems found, by aiding the machinist in the setting up of precision production machinery, particularly the practical and theoretical methods used during the screw-cutting of threads during the manufacturing stages of turned and screw-cut components. This particular setting-up information is provided specifically for use on the center lathe and is described in chapter 18 (for external screw -cutting), and in chapter 19 (for internal screw-cutting). The production processes being used for these operations are virtually identical to those performed in many of the United Kingdom's typical engineering and manufacturing establishments.

INTRODUCTION

Why you may ask, is there a need to publish yet another book on trigonometry, when the shelves of the high street book -shops are already bulging with books explaining how to perform theoretical trigonometry?

The author's main reason for writing this book is that it is based purely on the use of *practical trigonometry* in the work-shop. It has therefore been designed and written with the sole purpose of bringing to the notice of both the novice student and those who are undergoing engineering training, including those among the current population who secretly consider they are less able mathematically than their peers in the calculation of trigonometry problems, the existence of this completely new and easier to-perform *practical* method of triangle calculation that now allows all students, including the experienced workers in engineering, to partake in this new practical calculation method to solve all those difficult triangle calculation problems occurring in the school, work-shop office, and on the shop floor of the engineering work-shop.

This completely new trigonometry calculation method is designed for use by both students and their 'fully skilled' peers, for solving all of their triangle calculation problems that require trigonometry skills to be used to obtain those vital answers. It should be noted that unresolved trigonometry problems are often discovered while an engineer or student is referring to a sketch or an issued engineering drawing, and the student is often obliged to work out how to machine an exact angle on a problem workpiece, or the angle of a sheet metal bend, or a tube's overall 'developed length' while working on the bench, or in the engineering workshop office.

This book is therefore written in a very simplistic form, which makes it equally suitable for those students who may have previously found difficulty in coping with the basics of mathematics, and their trigonometry calculations frequently required during their college education, and this includes the fully skilled engineer, who will find the book's contents extremely helpful while working in their current employment in the engineering workplace.

This book will be found particularly useful to Students who may have previously decided that the whole subject of undertaking calculations that involve trigonometry to be much too difficult for them to get to grips with at the present time. The book also contains a host of important but required mathematical facts and engineering experiences that will be found to be extremely useful both to the novice student and to the budding engineer. This full and in-depth knowledge of the engineering workshop calculation practice will be found to provide a true awareness and grounding in the use of *practical trigonometry* for all triangle calculation requirements.

The author has used American spelling for many of the words used in this book, which then allows its contents to be easily read by a truly universal public. This book is specifically designed to meet the *practical* requirements of personnel who are currently studying in schools, colleges, and those who are presently employed in the United Kingdom and other countries that have manufacturing industries.

The reader will very soon discover that its written text contains a very practical approach to the calculation methods used when compared to the theoretical trigonometry currently used and taught in most engineering training establishments.

It is the author's considered opinion that there still exists a tendency for other books dealing with the teaching of trigonometry to cater mainly for the requirements of the examination test boards, in order to comply mainly with the requirements of their issued examination papers, rather than meeting the full educational and practical needs of the engineering student or the work-shop engineer undergoing training, or the basic requirements of the manufacturing company that employs them.

A thorough study of this book will, it is hoped, provide the opportunity for all students and skilled personnel to become fully aware, from the grass-roots level in the industry to the problems encountered while they are attempting to solve those awkward but necessary triangle calculation problems that currently occur in the engineering working environment.

By Inventing this innovative Probe and Prompt easier practical triangle calculation system it now makes it possible for the book's contents to be used by all those students who have long wished to have the opportunity of taking part in and using a much simpler system for solving their difficult triangle calculation problems.

As an additional aid to solving all those triangle calculation problems, the author has designed a miniature trigonometry aid (in fig. 50), that can be worn affixed unobtrusively (for example) to the wristwatch strap. This aid contains all six of the newly designed and necessary prompts required to solve all of the most likely-to-occur right-angle triangle problems occurring in the workshop.

The author considers that the practical contents of this book are unique if compared to the general run-of-the-mill mathematics books found on sale in the local high street.

The book's contents will be found to contain a full and detailed explanation of how components are designed, calculated, and manufactured in the work-shop, including many detailed and dimensioned sketches (some in cross section), used for solving many of the practical triangle problems encountered in the engineering workplace.

As previously stated, the actual solving of these triangle problems in a *practical* way in the industry is quite often glossed over, and the subject is often poorly explained in many of the currently published books dealing with workshop calculations.

Other publications tend to deal mainly with the more theoretical (i.e. 'complicated') aspects of trigonometry. The author considers that other writers on this subject seem to possess a marked reluctance to deal *specifically* with the most important *practical* side of the problems normally encountered on the shop floor by the student, when he (or she) is attempting to solve a triangle problem on the bench, or a machine, in the engineering workplace.

This book's main aim therefore is to allow student engineer to develop the necessary *practical expertise* in work-shop practice, and to obtain a thorough working knowledge in the use of *practical trigonometry*, to a sufficient level that will allow him or her to accomplish and solve the vast majority of triangle calculations, and practical problems, found in their day to-day workplace and training establishment.

Surprisingly, these triangle calculation problems tend to occur quite frequently in the school, college and in the engineering workplace. A frequent problem found in the engineering work-shop is the lack of some vital verbal instruction being received from either a supervisor or the engineering manager, regarding a particular angle required on a work-piece, or a missing or a possibly helpful dimension being omitted from an issued manufacturing or pre-production engineering drawing.

While working in the tool-room section of the engineering shop floor, the author became aware of a deliberate policy of the drawing office staff, deliberately omitting an important dimension from an issued drawing so that this omission of certain details could be used later to give prior warning to the office that production of the work-piece was about to be started; this warning would be heralded by the approach of a member of the production shop floor staff requesting a clarification of a particularly unclear dimension on the drawing. The use of this ploy then enabled the actual draftsman concerned to visit the workshop floor, provide the missing dimension, and (while doing so) take advantage of this golden opportunity to make last-minute modifications (that 'had just come to light') on a drawing that had an original issue date of several months before. (The author admits that there have been occasions during his engineering career where he has ventured to short-circuit this particular drawing office ploy, by actually calculating the unknown dimension himself, on the shop floor, by using the Probe and Prompt method of triangle calculation to find and resolve the problem and, by doing so, causing a certain amount of dismay among the drawing office staff on their discovery that production of the component was now well underway and on schedule as originally intended. The author now admits that pursuing this course of action also provided him with a certain amount of job satisfaction.

This book also explains in minute detail and in purely practical engineering terms, the wide scope of technical know-how that is actually required by the working personnel in both the design department office and the machine shop floor in many of the world's manufacturing industries.

The 'complicated to solve' triangle problems that do occur on the shop floor have previously required a complete working knowledge in the use of trigonometry, coupled with the necessary expertise to deal with the problem practically. This mental conversion from the purely theoretical way of performing the task and being assisted in a practical way to complete it is rarely found in the texts of other publications on the subject.

This book therefore contains a host of useful information on how one performs these practical triangle calculation tasks while using its main calculation aid called the Combination Probe, supplied (in page form) within the drawing (fig. 37f).

Also included (for the specific use of trainee sheet metal bending engineers) are the two new bend development tables (shown in drawings fig. 32 and fig. 32a). These charts will be found extremely useful in the calculation and accurate production of sheet metal components that have the need to possess very accurate bends, and extremely accurate dimensions and overall lengths. These new bending aids are designed to assist the trainee sheet metal fitter and the trainee engineer whose ultimate aim is to take up precision sheet metal engineering as a full-time career.

The two new (unique) metric bend development tables allow the student to calculate the exact length of material that is used up in the bend of a sheet metal workpiece. These charts can be used to calculate the *exact length of bend arc* required when calculating the important *developed length* of both sheet metal and small-diameter tubular components. The use of these tables will also allow the vital (but rarely specified on the drawing) length of arc and cut-off length of the workpiece to be easily established, prior to bending the component with the required accuracy.

The importance of these two new bend development tables can however be relegated to second place when compared to the importance and ease of use of the new and innovative Probe and Prompt aid, with its new practical system of triangle calculation. This system can now be used to solve all those right-angle triangle trigonometry problems that currently occur in the school, in the office, and on the bench of the actual work- shop. The use of these calculation aids provides a very accurate and *practical* method of obtaining the dimensions and angles required to solve a triangle calculation problem.

While the student is in the process of using either of these two new bend development tables, he or she will now realize that these charts actually point the way to a completely new and unique method of solving sheet metal bending calculation problems. They will of course need to use the Probe and Prompt triangle calculation system, combined with the use of a scientific calculator. This new practical method, currently used in the bending of sheet metal components is explained in full detail in the example drawings, shown in figures 20, 22, 23, 24, 25, 28, 29, 30, and 31).

The two new and unique metric bending charts (shown in fig. 32 and fig. 32a) can be considered a first in the sheet metal bending industry, for they now provide (for the machine operator and the bend design engineer) a much easier and more accurate method for use in calculating the length of material that is used up in the arc of a bend in a sheet metal component. These charts are unique, as they are currently unavailable in the sheet metal bending industry.

To reiterate, the use of these two bending charts now allows the student (and the skilled sheet metal bending engineer) to be absolutely precise in the calculation of the amount of material actually being used up in the precision bend of a sheet metal workpiece.

Of the two charts, fig. 32 a deals with stretched bends, and fig. 32 deals with normal bends. The use of either of the charts then enables the student to obtain without difficulty, the true and accurate practical developed length of the required 'stretched' bend's arc, or the 'normal' bend's arc that exists in the component. This will then allow the vital cut-off length of material to be accurately calculated in order to produce a completely accurate bend in the workpiece, and provide the vital and exact overall length of material being used, while the material is still *in the flat* prior to the bending process taking place.

In the past, an accurate but required dimension for the cut-off length of material, prior to the components manufacture, was rarely (if ever) stated on the drawing. The student should now realize that the actual cut-off length dimension required for the component needs to be extremely precise if it is being used in the bending of extremely accurate components intended for use in the aircraft industry.

When obtaining a precise stretched calculation dimension for a bend, it involves a minute difference in length between the 'normal' and the 'stretched' length of bend arc; this is achieved by using a minutely calculated adjustment being applied to the original basic calculated arc length, resulting in the exact amount of stretch being accurately assessed in the material, during the calculation and planning stage, prior to the final bending process being carried out (as shown in the bend of drawing fig. 28).

The two new bending charts are calibrated to metric dimensions, as opposed to the (now superseded) imperial inch- dimensioned charts that are still being used in the U-K's industries.

The author considers that this accurate system of arc calculation will be welcomed by bending engineers, for they can now upgrade their old (but still current) imperial-dimensioned charts, found (through experience), to be less compatible when used with the metric-gauge thickness, of stock sheet metal materials in current use.

This availability of two new and distinct sources of bend calculation data, (in the form of being for either 'stretched' or for 'normal' bends), should prove to be extremely helpful, (in terms of accuracy), to both the student and to the skilled engineer.

Students will now have the ability to calculate the length of a work-piece's *bend center-line arc* dimension with extreme accuracy, enabling them to calculate the important and critical overall length of the material being used, this then allows the material to be accurately cut to length while still in the flat, prior to the bending operation taking place.

This newly found so-called bend freedom will prove to be of major assistance to those on the production shop floor and to a host of practical workshop engineers who are currently working in the sheet metal industry.

This so-called simplification of the trigonometry calculations normally required when a student is attempting to solve a triangle calculation problem, will prove to be indispensable to all employees, whether on the factory shop floor, or to those in the various inspection departments, where practical triangle calculations are often needed to check the finished components for accuracy. A sound knowledge in the use of practical trigonometry will also allow the 'less able' student, who has the wish to eventually progress to becoming an inspector in the engineering shop, will now find that he or she is not barred from applying for this particular post by any lack of triangle calculation knowledge.

In the past, where a shop floor worker was presented with a difficult triangle calculation problem, he would often need to seek the advice of his colleagues in the use of the required trigonometry calculations, in order to allow him or her to perform the bend calculations necessary to carry out the work accurately.

As previously stated, this situation can arise when a certain lack of information has been discovered on an issued sketch or drawing of a component, or by insufficient or unclear verbal instructions being received from either the supervisor or the management of the company, being passed on 'through the appropriate channels' from the design engineer. This situation is often made rather more difficult by the discovery that a particularly important detail has been left out of the issued drawing (or sketch) of the work-piece about to be manufactured.

The current practice of asking colleagues for triangle calculation advice can also lead to a certain feeling of loss of face by the fact that he or she is forced to ask colleagues for this particular calculation advice.

This book is based to a large extent on the author's personal experiences gained in the research and development engineering industry, where it was quite soon discovered there was a *real need* for the introduction (throughout the whole industry), of a much simpler triangle calculation system, that could be used by the whole workforce, particularly when called upon, at short notice, to calculate a so-called awkward triangle problem. This situation does occur quite frequently in both the research and development and production engineering departments.

This problem seems to occur most frequently during the research and development stage of a project, and often before the engineering drawings have been 'officially approved' by the drawing office inspection staff as being correct and suitable for general release to the production shop floor.

However, there still exists in the industry, many so-called grey areas concerning the shop floor personnel's own interpretation of an issued research and production sketch or drawing. This grey area applies particularly when the drawings or sketches are referred to on the production shop -floor. This problem becomes more concerning when it is found that some vital dimension is missing from the drawing, or some vital detail has failed to be transmitted verbally by the management. This problem is often coupled with insufficient manufacturing information being shown on the working sketch or drawing, which inevitably causes questions to be asked via the charge-hand or the management.

It is quite often the case that this missing information fails, for various reasons, to be included as part of the drawing's manufacturing instructions.

These so-called omissions often require additional time-consuming enquiries to be made, (via the charge-hand or foreman) or from the original drawing office source, in order to verify that all the vital required facts for the manufacture of the component are available, before the prototype or the production work-piece can be manufactured.

It can be gathered from the foregoing statement, that there is generally a pecking order, based on what is called the need-to-know system, throughout the whole of the engineering manufacturing industry. For instance, there is often the situation where the shop floor personnel may not be permitted to go directly with a problem to the drawing office in order to seek their advice on the matter, without first going through the 'proper channels', namely, by asking one's supervisor or foreman to make the enquiries on their behalf.

Workshop personnel also find it extremely difficult to gain access to the drawing-office-issued general arrangement drawing (called the G.A). This drawing contains virtually all the information and details required for the components manufacture, and usually includes the fully detailed assembly of all the components needed to make up the complete design of the assembled project. In fairness, this is usually the last drawing issued by the drawing office to the engineering workshop, but the problem remains that, after issue, this drawing is quite often kept strictly in the office, and guarded to some extent by the charge hand or foreman, who consider their work-shop personnel should work on a strictly need-to-know basis. This situation seemingly makes the supervision somewhat reluctant to 'let the G-A out of their sight'. The seriousness of this problem is explained by quoting an overheard comment made by the production office management, which stated, "If we let them have it, they'll spend all day looking at it".

When questions are asked concerning the project being worked on, the often heard reply is "But you don't need to know that', or, "It is not your concern'.

These comments can be a source of irritation among members of the production and development workforce, who would naturally have preferred the information to be provided willingly and as quickly as possible. In this case, the production work-force would have preferred to seek out the answer firsthand by themselves via a drawing office visit, and not have received the answer to their query in a rather hurried watered-down and abbreviated form consisting of a few hurried words, given in haste, such as, "'Oh, while I'm passing, that problem you are having . . .'

To quote an example of this situation, the following is a snippet of conversation overheard in the drawing office, while the drawing office staff were being approached by a shop-floor machinist who was having a drawing omission problem, The overheard comment from the drawing office personnel was, "'Don't tell them too much or they'll know as much as we do".'

This overheard comment explains to some extent, the situation that does occasionally exist between the drawing office staff and the shop floor personnel when a machinist is in the process of seeking information regarding a so-called drawing office omission.

Bearing this problem in mind, it reinforces the absolute necessity for both engineering students and skilled machinists to develop the ability to sort out their own manufacturing problems (regarding workshop calculations) by themselves and for them to develop the necessary skill and the ability to perform all the accurate calculations required, in order to obtain the dimensions or angles seemingly left out of the issued 'worked on' engineering drawing, without the need to seek out the rather grudging advice from either the charge hand or the office management staff, who often have the tendency to keep any possibly helpful manufacturing information rather 'close to their chests', resulting in this important information being made available only in the latter and critical stages of the project.

This so-called 'omission of vital information' does also occur on drawings received from outside suppliers, where certain discrepancies are often discovered while the drawings are being studied in the research and development office or on the work- shop floor. These drawings are often contain requests for prices, and the availability of production-manufactured samples, which of course cannot be produced until all the relevant information is made available via discussions, by phone, email, fax, etc.

* * *

The main innovation found in this new and unique practical triangle calculation system is that it provides a suitable alternative method for the student to use in calculating their triangle problems. This new triangle calculation system has the distinct advantage that it can be used by the mathematically less able; this is mainly due to it possessing the ability to effectively bypass some of the normal conventional methods that are currently being used during their theoretical trigonometry calculations.

Many of the words used in the currently theoretical explanations available to the student on this subject are often found difficult to be fully understood. For this reason, the author has carefully selected the following very short list of typical words, theorems, and phrases that the book effectively bypasses.

The following conventional mathematical phrases, used in other current publications on the subject of triangle mathematics have the tendency to become a breeding ground for the student's uncertainty, particularly with the absolute beginner and the mathematically 'less able', who will surely welcome the arrival of this much simpler method to obtain the answers to their triangle calculation problems, by using just this *practical* method of approach to their triangle calculations.

A very brief list of these so-called bypassed words is as follows: secant, cosecant, versed sine; conversed sine; third quadrant; ambiguous case, trapezium, ellipse formula, and so on.

This book provides the student with a complete instruction course in the use of this alternative practical trigonometry system, where it is used to supplement the theoretical trigonometry that is currently being used in upper schools and colleges. It will be discovered that this new system enhances the knowledge and practicalism of those who are familiar with the time-honored and traditionally taught theoretical methods of triangle calculation. This so-called supplement to the theoretical trigonometry system in current use will be found to differ quite markedly in its interpretation, and ease of usage. The following chapters fully explain this unique *practical* system of trigonometry calculation in minute detail.

This complete calculation system includes three specially designed memory aids for universal use in the school, and in the engineering workplace. These aids will be found to assist the student engineer in solving all of the triangle problems likely to be encountered in the work-place.

The author's own practical trigonometry expertise in mechanical engineering and triangle calculation has been diligently gained over many years of working in the production, research, and development engineering industries that covered a wide scope of practical, theoretical, and technical areas found in this particular working environment.

As previously stated, this complete book package includes not only the full instructions in the use of this unique method of triangle calculation, but also explains the use of the three newly designed innovative memory aids, (or so-called memory joggers). The upper half of the fig. 37f drawing of this removable memory aid, (its side 3), is used for the more advanced calculations required for a triangle that does not contain a right angle, and this is now known as the Wonky

Gabled House non-right-angle triangle calculation method. The lower half of the fig. 37f drawing of this aid (its sides 1 and 2) is used mainly for calculating basic right-angle triangles and is also used in the final calculations of the fig. 37 series, to complete the full calculation of the non-right-angle triangle.

The fig. 37f trigonometry calculation aid is called the Combination Probe. It is designed to combine the two aids into just one aid, and it is intended that a duplicate of this aid, is positioned at the end of the book just before the index so that it can be cut out and removed from the page, followed by being folded accurately as instructed for its immediate use in trigonometry calculations.

The whole aid (contained within the fig. 37f drawing) is extremely useful, as its side 3 is designed to assist in the calculation of the more advanced non-right-angle triangle calculation problems that occur from time to time in the design office and on the shop floor.

It's lower sections (sections 1 and 2), are used for the calculation of the basic right-angle triangle, (its inscribed formulas will be found to be identical to those shown in the fig. 2 drawing. It could be said that using the Probe and Prompt aid is like having the key to an Aladdins cave full of wonderful answers.

The third aid is in miniaturized form and is contained within fig. 50, it contains all six of the necessary prompts required to calculate the right-angle triangle. This aid is used primarily as a memory jogger, to aid those students who have by now acquired considerable experience in the use of the Probe and Prompt system, and will now therefore find themselves not requiring the use of it's full triangular Probe's outline shape to obtain its correct orientation in the problem triangle. This aid has been designed to be cut out and attached to the strap of a wrist-watch (this is explained in the index script adjacent to fig. 50), for use as an easy reference aid, it has the advantage that it will be unobtrusive in use. (The author has found this miniature aid to be quite useful as a 'memory jogger' during right-angle triangle calculations, and has fitted one to his own wrist-watch strap for triangle calculation use.

These three triangle calculation memory aids could in the passing of time be re-named Triangle Calculation Helpers.

These three aids and their method of use are unique in the world of published triangle mathematics.

They are now used to give that much-needed assistance to the student, when he or she is confronted with a seemingly unsolvable triangle calculation problem in the work-place.

However, it is not advisable to use any of these aids in the *examination room* in the school or college, as their use in this situation could be excluded by the examination board's rules. However, the six miniature prompts displayed on the smallest aid can be quite easily memorized and their recalled contents legitimately used in an examination room situation.

The explanations and calculation routes the student will need to follow are shown in a simple, easy- to-read-and-understand manner.

This new and unique approach to a much simpler method of triangle calculation will be found to be much easier to use by the less able in the field of triangle mathematics, who may have previously considered the traditional theoretical way much too complicated to be fully understood, resulting in the subject being given up entirely.

In use, these aids provide that much-needed assistance to the student when he or she is confronted with a seemingly unsolvable problem that involves the calculation of 'unknown angles' or the unknown length of sides in a worked-on right-angle triangle.

The use of these triangle calculation aids allows the student, (who may be unsure of the correct way to tackle or solve a triangle problem), to overcome this uncertainty by using the aids given prompts, its key sequences, (and by using of course the assistance of a scientific calculator), to ultimately solve the triangle problem.

The use of the probe will also allow the more able student to take part in, improve upon his or her experience, and ultimately to shine in what was originally a complicated subject.

The author's ultimate aim in writing this book, is to simplify the triangle calculation problem *once and for all*, and for all its users to accomplish their intended goals, by using all three of the calculation aids containing the easily understood prompts, in order to obtain the required dimension of the length of a side or the exact angle of the triangle under calculation.

The prompts written on the probe's surfaces actually point the way through the calculation, by leading the student through the correct sequence of key operations, using a scientific calculator to obtain the required and correct answer to the problem being worked on.

Using this method of triangle calculation will be found extremely helpful to those who consider themselves to be less able than their colleagues (in their mathematical ability), who will no doubt welcome having this golden opportunity of conquering the fear and trauma previously associated with using the traditional theoretical trigonometry, in their previously failed attempts to solve their triangle problems.

The practiced student will, however, after using the Probe and Prompt system for a relatively short time reach a stage where he or she will be able to positively shine in the subject, and to ultimately prosper in the enjoyment of solving those previously complicated so-called advanced non-right-angle triangle problems', (shown in detail in the fig. 37 example calculations that are required to solve the problem).

As a result of the experience gained in upper schools, engineering drawing offices, and by working on the industry's shop-floor and tool-room environment, the author has discovered, after several relatively short office and shop-floor discussions with staff, that a surprisingly large number of employees (if they dare admit it), are still not really capable of using trigonometry effectively to solve their triangle calculation problems in the workplace.

An example of the comments received in reply to my question, 'Do you use trigonometry at work?' I received the following replies: 'No I did it at school but I've forgotten how to do it now', or 'No, that subject wasn't thought to be important at school so they didn't teach it', or 'That way of working out problems is much too complicated for me, I'd rather get someone else to do it for me'. These replies, finally spurred the author on to write this book, with the aim and hope that its contents will eventually provide a definite mathematical advantage to both new students and to those existing workers who at the moment appear to have missed out on learning the subject, but who are fully prepared to enter into and to take on board this new method of practical triangle calculation, in order to help them in their daily work-piece and triangle calculation problems.

The author has designed, drawn, sketched, and calculated all the practical drawing examples shown and has supplemented this important information with descriptive sketches and drawings (so-called figures), with the intention that this information is being portrayed in the most simplistic form possible, to produce the required result.

The author has attempted this simplification exercise in order to assist the unsure students, by providing them with a thorough understanding of the calculation processes, and the routes that need to be followed (while using a scientific calculator), to achieve their ultimate goal of success in all their triangle calculation problems. By adopting this course of action, it is hoped that students will become fully enlightened to many of the possible snags and pitfalls that can and do occur in their working environment, and hopefully many of the mysteries presently surrounding this supposedly complicated subject will be removed.

The author wishes to point out at this juncture that the in-depth explanations given during the calculation sequences that describe the calculation routes to be followed, will inevitably take considerably longer (in time) for the student to read and digest, than would normally be taken in practice on the shop floor or office, while performing the required calculation sequences on the problem triangle being worked on.

While absorbing all of the many practical experiences gained over many years working in the engineering industry, the author has, during this period, acquired a considerable amount of in-depth mathematical knowledge, due to constantly working on three dimensional triangle calculation problems that involved the design of multi-bend tube assemblies and the design and construction of their respective checking jigs used in both the workshop and in the drawing office environment.

This experience has included precision design drafting skills, acquired while working on the design of sheet metal components, and on extensive tube bending research and development projects, including the tool-room development and the machining of components that require the use of precision machine tools for their manufacture. The experience so gained has allowed the author to complete this revealing and most informative book based on the main subject given the name practical trigonometry.

These acquired experiences have included not only precision drafting, but also the practical, hands-on experience in the welding and brazing of steel and aluminum, the precision machining of hardwood, metal, and fiber-glass used in the wind tunnel and scale model testing departments of the aircraft industry, also precision bench fitting, universal grinding, center-lathe turning and screw-cutting, universal milling, the inspection of precision machined components, problem solving, and modification of machine design, including the manufacture of precision components while using precision engineering machinery.

These experiences have also included the hands-on solving' of a host of additional shop-floor calculations, found necessary to complete the finished work-piece when working from an issued (and what the author would call) a 'limited information' production drawing.

While these calculations were in progress, it was often necessary to discover unknown dimensions and angles (essential for the accurate manufacture of the work-piece), from the issued drawing. This was achieved by utilizing every minute scrap of the (often meager) information supplied on the issued drawing or sketch for the manufacture of the component.

These trials, tribulations, and mathematical experiences, have been absorbed into the author's memory as a result of long periods of study in the research, production, and development departments of the country's typical engineering and manufacturing companies.

A large amount of this experience has been obtained at the following workplaces, various colleges, and periods of study taken up during the past fifty-five years of work experience.

It is with regret that due to the security regulations in force in the U-K, and currently enforced by the Ministry of Defence, the author has been advised that he should not divulge the names or locations of many of his previous places of work, seats of learning, or the colleges he has attended,including the names or locations of the engineering training courses he has attended during his engineering career.

It is due to these Security restrictions, that the locations and names of the following engineering establishments and seats of learning, have not been divulged in order to comply with with these security guidelines.

1. Service in The Royal Air Force, employed as an engine/ airframe technician. Rank SAC, A/CPL, (R/O) Pd. (obtaining ONC Mechanical. Engineering) through studies at the R.A.F. Education Sections, where an instructor's course was completed, including classroom techniques, and chalk board use.

2. Employed at a U-K engineering establishment as a tool – room universal grinder, this experience included (internal, external, tool and cutter, drill sharpening and surface grinding), engaged in the manufacture and precision grinding of measuring equipment, air gauging comparators and cemented carbide 'setting gages' used for checking comparators, including the use of diamond-impregnated grinding wheels, combined with the use of special Diaform radius attachments (for shaping the individual grinding wheel, to enable it to grind radii and V form shapes on thread measuring equipment), while also working to very close, temperature controlled, 'slip gauge' limits and tolerances.

3. Employed at a local Training Establishment where a Course for Engineering Bench Fitting, and a Universal Grinding course was completed

4. Employed at a U-K Aeronautics University as an R & D tool-room fitter/machinist/universal grinder. Instructing and advising college students in the design and manufacture of thesis prototype experimental aircraft engineering design, including the experimental explosive forming of sheet metal, the manufacture of wind tunnel delta models / stings, also various modifications to aircraft including those required on a Smiths Vickers 'Varsity' aircraft, for the manufacture of experimental very high-pressure hydraulic pipe unions used for blind flying / take-off and landing systems equipped with very high-pressure stainless steel hydraulic control equipment used for the actuation of the aircraft's flying control systems etc.

5. Employed at a U-K Engineering Establishment as an R & D metal model maker, in the manufacture and testing of precision scale-model aircraft for tests in high-speed wind tunnels, also producing precision fiber- glass mock-up models for use in high-speed wind tunnel experiments). These included Concord, Airbus, Tornado, Jaguar, and the Kestrel/ Harrier, vertical take-off and landing jet aircraft. An example of this work included extensive intake modifications made to the original Harrier VTOL nacelle in order to provide critical extra intake airflow capacity for the engine on take-off and landing, by using a system of automatic air vents (a design innovation introduced to augment the air flow to the engine while reducing the need to enlarge the nacelle's diameter); this modification allowed extra air supply to be obtained for take-off and landing etc. These controlled experiments were under contract to various companies within the aircraft industry.

6. Attended a local College where the author passed a course in metal machining and machine safety

7. Attended a local College where the author,as a student, completed a training course in precision hard –wood machining and the safety requirements of wood working machinery.

8. Attended a College of Higher education as a student in welding, brazing, heat treatment of metals, also plumbing / lead burning, including a course on the 'technology and maintenance of woodcutting machinery.

9. Employed at a local High School, as an (Engineering Workshop Technician), Instructing pupils in engineering production methods, together with their associated machine safety requirements; involved in bench fitting, the setting up of lathes for screw-cutting. Giving instruction in the art of precision tool grinding and preparation, for use on production center lathes, milling machines, and pillar drills for pupil's use, while engaged on project work for A levels, potential degrees, etc.

10. Employed at a local lighting manufacturing factory as a design draftsman on the design, development, bend calculation, production, and modification of sheet metal lighting fittings.

11. Employed at a local Engineering Establishment as a fitter / machinist / R & D crucible furnace specialist technician, selecting the individual metal materials for use in the electric furnace, the heating and pouring of molten metal, to produce experimental high-speed and stainless-steel metal powder, to be later used in the research and development of sintered lathe, and various other high-speed steel cutting tools required by various engineering industries and companies.

12. Employed at a U-K tube bending factory as a tube bending engineer engaged in the production of three-dimensional research and development tube bending components, and their accurate bend calculation, with the requirement to produce both batch production and precision samples for outside customers, also the production of shop floor production worksheets, containing sufficient detail to allow the precision manufacture (by the shop floor personnel) of large quantities of multi-bend tube assemblies, while using hydraulically operated Pines tube bending machines; also the design, calculation, and drawing of precision checking jigs for checking the accuracy of the completed bent tube assemblies, including the production of development drawings for issue to outside manufacturing companies for the manufacture of three-dimensional precision checking jigs as and when required.

13. Employed at a U-K engineering establishment as a tool-room engineer, working on the research, development, and manufacture of prototype, experimental, can and cap sealing machinery, for the food and drinks industries. Customers included Unilever and various Japanese bottling /canning companies.

14. Employed at a U-K aircraft engineering establishment as an aircraft detail fitter, working on the calculation and the precision manufacture of sheet metal aircraft detail components. This work included the design and manufacture of the necessary bending jigs, (as required) for the manufacture of very accurate sheet metal components requiring extremely precise dimensions and bends.

15. Attended a local engineering company for a course as a mature student in a Practical Power Press Safety course (1981) and a previously attended practical 'Abrasive Wheel' Grinding Safety course in (1971).

16. Employed at a local engineering establishment as a Tool-room engineer, works mechanical and electrical engineer / production setter, engaged in the setting up of all production machinery, including plastic injection molding machines and their tooling, offset gravure printers, automatic wire tagging machines, and the setting up and maintenance of the Artos multi wire stripper and cutting machines, also duties as maintenance engineer / problem solver, involved in the design modification and manufacture of production line assembly jigs. This task included full responsibility for the serviceability of all production machinery used for the production of electronic and manual timer controls for domestic central heating systems.

17. Employed at a Teaching resource center as a tool-room fitter and mechanical engineer, this work included the duties as an R & D, fitter-machinist, oxyacetylene/electric, arc welder, surface grinder, drill, cutter, and tool sharpener, center-lathe turner, universal miller, carpenter / wood machinist.

    Also employed as a metal and wood machining safety instructor, engaged in teaching courses provided for Design and Technology teaching staff, employed county-wide in the local upper and middle schools.

18. Employed at a local engineering establishment as a Tool-room engineer, research, development, modification, and destructive testing of rack-and-pinion steering gear assemblies, power steering, and cam-and-peg-type steering gears used in the motor vehicle and allied industries.

19. Employed at a local municipal vehicle manufacturing factory as a Design draftsman / troubleshooter, engaged in the production of and modifications to existing engineering drawings required for tube bending and also the calculations required for the company's sheet metal and hydraulic tube manipulation section; cab, chassis, and sheet metal component

modification, also the design and development of the in-house manufacture of municipal vehicles. These included municipal dust carts, gully emptier vehicles, and special, all-terrain airfield / fire and rescue vehicles.

20. Employed at a local engineering company as a Center-lathe turner working on the precision manufacture of very heavy, large diameter (often internally tapered) marine clutch components, while using the author's calculation conversions from imperial taper per foot to metric dimensions by using practical trigonometry to obtain workable metric digital taper readout dimensions required for the center-lathe taper settings, allowing the precision center-lathe turning of taper-turned machine and marine clutch parts etc.

21. Employed at a local County Council School as a Mechanical engineering technician / instructor, for the repair and maintenance of all design technology engineering and woodworking machinery, including lathe tool/plane/chisel, metal cutter, shears/scissors sharpening, repairs to sewing machines and allied equipment, including oxy-acetylene welding / brazing and machining repairs to sports equipment, science, art, music, and canteen equipment while also providing practical engineering and safety instruction to pupils, including precision marking-out and benchwork. Assisting the student design team in the manufacture of virtually all the components required to produce the STM 01/02/03/04 and 05 series of battery-powered endurance race cars allowing them to compete successfully at Bedford Autodrome, Rockingham Raceway, and at the Goodwood car race circuit, (achieving the endurance of 87 miles in 4 hours). The STM 03 model competed in the national race competition held at the Goodwood race circuit, and achieved 46 th place out of the 76 competitors (in their first year) in the four-hour endurance race. In 2014, car STM 03 finished in 63rd place and car STM 05 finished in 57th place out of the total of 220 endurance racecars competing at the same venue. On the second visit to the Goodwood circuit in 2015, the cars were in 40th and 41st place, out of the 220 participating

racecars; these improvements in performance were mainly due to modifications made to both cars and to the employment of much lighter and younger drivers from years 7 and 8 to drive in the race.

Part-time instruction is now being given to the students in manufacturing techniques, in the design and safe use of center lathes, milling machines, pillar drills etc, also safety instruction is being given in the practical use of oxy-acetylene welding/ brazing equipment, the heat treatment / hardening of metals etc. to pupils engaged on metal/ wood projects, in the sixth form, A level, year 10,year 11, etc.

The author has now retired but helps out as a part-time engineering volunteer, working with pupils on the development and design of a new (and lighter), electric/battery-powered endurance race car (STM 06) at a local Upper School.

Chapter 1

The author's main reasons for writing this book on the use of practical trigonometry in the work- place.

The author is seeking to introduce this new and much simpler method of triangle calculation to the world's working (and studying) masses. This novel and practical method of performing trigonometry calculations will be found to simplify the normal theoretical trigonometry that is currently being used, by adopting it's easy to-follow series of prompts to obtain the required answer to the problem.

This new system is more accessible to the less able, because of its simplicity in use. Its use now allows the budding engineer, and the 'man in the street' to enter into a world of triangle mathematics previously thought to be completely inaccessible to them.

Have you the reader ever wished you had the ability to calculate an unknown angle, or an unknown length of side in a triangular figure?

If you have, then this book of practical guidance, instructions, drawings, memory aids, and sketches, will be extremely useful, especially if the reader's long-term aim is to develop the necessary skills that will allow him or her to use trigonometry to solve those problems previously thought of as complicated.

An extended study of this book will provide the reader with a new golden opportunity or (to some) second chance at producing very accurate trigonometry calculations.

This new and innovative practical system of triangle calculation uses the *Probe and Prompt triangle calculation system* for the calculation of virtually all triangle calculation problems.

If we investigate the meaning of the word *probe*, we discover that it means to explore, or to search out something previously unknown.

It follows that for the purposes of this book the author has chosen this particular word to provide a simpler way for the student to obtain the required answers to a triangle calculation problem.

The word *prompt* means to '*instantly assist*'. It follows that a similar word is used for the title of this book, namely *prompting*.

Prompting can be a major factor when instructing those in need of mathematical assistance in the art of triangle problem solving.

The use of the word *practical* describes precisely the course of action needed to search out and use this novel approach to the problem of finding the required answer by use of this friendlier, hands-on, searching, and practical technique.

This new practical approach in the use of trigonometry calculations, is assisted by using the probe and its prompts, a scientific calculator, and in some cases the 'Combination Probe's side 3', for solving the more advanced unequal-angle triangle problems. This system is of major assistance to the student, who in the past has become completely baffled at times while performing a particularly difficult triangle calculation.

By introducing this new innovative practical approach to triangle calculation, the author wishes to assure the student that there is little need for him or her to have undergone a trigonometry-filled education prior to the study of this book.

In fact it could be said that it is to the student's advantage if he has *not* gone through the full experience of using the traditionally taught theoretical trigonometry, for this will then make it easier for him or her to approach this new method of triangle calculation with a completely clear mind, without the need to relinquish any aspects of the traditionally taught theoretical trigonometry, which often requires the student to actually name and commit to memory lots of named and variously shaped triangles.

The practical methods described, will also allow the less-able students the ability to shine eventually in the use of this new and unique method of practical triangle calculation.

A careful study of the contents of this book will also help to eliminate some of the mystery found in the normally difficult formulas currently being used in theoretical trigonometry calculations. The author has achieved this easy-to-perform situation by substituting new and innovative practical trigonometry formulas that also assist when using the conventional theoretical trigonometry. The Probe and Prompt system of triangle calculation will be found extremely helpful when used by students who are 'afraid of mathematics', or think they are not very good at it. It should also be stressed that the use of the contents of this book, with its Probe and Prompt system of triangle calculation, should not be considered to be in any way cheating, as its use merely allows the less-able student to temporarily bypass some of the difficult-to-remember formulas that seem to occur quite frequently when performing traditional theoretical triangle calculations.

The Probe and Prompt aid, while being used for engineering triangle calculations, can also be used effectively in our schools, workshops, and offices. There is of course a proviso that these particular aids should not be *taken into* an examination room in a school/college or be used during a mathematics examination, because of the fact that the examination rules often prohibit the use of such aids during the sitting of an examination.

It is therefore prudent for the student, prior to sitting a mathematics examination, to learn by heart the probe's *six main trigonometry formulas* used for calculating the right-angle triangle (as shown detailed in fig. 2, and fig. 50.

To reiterate, it will also be found advisable for the *advanced* student to commit to memory the formulas shown in the fig. 37f drawing that are required for using the 'Wonky Gabled House' method of calculating unequal-angled triangles (i.e. those triangles that *do not* contain a *right angle*). (This particular method of triangle calculation is fully described in the fig. 37, a, b, c, d, and e examples, detailed in the fig. 37f cut-out aid, and in the fig. 37g drawing of the Wonky Gabled House method of unequal-angle triangle calculation.)

The use of this practical method of calculating triangles will allow the student to arrive at the correct answer without having to undergo the stress and anxiety usually associated with the difficult formulas contained within conventional theoretical trigonometry.

It is also hoped that a keen interest in this subject previously thought of as difficult can be resurrected by those who may have fallen by the wayside, or given up in despair due to the complex requirements of the mathematical formulas that have previously been needed while using theoretical trigonometry methods to obtain the answer.

To make full use of this novel approach to this previously thought of as being a difficult subject, the reader will of course also need to use a scientific calculator, the main key functions of which must contain the capability of computing and displaying the scientific symbols of *sin, cos, tan,* and *square root* (accompanied by its sign, √). (Note that these symbols are abbreviations used on the calculator's keypad for 'sine', 'cosine', 'tangent', and 'square root').

There is little need for the student, while getting to grips with this easier and practical method of triangle calculation, to know exactly what all the signs and words represent, but to know that he or she can access their use by merely pressing the appropriate key as detailed in the prompts.

A scientific calculator (of the type used for calculating the trigonometry examples in the preparation of this book), may be purchased for as little as £5.

A complete study of this book, its methods of calculation, and its written text (including of course its sketched calculation examples), will not be found expensive, and will therefore allow the student to gain the necessary triangle calculation experience without undue cost.

However, a limited amount of triangle calculation can of course be performed by using a *basic-model calculator* (provided of course it has the addition of a square root key), but sadly, this particular model of calculator's calculating potential is rather limited, as this basic calculator will only be capable of resolving the 'length of side' prompts situated in the 'off-center area of the probe (as shown in figures 2a and 2b) and will not have the ability to use the important angle-finding prompts situated facing outward at the extreme edges of the probe, used for *angular* calculations.

The keys of a basic calculator therefore cannot provide (in this particular case) direct 'angular calculation' capability. If one is forced by financial circumstances to use just the basic triangle calculation route (that is, using just a basic calculator with a square root capability), then the student will be unable to accurately measure the right angle triangle's newly named The Angle, or its newly named Other Angle, without quoting and using reference books that contain the sin, cos, and tan scientific tables (which would of course need to be purchased, studied, and used).

This fact alone proves beyond doubt, that the scientific calculator (which contains all the required information for all angle calculation stored within its built-in memory) is by far the superior tool to use, for speed, convenience, and accuracy in the task.

*   *   *

Long ago in the dim and distant past, (around the time the pyramids were actually being built), it is believed that the local workforce discovered (purely by trial and error), that if three lengths of a suitable material, (possibly pieces of slatted wood or similar material), were cut to the individual lengths of 3, 4, and 5 respectively, (in the measurements used at that time, (probably cubits), the resulting outline shape around its outside edges, (when the slats were placed together in triangle fashion), it's outline shape would possess, at one of its corners, a very accurate and useful right angle, and would be found to provide a perfectly square and useful corner for use as a checking gauge to help in the accurate building of stonework.

fig. 1.

THE OTHER ANGLE.

THE 3,4,5, TRIANGLE

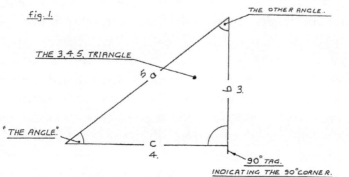

"THE ANGLE."

5 a

b 3.

c
4.

90° TAG.
INDICATING THE 90° CORNER.

A TYPICAL 3·4·5 TRIANGLE, SHOWING DESIGNATED SIDES a,b,c, THE 90° TAG,
AND 'THE ANGLE' (OR SHARP END), INDICATED. (ALSO SHOWN IS THE 'OTHER ANGLE'.)

BOTH SIDES OF 'PROBE' SHOWN.

fig. 2.

'THE ANGLE'
OR ∠= END OF PROBE.

'THE ANGLE,'
OR ∠=, END OF PROBE.

_2a._   90° TAG.   _2b._

THE DESIGN EVOLUTION OF THE 'PROBE'   figs. 2d, 2e, 2f, 2g, 2h.

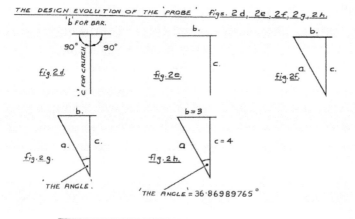

'b FOR BAR.

90°   90°

'c FOR CRUTCH

fig. 2d.

b.

c.

fig. 2e.

b.

a.   c.

fig. 2f.

b.

a.   c.

fig. 2g.

'THE ANGLE'.

b=3

a   c=4

fig. 2h.

'THE ANGLE' = 36·86989765°

The fig. 1 drawing shows a typical and similar 3-4-5 triangle, in this case with its individual sides designated as, a, b, and c. This triangle should be studied and its shape and designated side names memorized.

This drawing also confirms that the 'square' corner angle of this triangle will always contain exactly 90 degrees. This corner is now designated as its 'tag' corner.

Also shown in figure 1 is The Angle (this is positioned at the triangle's so-called sharp end'); it will also be found during later calculations, that the position of this particular angle in the triangle can be used in the calculation of angles as small as 0.01 degrees and can also be extended in a calculation up to a maximum of 89.9 degrees'. Similarly, it is also possible for its aptly named Other Angle (in this triangle) to also have the ability to expand from 0.01 degrees up to a maximum of 89.9 degrees in magnitude during calculations. It should now be realized that the Other Angle and The Angle of this triangle have the ability (during calculations) to become flexible in their magnitude (meaning that these particular angles can grow or shrink in one's mind's eye) during the calculation examples.

This flexibility in angle is shown in detail in figures 12 and 13, which show the full scope and magnitude possible to be reached when using it in its Expandable Probe mode.

The three included angles that make up this particular triangular figure (using in this case, the modern angular measurement of decimal degrees), do actually add up to a grand total of exactly 180 degrees.

Note that this particular number of degrees (180), will always be found contained within a right angle triangle (this fact also applies to any triangle of whatever shape). Remembering this fact will prove most helpful and will enable the accurate calculation of any later worked-on triangle problems.

The number 180 (degrees), refers to the total number of *included angles* found in a complete triangle, and most significant to the reader, is the fact that one of its angles, namely the newly identified 90 degree tag corner of the triangle (as shown in fig. 1.) will be found to contain exactly 90 degrees of included angle (this being a very important fact), and this too should be noted and remembered.

To establish the magnitude of the triangle's other two angles, (which incidentally when added together must add up to a total of 90 degrees), we must use the probe and follow the instructions given by its prompts. This allows us to discover the triangle's other individual angular magnitudes, contained within the outline shape of this triangular figure.

The side and angle prompts are displayed on all three sides of the probe (as shown in detail in fig. 4). These have been strategically placed facing outward at the probe's three outermost edges for easy reference, and apply to each individual side that is in turn identified by its individual a, b, or c, lettering. (Fig. 2 shows in detail both sides of the probe, The Angle of the probe, and its 90-degree tag).

As a practical example, a hand tool made to the (previously described), Egyptian simple right angle triangular shape was used quite recently on a building site, as a substitute for a conventional builder's square, which was unavailable at the time. It proved to be a very accurate measuring tool for checking the 90-degree included angle of the walls of a building under construction.

By adhering to this exact 3:4:5-dimensioned triangle, this homemade tool proved to be extremely accurate, and ensured that the corners and walls of the structure were being built squarely, and to the exact included angle of 90 degrees. The use of this tool ensured that the finished piece of work could be completed without the need to use a precision, professionally manufactured square, for this accurate angle-measuring job.

Other numbered combinations of a triangle's length of sides that provide this vital right angle, (with its very important 90 degree included-angled, square tag corner) are to be found with side lengths of 6:8:10, 12:16:20, and many others.

There are of course many other numbered combinations of triangles' side lengths, that will also provide us with a 90-degree square corner in this particular series, but the larger sizes tend to become rather too large to be of much practical use in the home, or the workplace.

Larger combinations in this number series are used mainly in land surveying and similar professions. They are used in calculations that involve the use of an instrument called a theodolite. This precision instrument is used for establishing the angles, the triangular side lengths, etc. of building sites and land areas, while using a calculation method called triangulation. (The student should note that the Probe and Prompt calculation system also uses this triangulation method of calculation on a much smaller scale during later calculated examples).

Triangulation is also used in the computing and manufacture of accurate world maps, and in the calculations required for space exploration, air and sea navigation, etc.

In these particular cases, the navigator of an aircraft or ship must calculate his proposed route, coupled with an obligation on his behalf to notify, in advance, the appropriate authority, by supplying them with a flight course (or sea route passage), detailing the proposed start time, (and in the case of aircraft, height to be flown), also the exact track (or intended course), while allowing for the effect of wind or current/tide on his aircraft or ship (in each case) during the proposed flight or passage.

This vital information is also obtained by using the so-called drift calculation method. This drift, (caused by the wind in the case of aircraft, or by leeway in the case of boats, is caused by the speed of flow of the tide coupled with the speed of the wind in the case of boats); this calculation process is called the triangle of velocities. The use of this method shows in graphic form on a prepared drawing, the angle and amount of drift or leeway, expected (over time), caused by the wind, or the movement of tide during the proposed journey. While this particular subject is only briefly mentioned in this book, beginners in this field of study, (who may possibly have doubts about their ability to solve their triangle problems using trigonometry coupled with triangle calculations), can obtain many of the required basic, and some of the more advanced trigonometry skills, not only for their preparatory examinations, but later on for charting their proposed course, by using 'The Probes' assistance in calculating their triangle navigation problems for use on land, sea, and in the air.

Chapter 2

Fig. 2 *The evolution of the Probe and Prompt system*

The author's design thoughts were finally developed during the evolution and planning stage of the original probe. These thoughts are now explored and explained in minute detail to allow their full potential to be practically understood by all levels of student. (This process is shown in figures. 2d, 2e, 2f, 2g, and 2h.). The probe and its prompts are fully explained in figures 4, 4a, 4b, 5, 6, and 6a.

The basic design of the original probe comprises a drawn right-angle triangle, printed on 80-gram normal typing paper. It has printing on both of its sides, (one side contains an approximate mirror image of its other side). These printed memory joggers are called prompts.

This original design has been given the name Probe and Prompt. It is designed to be used initially for the calculation of right angle triangles but will also be found useful in the latter stages of the final calculations required when dealing with the slightly more 'complicated' non-right-angle triangle' problems, (as shown in the latter stages of the fig. 37d and 37e series of calculation examples).

However, the student can now use just the cut-and-fold Combination Probe (designed to possess an identical Probe and Prompt aid in its lower section sides 1 and 2. It is now combined with the more advanced Wonky Gabled House method of triangle calculation contained in its side 3).

This cutout combination probe is available in fig. 37f, and is to be used as a perfect alternative to using just the original probe, after being cut out from the page of the book. This cutout aid is then folded as instructed, and used as the perfect alternative aid for use in all the following trigonometry triangle calculations.

There is also available a third trigonometry calculation aid that possesses unobtrusive miniaturized prompts. A copy of this aid is shown (immediately below its installation picture in fig. 50); it is intended that one of these miniaturized aids could be cut out and securely attached (by adhesive for example) to the reader's wrist-watch strap for unobtrusive use. The six formulas shown on this miniature aid are identical to those used on the original Probe and Prompt aid, but in this case, they do not possess the prompt's original triangular shape, originally provided to give a full triangular reference. This aid is duplicated below the installation picture in figure 50 in order to provide the reader with a generous supply of spares for later use. It will be found useful by students who have already familiarized themselves with using the original probe and its prompts, but who will now have become sufficiently experienced in using the shape of the original aid for orientation purposes, and now find there is little need to refer to its familiar shape when in use. The author now uses a copy of this particular miniature aid attached to his own wrist-watch strap for use as a memory jogger.

The following dimensions are provided (purely as a suggestion), in order to make this aid suitable for it to be attached to a standard gents' 18-mm-wide wristwatch strap. The external dimensions will of course need to be adjusted if using other widths of watchstrap.

To make this aid sufficiently durable, it has been found (through experience), to be best practice, (after removing the fig. 50 page from the book), to have its printed side heat-sealed in plastic while the page is still at its full size, followed by cutting each individual aid (as the need arises) to a convenient size of approximately 23 mm long by 16 mm wide (the extra outside material will be required to provide sufficient gripped overlap); the aid is then attached to the strap initially by using a short section of double-sided adhesive tape (after removing its protective backing), followed by a strip of black or other suitably colored duct tape, cut to a length of approximately 35 mm wide by 35 mm long, provided with a marked-out central window cut to the dimensions of approximately 16 mm by 11 mm (being careful not to over cut beyond the rectangle's corners). It will be found easier to cut out this central portion if this operation is carried out after being stuck back onto the roll of tape, and by the use of a very sharp craft knife, (care must be taken to avoid too deep a penetration into the next layer of tape). The width of the duct tape is then carefully wrapped around the strap, over the now-encased aid, and continued onto its underside by either overlapping the excess material equally, or applying another strip of tape to cover the inside raw edges. It will also be found convenient to have the strap already fastened (to the exact dimensions as worn, with its buckle's pin in its usual hole), and with the right-hand side of the duct tape placed abutting the buckles pin (the edge of the material being tucked under the buckle), in order to obtain the correct positioning of the six prompts and neatness of the aid during assembly.

There now follows a provisional explanation in the use of the 'cut and fold' Combination Probe, currently used for calculating both the Wonky Gabled House method of triangle calculation used for unequal angled triangles, (explained later) and also the basic Probe and Prompt method used for calculating just right-angle triangles.

The side 3 of the Combination Probe is used exclusively for the more advanced non-right-angle triangle (the so-called unequal-angle triangle) calculations.

This advanced memory aid (that incorporates the original designed probe in side 1 and 2 of its lower half), is unique, for its side 3 can now be used in the calculation process called the Wonky Gabled House method of triangle calculation, which is used for solving the unknown angles in the non-right-angle (or the unequal angled) triangle. This combination of the advanced and the basic aid is now used to solve virtually any triangle problem (with the proviso that the aids' written sequences are strictly complied with).

This more advanced aid is made from a similar paper material, but needs to be folded twice to allow for its multiple use; this design is produced in the fig. 37f drawing to enable the student to have the option of removing the page from the book (using scissors) to cut out the aid), and then for it to be folded into its required shape, followed by it being used in either of its folded forms to aid the required trigonometry calculations.

Its design is also featured in fig. 6b, and 6c, of the early drawings. The author intends that these originally drawn figure pages of the book should be retained for normal reference purposes. It is therefore for reasons of tidiness that this aid has been duplicated toward the rearmost pages, (a duplicated version of fig. 37 f) to enable it to be cut out and folded into its correct working shape for use, while still maintaining the neatness of the earlier main pages of the book. This folded aid is then used to assist in both basic right-angle triangle calculations (by using the Probe and Prompt sections contained within its numbered sides 1 and 2 for right-angle triangles), or by using its Wonky Gabled House section (numbered side 3), for use in the more advanced unequal-angle/non-right-angle triangle calculations.

It should be noted that if one is confronted with a non-right-angle triangle calculation problem, (where the triangle being worked on does not contain a right angle), then the student, during the latter part of this calculation, can introduce two 'new' right-angle triangles (with their orientations situated back to back), within the confines of the worked-on non-right-angle triangle.

By using the newly coined terms 'non right-angle triangle' and 'unequal-angle triangle', the author has (for simplicity purposes), avoided the need to use (and for the student to need to remember), the multitude of conventional names given to the various shapes of triangles that are not actually right angled.

However, when confronted with this particular calculation problem (as will be seen in the fig. 37 examples), one's aim *during this rather long calculation*, is first to find the magnitude of angle G in the problem, followed by finding the magnitude of angle H, followed by finding the magnitude of angle W; this can be followed by finding the vertical height of the triangle (if required) in order to complete the full calculation.

We do this by following *exactly* the sequences described on side 3 of the Combination Probe.

The full calculation is finally completed (if the vertical height of the triangle is required) in the calculation's latter stages by using (in its fully folded state), side 1, or its side 2 (if appropriate). The now-duplicated basic probe is used to complete the final calculations required in the problem by using two newly introduced right-angle triangles 'into' the original problem, thereby completing the final calculation required to obtain the vertical height of the non-right-angle triangle.

To reiterate, the Combination Probe is used in its partially folded state (folded just once) for calculating non-right-angle triangles, as shown in the calculated problem in chapter 14, fig. 37, examples a, b, c, d, and e.

Alternatively the aid can be used in its fully folded state (that is, with it folded completely twice), for use in the calculation of basic right-angle triangles (thereby adopting the identical shape and use of the original example shown in figure 2 of the basic probe.

It will be found that, using both of these novel calculation methods in turn, and following the further instructions given regarding the Combination Probe's use, the system will be fully capable of guiding one through the rather tortuous path needed to calculate virtually any shape of problem triangle, by merely using the instructions and the stated series of helpful prompts.

The design of the Combination Probe is shown at working size in fig. 6-b, and again in fig. 37-f, for easy and unobtrusive use (after being cut from the page).

It is also shown in fig. 6c (in this case enlarged to twice normal working size), to allow careful study and appraisal of its methods and formulas).

However, it will be found there is a distinct advantage in using for most of your triangle calculations, the fully folded version of the Probe and Prompt (sides 1 and 2 of fig. 37f). The advantages of using the fully folded Probe and Prompt is that it is small, unobtrusive, easy to use, and easily carried in the pocket, purse, or handbag.

In use, this particular aid can be moved freely around on the paper, and pointed (i.e. oriented), in its correct relative position in the same direction as the problem angle requiring calculation. This then allows the correct prompt to be easily read off, complied with on the scientific calculator, and used to solve the problem triangle.

The prime aim of the probe, therefore, is for it to be used for the easy calculation of angles and to obtain the unknown lengths of sides, in right-angle triangle problems as and when they occur in practice.

The combined design, given the name Combination Probe, is for use when the student is involved in the more complicated so-called non-right-angle triangle calculation problem. This situation occurs occasionally during this type of triangle calculation, where a 'convenient' right angle *cannot* be found in the worked-on triangle.

When arriving at this particular situation, there is the tendency for the student to become mildly despondent until a suitable alternative calculation route can be found to resolve the problem.

In order to deal with this particular problem, the student will need to call upon his or her reserves of ingenuity and brain-power to seek out (i.e. to contrive mentally) a way of solving this particular problem. This lack of a right angle in the problem triangle being worked on, can be accurately resolved by using the Combination Probe's side 3 to calculate its included angles of G, H, and W, followed by (if its vertical height is required)the introduction of two 'new' right-angle triangles, suitably placed in a back to-back situation, into the worked-on non-right-angle triangle problem. This move then allows the basic probe and its prompts to be used to complete the calculation to obtain the triangle's vertical height as shown in fig.37 d and e.

This resolving of calculation snags is often required when trying to solve a particular unequal-angle (non-right-angle) triangle problem. These problems are much easier to solve when one uses the formulas contained within the Wonky Gabled House calculation method.

To reiterate, the complete calculation process required when using this method of triangle calculation, is detailed in section 3 of the Combination Probe and is used in the calculation examples shown in figures 37, 37a, 37b, 37c, 37d, and 37e. (The Combination Probe is also shown in detail in figures 6b, 6c, and 37-f).

The trigonometry calculations needing to be completed to solve the non-right-angle triangle problem, must be carried out by following the given instructions in a strictly ordered and correct sequence, and no deviation from this particular sequence should be used.

These helpful instructions will provide the correct calculation sequence to follow, in order to obtain, firstly, the angle G (as shown in the fig. 37 example triangle problem), followed by calculating angle H, followed by calculating angle W.

Note that the last stages of this calculation sequence can be completed (if it is needed to find the triangle's vertical height) by using figures 37d and e, for the necessary prompts required while using sides 1 or 2 of the Combination Probe to complete the final vertical height part of the calculation.

As previously stated, this final part of the calculation is performed (if necessary) by using the assistance of two newly introduced right-angle triangles, these being placed strategically back to back (side b backed up to side b), into the sketch of the original problem. The calculation is then finally completed by using typically side 1, (or if the problem is in mirror image, by using side 2).

By using this sequential calculation method, it is possible to obtain the full angular magnitude of all the three corners (included angles), of the problem non-right-angle triangle'and (if necessary) obtain its exact vertical height.

In essence, the unknown angles required are calculated by utilizing the three known side lengths as shown in the fig. 37 sketch of the problem triangle, combined with using the 'Combination Probe's instructions. This is followed by using its sides 1 or 2, to finally complete the calculation.

It should be pointed out, however, at this stage, that this particular non-right-angle triangle problem, occurs only very rarely in the general run of workshop triangle calculations. The student should not therefore be deterred from using just the basic Probe and Prompt's sides 1 or 2 for the vast majority of basic calculation requirements involving the right-angle triangle.

This awkward problem triangle (a triangle with no known angles) will become apparent when the student is confronted with the rare situation where all the side lengths of the worked-on triangle are actually known, but it is then discovered that the problem triangle figure fails to contain any known angles with which to start the normal basic right angle triangle trigonometry calculations.

Therefore, by using the Wonky Gabled House method of non-right-angle triangle calculation', it is now possible to obtain all the included angles required to allow the problem to be finally resolved.

It is very important, however, when carrying out this particular calculation, that the correct sequences are followed exactly as indicated on side 3 of the Combination Probe; any deviation from this set instruction sequence, will ultimately prove to be incorrect. It is therefore very important that full attention is paid to the actual sequence of the instructions given. (This sequence is fully explained in the drawings fig. 37, 37a, b, c, d, and e, examples, used to solve the problem).

The Probe and Prompt aid is used to make accurate triangle calculations.

The fig. 2 drawing shows both sides of the probe and its prompts.

This aid has been designed to possess an inbuilt ability to be used on either of its sides, as dictated by the orientation of the problem found on the paper; this then allows the probe to be used pointing in any orientation, including upside down, where its written prompts will then, of necessity, need to be read and interpreted in this upside-down situation. In use, the probe can be pointed at the problem angle being worked on, exactly where it occurs on the paper, in any direction or orientation, thereby allowing all angle possibilities to be explored and calculated. The full calculation of the triangle can then be easily completed by following and acting upon the guidance of its prompts.

One advantage of using the Probe and Prompt aid is that it can be used in a discreet and unobtrusive manner in any workshop, office, or classroom situation to obtain the correct result, without actually drawing attention to oneself by the need to seek calculation advice from colleagues, or from the works supervision.

As previously stated, the probe can be used in its upside-down orientation, while still retaining the ability to display the correct prompt for use in each particular problem case.

There now follows a detailed explanation of the design and development of the 90,-degree tag (featured on the basic probe) as shown in fig. 2.

The design and development of this 90 degree tag is unique. This feature enables the user, while orienting the probe in its correct position on the paper to match the problem, to identify accurately and use the vital 90- degree corner's position for problem-orientation purposes,; even if the worked-on angle is drawn at less, or more than 90 degrees, (this occurs particularly if the problem triangle is drawn obliquely), it will still be found that this will indicate that the calculated object *does* possess three dimensions.

The probe is ideal for use on three-dimensional work, where angles are often distorted, and do not appear visually to be of any recognizable or 'normal' angular scale or shape (as shown in fig. 6, and shown without the tag in fig. 6a).

The probe is very versatile when used in a sliding motion across the surface of the paper, allowing it to arrive at its correct working position pointing at the angle to be calculated, for it now allows the student to concentrate all his attention and effort toward the pointed 'The angle' (or sharp end) of the aid, and of the problem angle requiring calculation.

Adopting this simple 'isolating the angle to be calculated' procedure, (by precluding the 'other angle' at this stage in the calculation), allows the student to ignore temporarily the newly named 'other angle' at this time. This has the advantage of simplifying the triangle calculation problem, by allowing only one calculation to be worked on at any one time.

Adopting this method of working avoids the mistakes which can and (as the author has found through experience) do actually occur, particularly when the student is forced to juggle with several formulas in his head at the same time, and while at this point in the calculation process, there is also the tendency for the student to give an enquiring 'glance around' for assistance (should he be working without the aid of the probe and its prompts).

The student, on reaching a certain stage in a calculation, is often uncertain as to which formula to use for each particular case, particularly if he is *not* using the probe and its helpful prompts for guidance.

The simple design of the probe allows even those who consider themselves to be less able in their mathematics ability, to understand its full potential quickly, allowing them to perform highly accurate work, with the aid of course of using a scientific calculator.

In order to allow the contents of this book, with its two main designs of probe, including the Combination Probe, to be used by everyone, every effort has been made to use simple everyday words in the text, and to make the prompts and the Combination Probe's instructions easily understood by all levels of student. It is also considered that the simplification of the actual wording contained within the written texts tends to enhance rather than detract from the accuracy of the system.

The author would like to point out at this early stage, that this book tends to focus more on providing the student with the required skills and abilities, by allowing him or her acquire the practical knowledge required by the engineering employers and for their employees, rather than for the demands of the academically qualified employed in the industry. This book has therefore been written for the benefit of the engineering industry's trainees, employed students, and machine shop floor workers, who may not necessarily be sufficiently educated in the mathematics required, but who feel they are fully capable of learning this new easier way of using trigonometry calculations to improve their ability, in order to complete their given work-pieces accurately.

It should be understood therefore that this book has not been written to suit the academically qualified, who may find the author's methods of triangle calculation, and problem solving, including the use of practical trigonometry and the easy-to-follow calculation instruction lists, to be a too simple a task for their higher intellect.

By the elimination of long, and difficult to-understand mathematical words phrases and formulas, this book now allows the ordinary person, (the man in the street') to participate in and effectively use this novel triangle calculation system to solve their triangle problems with considerable ease, thereby providing them with their ultimate goal of defeating their current triangle calculation problems.

Chapter 3

This chapter covers certain exceptions to the author's so-called simplification rule, followed by a full explanation of how one uses the abbreviated mathematical symbols written on the Probe and Prompt and explains it's easy to-follow symbols.

It has been found necessary to include a few longer words in the texts than was originally intended. The author has now included these in order to fully explain the functions used in many of the following calculations.

For example, the word 'sexagesimal' is a notation word used to describe the angular magnitude of an angle when it is written in the form of degrees, minutes, and seconds', (in this case, working to a base of 60, there being 60 minutes in a degree), and 60 seconds in a minute of a degree).

If an angle is written in the form of, say, 1 degree 30 minutes, it could also be written as $1^1/_2$ degrees in 'fractional notation', or 1.5 degrees if written in decimal notation.

However, all three ways of explaining the magnitude of this particular angle are identical.

The author has found, through experience, that it is much easier to calculate all the angles and side lengths found in a triangle problem by using solely decimal notation. It is therefore strongly recommended that students use this form of dimensioning at all times, in order to make their triangle calculations so much easier.

When using this innovative triangle calculation system, it will not be found necessary for the student to have had any prior in-depth knowledge of trigonometry, for, using the probe and its prompts enables students to perform highly accurate work by helping them solve their triangle calculation problems. It will also be found that there is little need for the student to possess any so-called expert mathematical knowledge.

During the author's introduction to mathematics and engineering in the 1950s, he found himself engaged in solving quite complicated triangle calculations. In the 1950s, it was necessary to refer to various reference books that contained the scientific tables including the log., sine, cosine, tangent, and square root mathematical constants; he was therefore obliged to go through a quite laborious process to obtain the necessary square roots to calculate the individual angles of triangles, (there being very few calculators available for use on the work-shop floor at that time in the U-K). Because of this, it took considerably more time and effort to perform these extremely necessary triangle calculations.

The introduction of the electronic calculator was therefore welcomed in the office and onto the shop floor, this being capable of providing instant square roots, and later, when the scientific models became available, made it possible to obtain instant Sin, Cos, Tan, and square root constants without having to refer to the old reference tables.

Having now acquired an easier way of obtaining these constants made the work involving two- and three-dimensional trigonometry calculations much easier and allowed the process to be speeded up considerably, with the added bonus that work with greater accuracy could be produced.

The 'Probe and Prompt's symbols

In order to make these calculations simpler, quicker, and less complicated, the author gradually evolved his own calculation system over the years, to help in the various decisions that were needed regarding the correct sequence, method, and formula to use when confronted with a particular triangle calculation problem. This development work gradually evolved into the author's own design of mathematic symbols (originally intended just for his own use). However, with continued use of these 'newly developed' symbols', it was realized that their use could also become extremely helpful to the modern-day student who encounters similar triangle calculation problems nowadays, hence the development of this book that now explains to everyone the whole subject in meticulous detail.

If the student now refers to fig. 2 and studies for example the symbol $\angle$ on the probe, this is an abbreviation symbol designed to identify The Angle in the right-angle triangle. This is interpreted as such on being spoken aloud (for practice purposes). If we then add the equals sign to this symbol, it is then interpreted as 'The Angle' equals' ($\angle$ =). It follows that the symbol '$\angle$ a', should be mentally interpreted as 'I know The Angle and a.' It also follows that the symbol '$\angle$ b' should be mentally interpreted as 'I know The Angle and b', and the symbol '$\angle$ c' should be mentally interpreted as 'I know The Angle and c.'

These are the known symbols and are immediately followed by the Prompt symbol, (this being the unknown or sought-after part of the Prompt sequence). The use of the full sequence will therefore provide the complete formula for use in each particular case.

It should be understood that in order to perform right-angle triangle calculations, it is vital that the student is given, or in possession of (or actually knows) either

1. the lengths of two of the triangles sides (when using a basic calculator with just the additional square root key), or

2. the magnitude of The Angle in decimal degrees, and the length of at least one of the triangle's sides, (when using a scientific calculator equipped with sin, cos, tan, and square root keys).

If we now refer to fig. 1 and wish to calculate its The Angle in degrees, we should study the sketch of the probe shown in fig. 2a and mentally place this and its tag corner in a similar *position* and *attitude* relative to the triangular figure in the drawing or sketch of the problem being worked on.

The probe should be positioned with The Angle end pointed at the angle needing to be calculated, while making sure that its 90-degree tag corner is accurately oriented to match the 90-degree square corner of the problem triangle.

In fig. 2a, we can see detailed at the probe's The Angle end (its sharp end), the prompt $\angle = b/c$ INV TAN.

This is a very important prompt and is used very frequently and usually at the beginning of a calculation to obtain the magnitude of The Angle in decimal degree notation; this is therefore very often used as one's first calculation move.

It will be found to advantage therefore if a serious effort is made on the part of the student to memorize this abbreviated prompt. This can be broken down further, for easier understanding as follows:

The Angle equals the length of side b, divided by the length of side c, followed by the calculator entered sequence, INV. TAN; this particular sequence can be abbreviated still further, (when experience has been gained), and it will be found an advantage if it is whispered (under one's' breath), as 'The Angle equals b over c, INV, TAN.' (The Angle equals b divided by c, INV, TAN.) Note that in mathematics the phrase $b \ / \ c$ is identical to the phrase $b \div c$ or b divided by c; this prompt is used very frequently in the later calculation problems found in the book.

In use, the first numerical part of this calculation sequence is entered into the calculator; this is followed by the division sign, followed by the second numerical part, followed by the key sequence, INV followed by TAN. The numerical result of the calculation is then shown in the calculator's display (in decimal degrees).

In the following fig. 4a example, the actual sequence of key operations used for this particular calculation on the calculator is underlined for clarity as follows, we enter 3 divided by 4 = (display shows 0.75), followed by INV, TAN. This sequence then gives us the answer to the triangle's The Angle, which calculates out to 36.8698989765 degrees. It is advisable to sketch your own triangle and repeat this sequence several times using the calculator, and to note the result down on paper for practice purposes; the aim of course when doing this is to obtain an identical result each time.

We should now use this identical triangular figure to illustrate a different triangle calculation example.

We can now do this by using one of the other prompts on the probe in order to demonstrate the probe's versatility in obtaining a similar result by using its now known The Angle and one of the triangle's other length of sides in the calculation.

Because we have completed the first task, we now have the advantage that (by referring to the first calculation), that The Angle part of this particular triangle is 36.8698989765 degrees; we also know the length (taken from the drawing), of at least one of the triangle's known sides, in that side *b* is 3 (and also incidentally noting that the length of side *c* is 4).

It is now possible, because we now know at least two vital pieces of information, (i.e. that of The Angle in decimal degrees and at least the length of one side), to calculate the length of the other unknown side, in this case the length of side *a*.

If we choose to use for this particular example the known length of side *b*, and the now-known The Angle, this being 36.8698989765 degrees in the calculation, we therefore use the prompt ∠ *b*, *a* =, *b* ÷ SIN ∠. This particular prompt is shown on the relevant outer edge of the probe) and its relative position on the probe should be noted. (We will also be later using the calculator's memory by using the X-M key and the return memory key R-M.

We therefore enter into the calculator, the sequence 36.8698989765, SIN, (display shows 0.6), XM; we then enter the sequence 3, ÷, RM, now giving us the length of side *a* as 5 (shown in the display).

Let us now suppose (for practice purposes) that we do not know, but wish to find, the length of side *b* (bearing in mind of course that we already know the length of side *c* and The Angle) which we will now use.

From a careful study of the probe we can now deduce that we need to use the prompt '∠ *c*, *b* = *c*, ×, TAN ∠'. For practice purposes, it will be found advantageous if we actually repeat this sequence (in a whispered tone to ourselves, for familiarization purposes).

The first part of the prompt ∠ *c*), explains to us that we already know The Angle and side *c*. The second part of the prompt (*b* =) informs us that it is the length of side *b* that we wish to find. The third part of the prompt (*c* × TAN ∠) explains the calculation route we must follow to obtain it.

This whole prompt is best remembered if it is said, preferably under one's breath, and committed to memory, as 'I know The Angle and c, b equals c × TAN The Angle.'

To reiterate, when starting the calculation to find the length of side b, it will be found much easier to enter the magnitude of the known angle into the calculator first. This must be in decimal degree notation and written in this case as 36.86989765, (degrees) and *not* in the sexagesimal notation of (36-52-11 degrees, minutes, and seconds, its angular equivalent). This is followed in sequence by pressing the TAN key. The tangent of the angle will now be displayed as 0.75.

This sequence is then carried onward by pressing the multiplication key, followed by entering the length of side c (which in this case is 4), followed by = (the equals key). The answer will then be displayed, that shows us the length of side b is 3.

Again for practice purposes, let us now suppose that we 'know The Angle and a, and wish to find c. We consult the probe and select the appropriate prompt: $\angle$ a, $c = a$ × COS $\angle$. We enter the key sequence 36.86989765, COS, (display now shows 0.8) ×, 5, =, giving us the answer 4.

It should be noted that some calculators have the usual INV key substituted by a 2nd F key. This key performs exactly the same function, and is used in place of the INV. key when using this particular design of calculator.

The main advantage of using trigonometry skills to find unknown angles, or the unknown length of sides in right angle triangular objects is that we can use the facts we already know about the existing angles, their length of sides, etc. and reuse them in further calculations to establish the unknown side lengths or the angles required.

It is possible, by adopting this practice, to build up a graphic picture of the problem in our mind's eye; we can then cement this experience into our memory, by carefully preparing a dimensioned sketch, correctly oriented (in a similar fashion to the problem triangle being worked on). In this particular case, the sketch need not necessarily be to scale, while using a separate piece of A4 paper.

It will soon be discovered that it is a combination of knowing the distances between the points of the problem triangle (in other words, its side lengths), coupled with the known included angles that occur between these points, that will enable us to calculate all the vital dimensions and angles needed to solve the problem.

The following notes explain the function of the important 90-degree tag. This has been specifically designed into the probe to assist in the accurate orientation of the probe's right-angled triangular shape relative to the shape of the problem triangle being worked on.

While it will be seen (and noted) that this tag is just a short extension of the *b* side of the probe, this extended symbol must also be purposely introduced into the worked-on problem triangle on the paper. This enables the reader to identify the exact position of the problem's 90 degree corner relative to the probe's 90-degree corner; this in turn also identifies the correct position of The Angle and the probe's individual sides relative to the problem triangle.

By adopting this simple orientation and checking procedure, it will be found quite easy to identify the position of The Angle (the pointed, sharp end of the probe) correctly in all the following calculations.

Prior to the invention of the Probe and Prompt system, the student would often find it necessary to expend considerable mental agility when juggling with all of the complicated trigonometry formulas in his mind, in order for him (or her) to arrive at the correct one to use for each particular case. The realization of this fact persuaded the author to look into the possibility of developing to a greater extent, this simple method of solving triangle problems in order to provide, ultimately a more straightforward easy-to-access system, requiring much less mental agility and effort on the part of the student, to enable him (or her) to calculate accurately the triangle problems found in a sketch, a drawing, or an actual work-piece.

It was ultimately decided to simplify the generally used theoretical methods traditionally used in triangle calculation, (that involve the use of Greek letters, and a rather confusing series of multiple shaped brackets), by guiding the student temporarily away from the old formulas being used in the originally calculated example shown in fig. 3, (now considered outdated), to be superseded by the new method used in fig. 3a and fig. 24, by using the author's new method of triangle calculation that involves just the joining up of triangles in a method called triangulation.

This change of method has also been achieved by the introduction of the basic 'Probe and Prompt' system, into the new Wonky Gabled House, a so-called advanced triangle calculating system' now used for calculating unequal-angled triangles, in order to solve those awkwardly shaped triangle problems that fail to contain a right angle, by using this much easier and straightforward method to resolve a particular triangle calculation problem.

It was realized that now a simpler alternative system has been designed and developed for those who are working on the basic right-angle triangle problems, then the introduction of the Wonky Gabled House advanced version, with its added instructions, would be beneficial to those students who may find themselves having to calculate the unequal or non-right-angle triangle' problems, featured in the fig. 37 series of calculations, or in the fig. 19 example, with the need to calculate the three-dimensional triangle problem that now includes a tube bending calculation requirement.

In order to develop this simpler calculation process fully, it became necessary to make a few changes to the current calculation system by the introduction of certain modifications to the normally accepted trigonometry formulas in current use, and to base these new designs on more suitable formulas that could be used directly (and be more easily acceptable), to the latest design format of the electronic circuits used in the currently available scientific calculators.

Chapter 4.

Explaining the advantages of using the Probe and Prompt system for the calculation of right-angle triangles

This new method is used to either substitute or augment the normally accepted formulas presently used for the calculation of the right-angle triangle's side lengths and angles.

Its main advantages lie in it providing the correct choice of formula for use in each problem's case. This choice is actually made by the student firstly identifying the problem area, then orienting and consulting the probe's prompts correctly, followed by reading off and complying with the appropriate prompt displayed for this particular situation.

All the student is required to do is to carry out methodically the prompt's instructions, while using a scientific calculator to complete that particular part of the triangle calculation.

If we now refer to figures 2d, 2e, 2f, 2g, and 2h, and use these particular made-up progressive triangular shapes as an aid to the recognition of the problem's right-angular shape, it will be found that it will aid the reader's memory in the required positions of its identifying lettering, The Angle, and its 90-degree tag.

To make this unique 90-degree tag more easily remembered visually in its correct position, its design, together with that of the actual probe, will be seen to be based on the shape of a slightly modified conventional capital letter T.

This design shape was specifically chosen because all the junctions of a T's straight lines actually form an angle of exactly 90 degrees at their respective junction points. (It is also the first letter of the word 'tangent' (abbreviated to TAN in the book), a word we will be using quite frequently in later calculations); it is therefore a good idea to become completely familiar with this similarity at this early stage of the book.

To allow the outline shape of this modified capital letter T to be easily remembered, the top line of this figure (with its 'T' shape oriented vertically), is called *b* for 'bar', and its leg (or support), is called *c* for 'crutch' (similar to the word used for a walking support appliance used in hospitals to aid a leg injury while walking). This is shown detailed in fig. 2-d.

It will also be found much easier for the student to remember this important T-shaped figure, if we consider the planning ideas, and the process that needed to be pursued in order to complete the design of the probe and its functions.

In fig. 2-e, we see that the top bar (*b*) has been moved along to its left to leave just a small overhang at its junction with (*c*). This actual overhang now forms the shape of the important 90-degree tag (which now indicates its vital and square 90-degree corner).

By making this sliding move, the two drawn lines, *b* and *c*, can now be converted (in one's mind's eye) into the two sides of a right-angle triangle as shown.

Now, adding line *a* (as seen in fig. 2 f), makes it possible to complete the outline shape of a right-angle triangle, together with its' 90-degree tag situated in its correct corner position.

It should be noted that this diagonal line (*a*) is always the longest side in a right-angle triangle, and remembering this fact will prove to be extremely helpful during our future calculations.

The correct positioning of this tag, in a sketched triangle calculation, is now used to identify the position of the 90-degree corner positively in any worked-on problem triangle.

In order to use this Probe and Prompt triangular memory aid correctly in trigonometry calculations, we must first positively identify each of the worked-on problem right-angle triangle's three sides, by using the letters *a*, *b*, and *c* respectively, positioned in their correct and permanent locations for its side identification purposes.

In fig. 2-f, the line designated to be at the top, (while being viewed at this particular vertical orientation), will always be called *b* for 'bar'. The vertical support leg (while also being viewed at this particular orientation) will always be called *c* for 'crutch'. The longer diagonal line that completes the triangle will always be called *a*, and will therefore always positively identify this particular side of the triangle as being the longest side of any right-angle triangle.

We now refer to fig. 6.

[We have now noted that side *a* of a right-angle triangle will always be the longest side of any right-angle triangle, when it is drawn in the plane of the paper (IPOP). This is an important fact and should always be remembered. However, if calculating triangles that have been purposely drawn to simulate an obliquely viewed triangle, then these triangles will now have the ability to represent all three dimensions of length, depth, and height, and provide a view of a triangle where its side 'a' does *not always* appear to be its longest side. Typical examples of these distorted three-dimensional views of triangles are shown in fig. 6, where, with its tag now positioned at its 90-degree corner, it gives a true impression to the viewer that side '*a*' now possesses a length that will differ from its true length if drawn 'in the plane of the paper'].

It will be seen in figures. 2d, e, f, g, and h that *b* is positioned over *c*. This is written as *b/c* (*b* over *c*); in practice this is '*b* divided by *c*', therefore, by using mathematical convention, it signifies that *b* will need to be divided by *c* in the calculation, in order to find The Angle in decimal degrees.

We have now arrived at the beginning of the prompt that is used in the first part of the sequence to find The Angle, this being, $\angle =$, *b/c*.

Having now described the first part of this calculation sequence, this is followed by pressing (in sequence) the INV and TAN keys; this will then complete the full prompt sequence used to find The Angle in decimal degrees in any right-angle triangle.

Note that this particular prompt will be seen written diagonally across the sharp end (The Angle end) of the probe in figures 2a and 2b as $\angle =$ *b/c* INV. TAN.

A study of the fig. 4 drawing explains the probe's prompts in detail.

A study of fig. 4 a shows how we can find the unknown length of side *a* by using, (via the probe's prompts), its known The Angle (shown in decimal degrees as 36.86989765), and the length of its side *b* (this being 3), by using the prompt $\angle\ b$, $a = b \div$ SIN $\angle$. We therefore enter the sequence 36.86989765, SIN (display shows 0.6), XM. We then enter the sequence 3 divided by RM (return memory), = 5, giving us the length of side *a*.

Fig. 4b explains how we can use the block of formulas displayed in the off-center block of the probe to find (in this case) side *a* of the problem triangle, when the length of this side is unknown. We can do this, by using the off-center prompt, which states that the length of *a* equals the square root ($\sqrt{}$) of, *b* squared, + *c* squared. We therefore enter into the calculator the sequence 3 × 3 = 9, followed by entering the sequence 4 × 4 = 16. We then add these two numbers together by entering the sequence 9 + 16 =, to obtain 25. We then find the square root of 25 by pressing the INV key, followed by pressing the SQUARE ROOT key, to give us 5, the length of side *a*.

Figures 4a and 4b therefore explain that there are two ways of finding the length of line *a* in a right-angle triangle.

Fig. 4a shows the method of finding the length of *a*, by utilizing the already-known The Angle, combined with using the known length of one of the triangle's other sides. Note that The Angle calculating prompts are shown facing outward on all three sides of the probe, and the length-of-side prompts are shown in the off-center block of basic prompts displayed within the probe.

The off-center-block prompts are used to find a triangle's unknown side lengths, by using a calculation involving solely the known side lengths (without in this case the student needing to know The Angle to complete this calculation). The position of these prompts is shown by arrow in fig. 4, and is explained fully in the following very detailed description of their use. The following details preferably should be repeated verbally in order to cement them firmly into the student's memory. Therefore, we must use, for example, the correctly chosen prompt from the off-center block to find the length of *a*, by now referring to fig. 4 b, where it shows the correct formula to use to obtain the length of side *a*.

This calculation to find the length of *a* is not as complicated as it first appears. It will be found easier to first make a sketch of a right-angled triangle with its tag in its correct position, on a sheet of A4 paper, with the (moveable) probe positioned at a similar orientation to the problem triangle; this is followed by introducing into its sides the identifying letters for side *a*, side *b*, and side *c*, followed by entering the known and unknown lengths of sides as *a* =?, *b* = 3, and *c* = 4.

A close study should then be made of the group of off-center prompts; select the one necessary for finding the length of line *a*, which states, the length of side *a* equals the square root ($\sqrt{}$) of *b* squared, plus *c* squared. This formula should be memorized, as it informs us that the length of line *a* can be found by the result of adding side *b* (multiplied by itself), to side *c* (multiplied by itself), followed by finding the square root of the resultant number by pressing in sequence the INV key followed by the square root key ($\sqrt{}$), to give us the answer 5, the calculated length of *a*.

In practice, therefore, to find the length of line *a* in a right-angled triangle, we need to multiply the length of line *b* by itself, by entering the sequence, 3 × 3 =, -to give us 9, followed by pressing the X M key (which transfers this figure into the calculator's memory). We then enter the sequence 4 × 4 =, giving us 16. We then enter the sequence + RM (which returns the memory), =, giving us 25, We then press the INV key, followed by the square root key ($\sqrt{}$), which then gives us the answer 5, the length of line *a* (also shown in the calculator's display) . This calculation sequence should be practiced several times on the calculator until it is fully understood, followed by entering the result in its correct (line *a* position) in the previously sketched triangle.

For practice purposes, we can now find the length of side *b* by using the known-length of sides *a* and *c*. We now multiply the length of side *a* by itself, i.e. (5 × 5 = 25), and subtract from this number, the result of the length of side *c* multiplied by itself, i.e. (4 × 4 = 16), by entering the sequence 25, – 16, =, to give us 9. We then press the INV key followed by the square root key ($\sqrt{}$), completing the calculation, to give us the answer 3. (This calculation should also be practiced several times until it is fully understood) and the result entered correctly into side *b* of the sketched triangle.)

Should we now wish to find the length of line *c* (and know the length of sides *a* and *b*), we multiply the length of line *a* by itself, by entering the sequence 5 × 5 =, giving us 25. We then subtract from this number the length of side *b*, multiplied by itself, by entering the sequence 3 × 3 = 9. We now enter the sequence 25, −, 9, =, giving us 16, followed by pressing the INV key then the square root key (√), to give us the correct answer 4. (This sequence should also be practiced several times until it is fully understood) and this result should also be entered into side *c* of the sketched triangle.)

For familiarization purposes, the author now includes more examples in the use of the probe when it is being used to solve a right angle triangle where *The Angle is known*). These examples are shown in greater detail to familiarize the student further with their complete method of use.

Referring to figures.1, 2, 2a, 2b, 4, 4a, we can now extend the scope of our calculations to include the angle prompts (found on the outside of the probe, facing outward). In fig. 4a, we use the angle prompt to find the length of *a*. We first study the problem triangle and select the known The Angle and the known length of side *c*. In this case, we choose to use the known length of side *c* (4), and the known The Angle of 36.86989765 degrees.(note that this angle is calculated in decimal degrees), and we are using the outside displayed prompt on the probe that shows ∠ *c*, (The Angle and *c*), followed by the sequence, *a* = *c* divided by COS, The Angle. The abbreviated prompt formula for this is, ∠*c*, *a* =, *c*/COS. ∠, which actually states that we already know The Angle and *c*, and the length of *a* equals the length of *c* divided by COS The Angle. (This formula is interpreted in the mind's eye as 'We already know the angle and *c*, therefore *a* equals *c divided* by the cosine of The Angle.')

We therefore need to calculate the result of 4, divided by the COS. of 36.86989765 degrees.

At this stage in the calculation, the student should now be aware that it is much easier to enter into the calculator the known angle (in decimal degrees) first, by entering 36.86989765, followed by using the COS. key, giving us 0.8 (in the display), followed by pressing the XM key (storing this number into the calculator's memory), followed by entering the key sequence, 4 ÷ by, RM (which returns the memory to the display) followed by the equals (=) sign, which finally gives us the answer 5, (shown in the display), the calculated length of side *a*).

We have now obtained this answer by using the known angle of 36.86989765 (in decimal degrees), and the known length of side *c* (4), by using the trigonometry formulas contained within the calculator, while also using the guidance of the probe and its prompts.

This calculation should be repeated several times in order to allow the student to become familiar with this method of triangle calculation.

While in use the probe can be used anywhere on the surface of the paper in the problem triangle being worked on. It can be rotated to any attitude' and can be turned over, allowing it to be used on its reverse (mirror-image) side, in order to calculate other sought-after problem angles. The student should aim to develop the ability to read off the prompts wherever they occur, (even if the actual written words on the probe are found to be upside down). (This skill can soon be learned through practice.)

The probe can be used to calculate both two-and three-dimensional triangles, (see also figures. 5, 6, and 19 as examples of this proof). Also note that if the problem triangle 'being worked on' is drawn containing just two dimensions, (i.e. flat on the surface of the paper), this situation, by convention, is said to be 'in the plane of the paper' or IPOP (for short).

Chapter 5

Three-dimensional triangle problems solved

The Probe and Prompt triangle calculation system can also be used to solve a three-dimensional triangle problem' that can quite often be found in work-pieces, sketches, or issued drawings.

When a triangle problem being worked on is discovered to possess all the three dimensions of length, width, and height, the possession of this additional depth dimension, will make it a three-dimensional trigonometry problem). In this case, and in order for the student to get a true feeling of the object's depth, it is considered best practice to sketch the problem triangle on a separate piece of A4 paper, with the actual problem figure located within a three dimensioned box that is angled toward the viewer by approximately 30 degrees, (as shown in fig. 19).

The perfectly square corners and straight sides of this box can then be used as reference datum lines that will allow their junction points to be visualized as being true 90-degree angles for the ease of further trigonometry calculations, (as shown in fig. 19).

The use of this angled and dimensioned box method of calculation, will provide us with the ability to scrutinize all of its three dimensions quite clearly, and will also open up to students a new area of previously unexplored three-dimensional calculation possibilities. The accurate use of the 90-degree tag to mentally and accurately position the probe relative to the worked-on problem in the box while this problem is a three-dimensional situation will then be fully realized, for its 90-degree tag corner will now be capable of identifying what is actually a true 90-degree corner in the problem triangle while it is not being accurately illustrated as a true 90-degree right angle.

When using this box method of triangle calculation, the student will be assisted in positively identifying the position of The Angle, and the triangle's sides *a*, *b*, and *c*, by identifying the position of the 90-degree tag to accurately place the triangle in the problem. Care should be taken, however, during the initial positioning of the tag, to ensure that it is correctly located at the (now-distorted) 90-degree corner within the problem triangle being worked on.

If we now refer to fig. 5, (which shows eight rotated right-angle triangles, with their 90-degree identifying tags), it will be seen (in this case) they have been drawn flat on the paper and therefore possess just two dimensions. These triangles are, by convention, said to be 'in the plane of the paper', or IPOP (for short). These triangular figures have been deliberately drawn rotated into a series of random attitudes in order to demonstrate clearly how it is still possible for the 90-degree tag to perform its full identification task while at any attitude, provided the tag is oriented accurately into its correct position relative to the sketch of the problem triangle being worked on. It therefore accurately identifies the position occupied by The Angle and its sides *a*, *b*, and *c*.

In fig. 6, the illustrations feature eight three-dimensional (obliquely viewed) and *tumbling*, triangular shapes. It will be seen that although each one actually possesses a right angle, they actually appear to the viewer to be distorted in shape, particularly at their tag corner. Because of this obliqueness, they at first do not appear to represent true right angles, due to them *not* being in the plane of the paper (IPOP). Therefore in these particular cases, it should be noted that the tag (which is actually a short extension of the side *b* of the probe and of the object being viewed), still possesses the ability to accurately identify the 90-degree corner of the right-angle triangles, and the true position of The Angle and its sides *a*, *b*, and *c*. As previously stated, these particular shapes are all right-angle triangles, but because they are viewed obliquely, it causes their side lengths and angles to appear visually untrue in both angle and scale.

In fig. 6-a, these eight tumbling right-angle triangles are positioned identically to those in fig. 6, but are drawn *without* their identifying 90-degree tags, and doing so makes it virtually impossible to recognize their true shapes; this reinforces the importance of introducing the 90-degree tag into all triangle calculation sketches.

Up to this point in the book, a right-angle triangle's longest side has always been considered to be its side *a*. However, after viewing the previously mentioned oblique views of right-angle triangles, (when displayed without their tags), it will now be realized that the sides of at least four of these particular triangles do not now appear to have side *a* as their longest side, due to them being viewed in this oblique manner. This fact should be noted down and remembered, particularly when one is studying 3D sketches or performing any of the triangle calculations featured later on in this book.

To reiterate:

Fig. 5 depicts eight rotating right-angle triangles, viewed in the plane of the paper (IPOP): and showing their identifying 90-degree tags.

Fig. 6 depicts eight tumbling right-angle triangles, drawn obliquely, but still having their identifying 90-degree corner tags positioned correctly. These distorted triangular figures show how important it is that the position of the 90-degree triangle's tag corner is correctly identified and positioned in the calculation, to allow its sides *a*, *b*, and *c* to be truly recognized, and to identify the position of The Angle in the sketched distorted figure.

Fig. 6a depicts eight tumbling right-angle triangles, but they are shown *without* their right-angle, 90-degree corner identifying tags (previously shown in fig. 6); these figures are also drawn obliquely. A study of these right-angle triangles, while being made without the 90-degree tag corner indication being used, actually demonstrates how important the tag is, in identifying the 90-degree corner of the problem effectively, to allow further accurate calculations to be made.

It should be noted that without this new 90-degree corner tag identification, the perceived shape of these figures renders them largely unrecognizable as true right-angle triangles and, as a result, can be the cause of potentially large errors being made in any resulting calculations. This distorted view of a potential problem must be spotted, noted, and taken into account before making any further calculations.

*   *   *

Fig. 6-b is a working size drawing of the Combination Probe. This drawing contains full instructions for its use in the calculation of both basic right-angle triangles and for the calculation of the so-called non-right-angle triangle.

Fig. 6c is an enlarged version of the Combination Probe shown in fig. 6b, and is to be used only for reference purposes. This drawing contains the working instructions and duplicates the cutting and folding instructions needed when using the cut-and-fold enlarged version shown in the ENLARGED FIG 37f located near the index, and specifically designed for practical use in the office or in the workshop environment.

The full details that describe the usefulness of this Combination Probe for the calculation of right-angle and non-right-angle triangle' problems, are shown in Chapter 14 figures, 37, 37a, 37 b, 37c, 37d, 37e, and in figures 37f, and 37g).

fig. 1.

THE OTHER ANGLE.

THE 3,4,5, TRIANGLE

5 a

b 3.

" THE ANGLE."

c
4.

90° TAG.

INDICATING THE 90°CORNER.

A TYPICAL 3·4·5 TRIANGLE , SHOWING DESIGNATED SIDES a,b,c, THE 90°TAG,
AND ' THE ANGLE ' (OR SHARP END), INDICATED. (ALSO SHOWN IS THE 'OTHER ANGLE'.)

BOTH SIDES OF 'PROBE' SHOWN.

fig. 2.

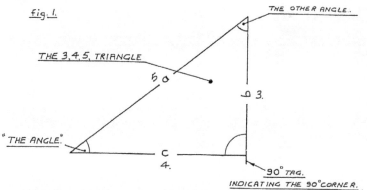

'THE ANGLE !
OR ∠= END OF PROBE.

'THE ANGLE,'
OR ∠=, END OF PROBE.

2a.   90° TAG.   2b.

THE DESIGN EVOLUTION OF THE 'PROBE' figs. 2d, 2e, 2f, 2g, 2h.

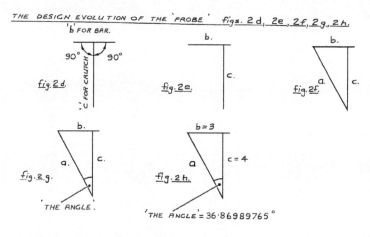

'b' FOR BAR.

90° 90°

fig. 2d.

b.

c.

fig. 2e.

b.

a. c.

fig. 2f.

b.

a. c.

fig. 2g.

' THE ANGLE '.

b = 3

a c = 4

fig. 2h.

'THE ANGLE' = 36·86989765 °

Fig 3.

fig 3.

MY ORIGINAL FORMULA, USED IN CALCULATIONS TO FIND THE VERTEX POINT, AND THE UNKNOWN ANGLE OF BEND ϕ, GIVEN X, Y, AND R ONLY. THIS IS NOW SUPERSEDED BY THE "PROBE AND PROMPT" SYSTEM, WHICH PROVIDES ALL SIX OF THE SOUGHT AFTER UNKNOWNS, REQUIRING MUCH LESS MENTAL DEXTERITY. SHOWN IN fig. 3a.

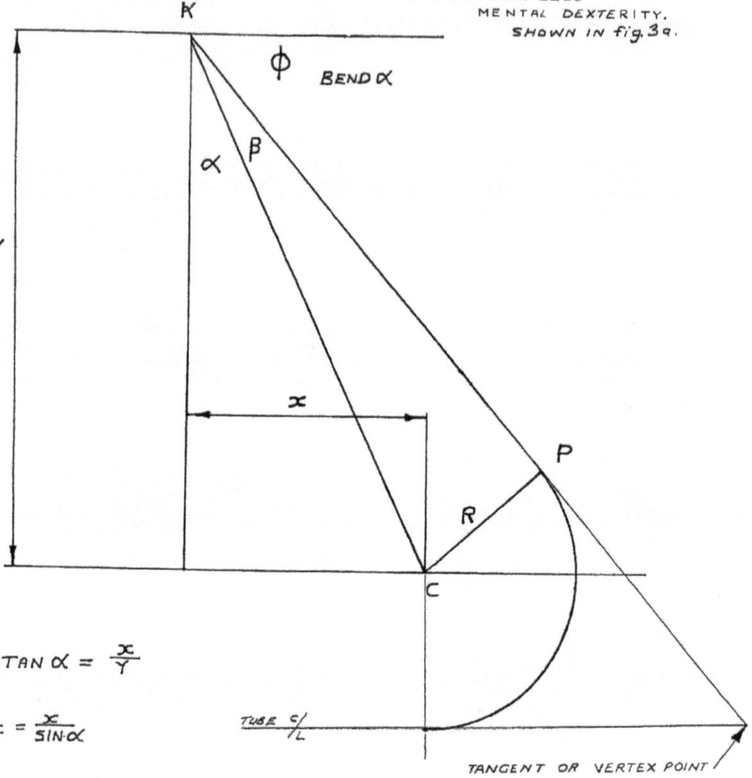

BEND α

$$\tan \alpha = \frac{x}{Y}$$

$$Kc = \frac{x}{\sin \alpha}$$

TUBE C/L

$$\sin \beta = \frac{R}{Kc} = \frac{R \sin \alpha}{x}$$

TANGENT OR VERTEX POINT

$$\therefore \phi = 90 - \left[\left(\sin^{T} \frac{R \sin (\tan^{T} \frac{x}{Y})}{x} \right) + \tan^{T} \frac{x}{Y} \right]$$

$$\alpha = \tan^{T} \frac{x}{Y}$$

$$\therefore \phi = 90 - \left[\left(\sin^{T} R \sin (\tan^{T} \frac{x}{Y}) + \tan^{T} \frac{x}{Y} \right) \right]$$

ONE SHOULD NOT BE TOO CONCERNED BY THE ABOVE COMPLICATED, BUT CONVENTIONAL FORMULA. THIS COMPLICATION HAS NOW BEEN SIMPLIFIED BY USING THE 'PROBE' IN fig. 3a/24 AND ELSEWHERE IN THE BOOK.

fig 3a.   <u>CALCULATIONS TO FIND 'ANGLE OF BEND' AND 'TANGENT POINT' (x DIMS.),</u>
(ALSO SHOWN   <u>ON SHEET METAL BRACKET fig 23.</u>
LATER AS
fig 24.)

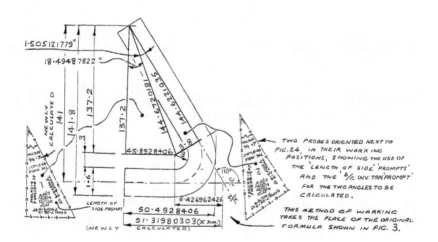

1·505121779°

18·4948 7822°

137·2

137·2

NEWLY CALCULATED D

141

141·8

3

144·6720181

144·6221035

45·8928406

1·6

2·3

2·3

110°

5·426962426

LENGTH OF SIDE PROMPT

50·4928406

51·31980303 (x DIM.)

(NEWLY CALCULATED)

TWO PROBES ORIENTED NEXT TO
FIG. 24, IN THEIR WORKING
POSITIONS, SHOWING THE USE OF
THE 'LENGTH OF SIDE' PROMPTS'
AND THE 'b/c INV. TAN.' PROMPT'
FOR THE TWO ANGLES TO BE
CALCULATED.

THIS METHOD OF WORKING
TAKES THE PLACE OF THE ORIGINAL
FORMULA SHOWN IN FIG. 3.

fig. 4. EXPLAINING THE 'PROBES PROMPTS' IN DETAIL.

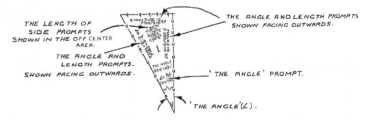

THE LENGTH OF
SIDE PROMPTS
SHOWN IN THE OFF CENTER
AREA.

THE ANGLE AND LENGTH PROMPTS
SHOWN FACING OUTWARDS.

THE ANGLE AND
LENGTH PROMPTS.
SHOWN FACING OUTWARDS.

'THE ANGLE' PROMPT.

'THE ANGLE' (∠).

THE LENGTH OF SIDE 'a' FOUND BY USING THE 'PROBES' PROMPTS.
TWO SIDES AND 'THE ANGLE'. ARE KNOWN IN THIS CASE.

fig. 4a

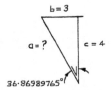

b = 3

a = ?   c = 4

36·86989765°

---

fig. 4b.   CHOOSING THE 'INSIDE PROMPT' FROM THE 'OFF CENTER' AREA TO FIND 'a'.

$$a = \sqrt{b^2 + c^2}$$

---

fig. 5 SHOWING EIGHT ROTATING RIGHT ANGLE TRIANGLES WITH IDENTIFYING 90° TAGS.

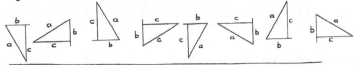

---

fig. 6. SHOWING EIGHT TUMBLING RIGHT ANGLE TRIANGLES WITH IDENTIFYING 90° TAGS.

IT WILL BE SEEN THAT THESE ARE DRAWN OBLIQUELY.

fig 6a. SHOWS   EIGHT TUMBLING RIGHT ANGLE TRIANGLES WITHOUT IDENTIFYING 90° TAGS;
OR DESIGNATED SIDES. THESE ARE SIMILAR TO fig.6 AND ALSO DRAWN OBLIQUELY.

CUTTING AND FOLDING INSTRUCTIONS BEFORE USING THE 'COMBINATION PROBE'.

FIG 6B.
1. USING SCISSORS, CUT OUT TO THE OUTSIDE DASHED LINE.
2. TO USE SIDES 1 OR 2 OF THE 'PROBE AND PROMPT', FOLD ALONG LINE 'A' AND FLATTEN, LEAVING THE TEXTS FACING OUTWARD. FOLD ALONG LINE 'Z', FLATTEN, LEAVING SIDES 1 AND 2 EXPOSED IN A POSSIBLE STANDING POSITION.
3. IF REQUIRED FOR THE CALCULATION OF 'WONKY' ANGLES, LINE 'Z' IS UNFOLDED TO EXPOSE THE FORMULAS ON SIDE ③, (LYING RELATIVELY FLAT FOR USE).

THE COMBINATION PROBE IS DESIGNED FOR USE ON BOTH RIGHT ANGLE AND NON-RIGHT ANGLED TRIANGLE CALCULATIONS. (THIS CAN INCLUDE ANY FIGURE MADE UP OF STRAIGHT LINES). SIDES 1 AND 2 ARE USED FOR RIGHT ANGLE TRIANGLE CALCULATIONS. ALL THREE SIDE LENGTHS ARE NEEDED TO BE KNOWN WHEN CALCULATING 'NON-RIGHT ANGLED TRIANGLES'. FOR THIS PURPOSE THE CALCULATION SEQUENCES SHOWN IN D, E, F AND G BELOW, ARE TO BE FOLLOWED. (THESE ARE SHOWN ON SIDE ③ OF THE 'COMBINATION PROBE'.)

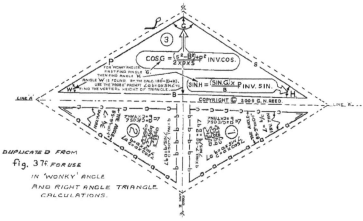

$$\cos G = \frac{(s^2 - B^2) + P^2}{2 \times P \times S} \quad \text{INV. COS.}$$

$$\sin H = \frac{(\sin G) \times P}{B} \quad \text{INV. SIN.}$$

COPYRIGHT © 2009 G.N.REED.

LINE A

DUPLICATED FROM
fig. 37f. FOR USE
IN 'WONKY' ANGLE
AND RIGHT ANGLE TRIANGLE
CALCULATIONS.

REFER TO TRIANGLE 3.

Ⓓ FIRST FIND ANGLE 'G' BY USING THE FORMULA $\cos G = \frac{(s^2 - B^2) + P^2}{2 \times P \times S}$ INV. COS.

Ⓔ THEN FIND ANGLE 'H' BY USING THE FORMULA $\sin H = \frac{(\sin G) \times P}{B}$ INV. SIN.

Ⓕ ON COMPLETION OF THE ABOVE TWO SEQUENCES, 'W' IS FOUND BY CALC. $180° - ('G' + 'H')$.

Ⓖ THE VERTICAL HEIGHT OF TRIANGLE 3 IS FOUND BY THE 'PROMPT' $\angle a$ $b = a \times \sin \angle$. (SEE SIDE 1).

SEE fig. AND TEXT. 37. a, b, c, d, e, f, FOR FULL DETAILS AND WORKED EXAMPLES, OF 'WONKY GABLED HOUSE', OR 'NON-RIGHT ANGLE TRIANGLE' CALCULATIONS.

NOTE. fig 6c SHOWS THIS DRAWING AT TWICE NORMAL WORKING SIZE.

AN ENLARGED VERSION OF THE 'COMBINATION PROBE', SHOWN IN fig. 6b.

THE CUTTING AND FOLDING-OPERATION REQUIRED BEFORE USING THE COMBINATION PROBE

1. USING SCISSORS, CUT OUT TO THE OUTSIDE DASHED LINE.

2. FOR 'PROBE AND PROMPT', USE, FOLD ALONG LINE 'A' AND FLATTEN, LEAVING THE TEXTS FACING OUTWARD.
FOLD ALONG LINE 'Z', AND FLATTEN, LEAVING SIDES 1 AND 2 SHOWING, (IT CAN BE LEFT IN A STANDING POSITION FOR IMMEDIATE USE)

3. IF REQUIRED FOR THE CALCULATION OF WONKY ANGLES, LINE 'Z', IS UNFOLDED TO EXPOSE THE FORMULAS ON SIDE 3,
AND USE D (IN THIS CASE), LYING FLAT.

LINE 'A'

ALSO SHOWN AT
NORMAL WORKING
SIZE IN fig 6 b.

REFER TO TRIANGLE 3.

(D) WE FIRST FIND ANGLE 'G' BY USING THE FORMULA
COS 'G' = $\frac{S^2 - B^2 - P^2}{2 \times P \times S}$ INV. COS.

(E) WE THEN FIND ANGLE 'H' BY USING THE FORMULA
SIN 'H' = $\frac{(SIN G) \times P}{B}$ INV. SIN

THIS 'COMBINATION PROBE', IS DESIGNED FOR USE IN BOTH RIGHT ANGLE AND NON-RIGHT ANGLE TRIANGLE CALCULATIONS.(THIS INCLUDES
ANY FIGURES CONSTRUCTED SOLELY OF STRAIGHT LINES. SIDES 1 AND 2 ARE USED FOR RIGHT ANGLE TRIANGLE CALCULATIONS. THE
LENGTH OF THREE SIDES ARE NEEDED FOR THE CALCULATION OF NON-RIGHT ANGLED TRIANGLES WHEN USED IN CALCULATION
SEQUENCES SHOWN IN D.E. AND F ABOVE.

W → FIND ANGLE

C
C
C
C
FOR 'WONKY' ANGLES'
FIRST FIND ANGLE 'G'.
THEN FIND ANGLE 'H'.
ANGLE 'V', IS FOUND BY THE CALC.180°-(G+H).
USE THE PROBE PROMPT $\angle a.b = a \times SIN \angle v$.
FIND THE VERTICAL HEIGHT OF TRIANGLE.

P.

INV. TAN
$\angle 7 = b/c$
89.9° TO 0.1°
THE ANGLE

(3)

$COS G = \left(\frac{S^2 - B^2 + P^2}{2 \times P \times S}\right)$ INV COS.

$SIN H = \frac{(SIN G) \times P}{B}$ INV. SIN

G

B

S

H.

COPYRIGHT © 2009 G.N REED

(G) THE VERTICAL HEIGHT OF TRIANGLE 3 IS FOUND
BY USING THE PROMPT $\angle a.b = a \times SIN \angle L$, USING
SIDE 1, OF THE COMBINATION PROBE.

(F) AFTER COMPLETING THE D AND E SEQUENCES, 'V' IS FOUND BY
THE CALC.180-(G+H)

FOLD
LINE 'Z'

THE ANGLE
89.9° TO 0.1°
$\angle 7 = b/c$
INV TAN

C

D D D D D D

b — b — b — b — b

B

b — b — b — b

D D D D D D

LINE 'A'

Fig 7. Fig 8, and Fig 9.

fig 7.   USING THE 'PROBE' AND 'PROMPT' TO CATCH A FLY.

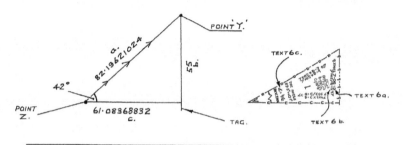

POINT 'Y.'

a. 82·19621024

55 b.

42°

POINT Z.

61·08368832
c.

TAG.

TEXT 6c.

TEXT 6a.

TEXT 6 b.

fig. 8.   THE CAPTURE SCENE.

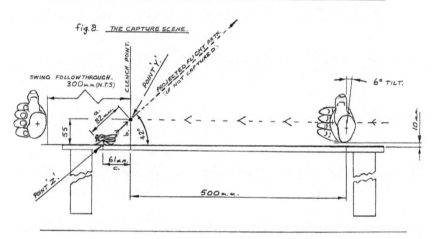

SWING FOLLOW THROUGH.
300 m.m. (N.T.S)

CLENCH POINT.

POINT 'Y.'

PROJECTED FLIGHT PATH.
(IF NOT CAPTURE D)

6° TILT.

10 m.m.

a. 82 m.m.

55

b.

42°

61 m.m.
c.

POINT 'Z.'

500 m.m.

fig. 9.   THE 'ANGLE OF PALM TILT'

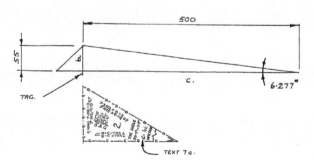

500

55

b.

c.

6·277°

TAG.

TEXT 7a.

Chapter 6

There follows the author's tongue-in-cheek calculation exercise, using practical trigonometry *to catch a fly*.

This trigonometry task is performed by using practical trigonometry, a scientific calculator, and side 1 (or 2), of the Probe and Prompt right-angle triangle calculation aid.

The following item has been written specifically to allow the student to become fully aware of and conversant with the methods and sequences used by the Probe and Prompt aid and informative sketches to obtain all the answers required to the following problems.

It is important that the student should now just relax and visualize the full situation as it unfolds in figures 7, 8, and 9. The following examples are considered a novel way of solving an age-old problem that has puzzled man for centuries.

It is not generally realized that the common housefly uses a rather devious tactic to avoid being killed or captured.

When disturbed from rest, it takes evasive action by jumping backward and upward away from the supposed threat. This is, no doubt, a reflex action developed to improve its survival strategy over the many millions of years of its development. This latent survival strategy must have been inherited through its countless offspring and passed on in its genes.

It is also suspected that the fly at rest suffers quite a major disadvantage in that its rearward vision is partially restricted by the masking effect of its own abdomen.

When one studies this alarm behavior, it would appear that it possesses a formidable set of 'spring-loaded' legs; these being capable of propelling it instantly backward and upward, to evade the attacking hand or paw of its predator.

In modern times, of course, this could well be a rapidly descending palm of the hand, or a fly- swatter in a clutching hand. This evasive behavior does explain, to some extent, why the fly so often escapes capture.

During a whole series, of development trials, extensive measurements were taken to identify and accurately establish the average angle of the flies'evasive backward jump. After exhaustive tests, this particular angle was narrowed down to an average of approximately 42 degrees, relative to its starting surface.

Now, by using this newly discovered angle information, it is now possible, with the aid of the probe and its prompts, to anticipate the fly's actual position in mid-air, during its initial lift-off.

To actually calculate this crucial point accurately, we need to consider the facts already known about the problem.

It is estimated, (after taking a series of accurate measurements), that the width dimension across the cupped hand of an average person is approximately 90 mm. This dimension also confirms that the actual tilted palm's centerline (with the thumb held vertically upright) must be 45 mm. If we now refer to fig. 7, and allow a working clearance of 10 mm. between the working tabletop and the inclined lower underside surface of the hand, then the cupped palm's center-line point will be 55 mm above the surface of the table during the forthcoming accelerated catching operation. It will also be seen in the sketch (fig. 7), that by using these known dimensions, it is now possible to calculate the distance the fly will actually move backward and upward from its starting point, (now called point Z), in order to reach a theoretical midair point, (now called point Y), that occurs during its 42-degree, (sloping) airborne trajectory, backward from point Z.

From our recently obtained figures, we can now establish that we now know The Angle to be 42 degrees, and the length of probe side *b* (the palm's center-line), is 55 mm.

If we now study the probe again with reference to fig. 7, it will confirm that the correct prompt to use for this calculation is, $\angle b$, $a =$, b/SIN $\angle$.

The figures for this are now entered into the calculator using the sequence, 42. SIN (display shows 0.669130606) X-M; we then enter 55 divided by R-M = giving us 82.19621024. This gives us the length of the triangle's side a (the length of the fly's anticipated flight path, from point Z to point Y). This is also shown in fig. 7.

To find the length of c (the theoretical tabletop distance that point Y is actually behind point Z), we use the prompt $\angle b$, $c =$, 90 minus $\angle$, TAN $\times$ b.

For this we enter the sequence $90 - 42 = (48)$, TAN (display shows 1.110612515) $\times$ by 55, =, giving us 61.08368832, the total length of c.

It is important at this stage to check that our figures are correct; it is therefore a very wise precaution, to choose a different prompt and calculation route for this checking calculation; We do this by using the new figures gained by using the previous prompt source and making a check for possible calculation errors.

For this task, it is best to use a completely different prompt sequence, by using the length of a different side, with the expectation of reaching the same dimensional result, to confirm that the previous result was correct. We know The Angle and c, therefore in this case we can use the prompt $\angle c$, $a =$, c / COS $\angle$.

We therefore enter the sequence 42, COS, (display shows 0.743144825) XM; we now enter 61.08368832, ÷ by RM, =, giving us 82.19621025, thereby confirming that the original dimension obtained for the length of a (the fly's flight path) was correct.

We must now concentrate on the details of the arm's swing, (shown in fig. 8), and the angle of 'palm tilt' shown in fig. 9.

Referring to fig. 9, the arm's swing is simulated in the sketch by the triangle's 500 mm, side c. The inclination of the hand from the vertical is simulated by the calculated angle of 6.27729849 degrees (shown in the sketch). This has been obtained by using the prompt $\angle = b/c$ INV TAN.

To obtain this angle, we enter the sequence 55, ÷, by 500, =, (display shows 0.011) INV TAN, giving us the displayed figure of 6.27729849 degrees, (answer).

This calculation proves that, to comply with the sketch, we must tilt the cupped hand backward during the swing, to an inclination of approximately 6 degrees.

To actually catch the fly in the hand (the whole point of the exercise), we will need to summarize all our usable working data as follows: With the cupped hand tilted backward to 6 degrees, positioned at a point 500 mm, behind point Y, and poised at 10 mm above the table's surface, we begin an accelerated swing, aiming for point Y, the estimated mid-air position of the fly (the actual point of contact), calculated to be 61 mm.behind, 55 mm above and 82 mm. in advance of the fly's anticipated 42 degrees angled flight path, (at the final contact point Y).

Of course, the table must be cleared of anything in the line of fire, such as sugar bowls, sauce bottles, milk bottles, candles, drink cartons, and any other table impedimenta, including breakable drinking glasses etc.

It will also be found advantageous, if a few grains of sugar (used as an attractant), are sprinkled on the table's surface at the flies' expected launch point' (now called point Z). However, one does have to wait until the fly is fully absorbed in sampling the sugar, and most importantly, it is facing away from the soon-approaching cupped and angled hand of the predator.

This sweeping cupped hand should be commenced at a point calculated to be 500 mm. behind point Y (see figures 8 and 9), and should be allowed to carry on in this movement throughout the swing, while still maintaining a 10-mm. clearance above the table surface, until contact with the fly is made (at the expected point Y) where a follow-through past this point will be needed to finish the swing.

During the swing, the cupped hand is held inclined backward with a slope of approximately 6 degrees. The arm is then swung, in table-tennis fashion, (with the cupped hand inclined at a similar angle to that used for a table-tennis forehand chop), during this phase of its acceleration, while progressing on through and past the intended point-of-contact position, (now called point Y). The position of this point is calculated to be 82 mm. behind the fly's start point.

At the instant point Y is reached, the hand is forcibly clenched, and a follow-through is carried onward in the extended swing for a further 300 mm to allow the hand and arm to slow down progressively.

Having fully adopted this procedure, and according to our trigonometry calculations, the center of the hand should now contain the fly. This can of course be released (if desired), by opening the clenched hand, or alternatively the hand's contents can be disposed of in a suitable receptacle, (should this have been provided for the purpose).

If the level of actual captures turns out to be less than expected, one should be made aware of certain variable factors that must be seriously considered, for they can have a negative effect on the expected results. These are listed as follows.

1. The non-cooperation of the fly
2. An unsuitable grade of sugar.
3. The wrong type of fly.
4. Air turbulence caused by previous unsuccessful attempts.
5. Flies' inability to jump correctly.
6. Flies' refusal to face in the desired direction.
7. Predator's inability to estimate point Y correctly.
8. Inadvertent contact with table-ware during the swing.
9. Non-belief of the (by-now) accumulated audience.
10. Premature snatching of the hand, during the final milliseconds of the approach swing.
11. Predator impatience.
12. Reduced hunger level of fly.
13. Excessively high room temperature, caused by the combined effect of body heat of the predator and the now-extended audience.

It will be discovered that with the acquired expertise, (gained over time), the capture rate will improve; that is, of course, until the table's surface is urgently required for the serving of tea.

Chapter 7

Explaining the method of calculation used when dealing with so-called shallow and steep angles, by using the examples shown in figures. 10 and 11.

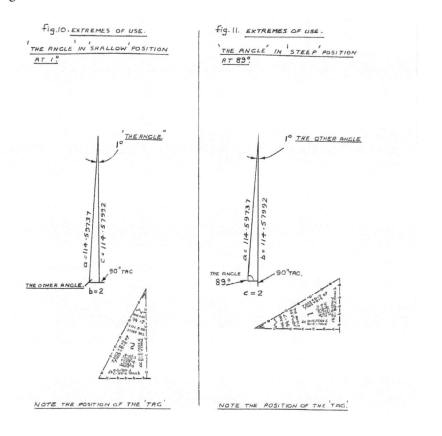

fig.10. EXTREMES OF USE.

THE ANGLE IN SHALLOW POSITION AT 1°

fig. 11. EXTREMES OF USE.

THE ANGLE IN STEEP POSITION AT 89°

THE ANGLE. 1°

1° THE OTHER ANGLE

a = 114·59737   c = 114·57992

a = 114·59737   b = 114·57992

90° TAG

THE OTHER ANGLE.   b = 2

THE ANGLE 89°   90° TAG.

c = 2

NOTE THE POSITION OF THE 'TAG'

NOTE THE POSITION OF THE 'TAG'.

51

The probe being used to calculate a right-angle triangle when The Angle is in its shallow angle position.

The Expandable Probe explained

The probe being used to calculate a right-angle triangle when it's The Angle is in its steep angle position.

It will be seen (for identification purposes and for the student's easier understanding), that a correctly oriented print of the probe is prominently displayed next to the individually calculated triangle examples in this book.

This orientation of the printed probe is deliberately placed with its tag in a similar attitude to the example triangle being worked on and calculated in each particular case.

However, the student should note that the tag and The Angle in the fig. 10 and fig. 11 example drawings, are being shown positioned at completely different orientations, (the amount of rotational difference between the two drawings can be mentally achieved by turning over the fig. 10 probe *in one's mind* (from its previously displayed probe side 2), followed by rotating it in the plane of the paper (IPOP) by 90 degrees anti clockwise) to allow it to arrive at its new position shown in fig. 11 (showing its probe side 1). This move has been demonstrated in order to show how the probe and its multitude of prompts can be used at widely differing orientations and angular magnitudes after going through the process of being mentally and practically turned over and rotated, in order to allow the triangle calculation to be completed at any desired orientation.

The calculated lengths of the triangle's sides are identical in both of the examples shown but the magnitude of their included angles will be seen (in this case) to have been completely changed, in order to arouse and cement the reader's full mental awareness.

These two examples illustrate that the so-called The Angle element of the problem triangle, can be calculated by using the probe as in fig. 10, where it is shown nearing its minimum calculation angle, of 1 degree, contrasting with that shown in fig. 11, where its The Angle is approaching its maximum calculation angle of 89 degrees.

In order to explore and calculate a typical so-called shallow angle (as shown in fig.10), it should be noted that in this case there is very little dimensional difference between the lengths of sides a and c, but their positions and actual lengths can now be positively identified by noting the position of the short extension of the triangle's side b (which forms the 90-degree tag), for this identification.

If we now compare this with fig. 11,(the steep-angled example), we can see that, similarly, line 'a' a is always longer than b (even if only by a very small amount).

This tag extension (of line b) will still enable the identification of the 90-degree corner in its correct position for further calculations to be carried out.

If we now refer back to fig. 10, we are given that side c = 114.57992 mm. and side b = 2.

The method used to calculate this triangle example is as follows:

With the probe placed on the paper with its The Angle end placed pointing in the direction of the shallow angle, and by using the 90-degree tag to confirm this orientation, The Angle (in decimal degrees), is found by using the prompt $\angle = b/c$ INV TAN.

To calculate this, we enter the sequence, 2 ÷ 114.57992 (display shows 0.0174551), INV. TAN, giving us the answer 1 degree.

We now know that The Angle is 1 degree, and should we wish to find the length of a, we again consult the probe and select the appropriate stated prompt for this, which is, $\angle c$, $a = c$ / COS. $\angle$. We therefore enter the sequence 1, COS. (display shows 0.999847695), X-M, followed by entering the sequence 114.57992 ÷ by R-M =, giving us 114.5973737, the length of a.

If we consider all three of the angles that make up the outline of the triangular probe's shape, we find that we have

1. The Angle,
2. the 90-degree tag corner, and
3. the Other Angle.

All three of these included angles, when added together, will total 180 degrees (this fact will be found very useful in later calculations and should be remembered).

The actual angle of the Other Angle is obtained by the calculation sequence, 90 minus The Angle = the Other Angle.

If we now refer to fig. 11, and wish to find its steep angle, we are given that side b is 114.57992 and side c, is 2. In this case, we orientate (or position) the probe on the paper, to point to The Angle, and read off the appropriate prompt, which in this case is, $\angle = b/c$ INV TAN.

We enter the sequence 114.57992, ÷, by 2, = 57.289962, followed by operating (in sequence), the two keys, INV and TAN. The display will now provide the answer, namely 89 degrees.

Now, knowing that The Angle is 89 degrees, and wishing to find the length of a, we use the prompt $\angle b$, $a = b$ / SIN. $\angle$. We therefore enter the sequence 89, SIN (display shows 0.999847695), X-M; we then enter the sequence 114.57992, ÷ RM =, giving us 114.5973737, the answer required.

You may notice that some of the calculations in this book are displayed using ten figures, while some are using just eight figures. This difference in length of these numerical displays is the result of using a different model of calculator for some of the calculations. The particular eight-figure calculator being used in this case possessed a smaller number of displayed figures. However, the end result is virtually identical and both can be accepted as being correct.

It is wise to point out at this juncture, that it is always the best course of action to use all of the available displayed figures in a calculation to ensure there is complete accuracy in the final result.

In other words, one must not, at an early stage in a calculation, round up or round down the figures, until the calculation is finally completed. Observing this important fact will ensure that the necessary accuracy is retained in the final result.

Your attention is now drawn to the note regarding the calculation of steep angles (explained in the text and shown in fig.11). This may indicate to the reader that there is a possible limit to the calculation magnitude of The Angle achievable, when one is approaching or exceeding 89 degrees of magnitude, and how this possible limiting problem can be resolved.

In practice, the extreme angles possible to be reached when approaching the probe's virtual angular calculation limit of 89. 9

degrees, will not in practice be found necessary to be used, because the probe can now be physically rotated on the paper, to allow The Angle end to point to the smaller, shallower angle's position in the problem triangle (as shown in fig.10). This rotational move of the probe allows the student to calculate accurately the shallower angle first. This new calculated angle is then entered into the calculator using the key sequence X M, followed by the sequence, 90 - R M =, giving us now the original problem angle's magnitude in decimal degrees for use in further calculations.

If we now refer to fig. 10, (which shows The Angle's magnitude to be 1 degree), and we now have the need to find the Other Angle, in this particular case, we use the calculation, 90 minus 1 = 89 degrees, to give us the answer.

It follows that when we refer to fig. 11, (which shows The Angle to be 89 degrees), then to find the Other Angle, we would use the calculation 90 degrees – 89 degrees = 1 degree.

This course of action can be taken because (as previously stated) a triangle contains a total of 180 included degrees of angle, of which 90 degrees are used up at the tag corner, leaving the remaining 90 degrees, to be shared between the two remaining angles, called respectively The Angle and the Other Angle. This simple calculation is used quite often in later calculation examples and should be remembered.

The fig. 12 and fig. 13 drawings of the Expandable Probe show the basic shape of the 'probe in outline, being superimposed by a series of (dashed) triangular figures designed to simulate the extent (in one's mind's eye) of the possible expansion or contraction of 'The Angle' during a calculation; this then enables the probe's' shape to simulate the problem triangle being worked on.

These particular drawings will be found slightly easier to interpret if the dashed lines at *a* in fig, 12, can be imagined (in one's mind's eye), as a side view of the open lid of a grand piano, with its hinge at its The Angle position and its support prop being held vertical in its *b* position on the piano, with its lid not allowed to exceed 89.9 degrees relative to the actual piano's flat top.

It should also be noted that the relative position of the dashed lines at *b* and the full line at *c* are always forming a 90-degree corner at their tag junction. The dashed line of *b* and the full line of *c* will also need to vary in length in order to simulate the shape of the calculated triangle.

The use of the two sketches at fig. 12 and fig. 13, and the use of our imaginative powers, will allow the probe's shape (in one's mind's eye) to cope with all the possible angular magnitudes likely to be encountered when dealing with either the very shallow angles of, say, 0.1 degree, up to the very steep maximum of, say, 89.9 degrees of magnitude.

One must therefore be mentally prepared to allow the physical shape of the actual probe to be distorted in one's mind's eye, to allow its outside shape to simulate the problem triangle being worked on. We do this to allow the probe to deal with the wide range of angle possibilities encountered while calculating worked-on problem triangles. 'The problem angle' encountered may, of course, be much steeper or shallower than the fixed triangular shape of the actual probe. Therefore by simulating the possible change in its angular magnitude, as shown in the fig. 12 and fig. 13 sketched examples, we are indicating (by the use of dashed lines), the theoretical possibilities imaginable while one is using the-called Expandable Probe.

Fig. 12 shows the Expandable Probe being used horizontally, in this case for the triangle calculations being made in the horizontal plane of the paper.

Fig. 13 shows the Expandable Probe being used in a vertical plane for the triangle calculations being made in the vertical plane of the paper.

When placed at any chosen orientation in the problem triangle, it will be found that the probe's tag still provides the ability for the student to identify the 90-degree corner of the problem triangle; it also identifies the position of The Angle and each of the triangle's sides correctly. The prompts will also be found to perform their dedicated function at any orientation.

Fig. 14

The method used to convert degrees and minutes into decimal degrees, by using the 'Probe and Prompt'. Also included is a separate calculation of a so-called shallow angle.

If we now refer to fig. 14, with the requirement to find the magnitude of The Angle depicted, we will see (from studying the probe) that the prompt to use in this case will be, $\angle = b/c$ INV TAN; we therefore input into the calculator the sequence 3 divided by 300 (3/100) as shown;.

We enter the sequence, $3 \div 100 =$ (display shows 0.03), followed by INV. TAN; this gives us 1.718358002, degrees (answer). (Note that the angle required is now in the preferred decimal degree notation.).

Alternatively, if we require the answer to be in degrees minutes and seconds (i.e. sexagesimal notation), we will need to select the calculator keys: INV and DMS (note that this method of conversion only applies if these particular keys are available on your calculator's keyboard); if available, they will give us the answer 1 degree 43 minutes and 6 seconds.

SHOWING THE SCOPE OF THE 'EXPANDABLE PROBE', FROM
'SHALLOW' TO 'STEEP' ANGLES, IN HORIZONTAL AND VERTICAL SITUATIONS.

fig.12.

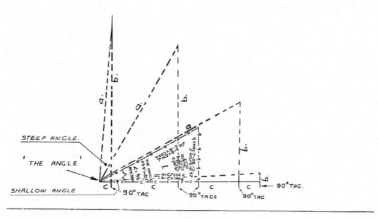

STEEP ANGLE.

'THE ANGLE'

SHALLOW ANGLE

90° TAG

90° TAGS

90° TAG

90° TAG.

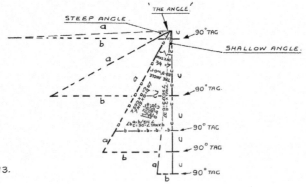

THE ANGLE

STEEP ANGLE.

SHALLOW ANGLE.

90° TAG

90° TAG.

90° TAG

90° TAG

90° TAG

fig.13.

Fig 14. Fig 15, and Fig 16.

CONVERTING DEGREES AND MINUTES INTO DECIMAL DEGREES, USING 'PROBE' AND PROMPT.

fig. 14.

IN THIS EXAMPLE SIDES 'b' AND 'c' ARE KNOWN.

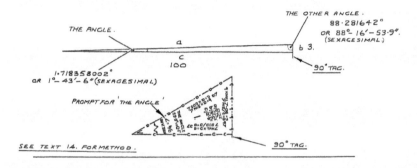

THE OTHER ANGLE.
88·281642°
OR 88° 16'–53·9".
(SEXAGESIMAL.)

THE ANGLE.

a

b 3.

c
100

90° TAG.

1·718358002°
OR 1°– 43'– 6"(SEXAGESIMAL)

PROMPT FOR 'THE ANGLE'

SEE TEXT 14. FOR METHOD.

90° TAG.

FINDING THE LENGTH OF SIDE AND THE UNKNOWN ANGLE
IN A RIGHT ANGLED TRIANGLE.

fig. 15.

? SIGNIFIES UNKNOWN SIDE AND ANGLE:

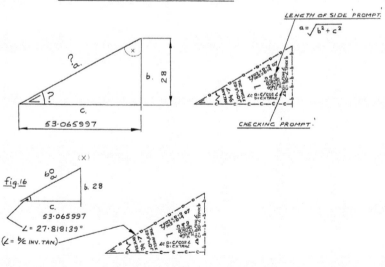

LENGTH OF SIDE PROMPT.
$a = \sqrt{b^2 + c^2}$

? a.

b 28

?

c.
53·065997

CHECKING PROMPT.

(X)

fig. 16.

60°

b. 28

c.
53·065997

∠ = 27·818139°

(∠ = b/c INV. TAN.)

SEE TEXT 16 FOR METHOD.

fig 17.   THE PYRAMID, PRIOR TO THE 'STEPPING STONE' APPROACH.

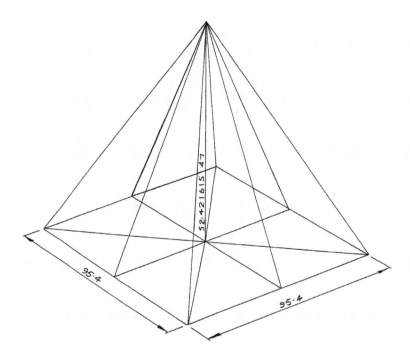

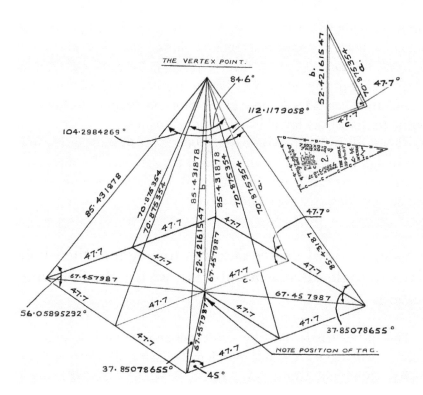

fig17a. THE PYRAMID AFTER THE 'STEPPING STONE' APPROACH.
USING FIGURES FROM fig.17.

THE VERTEX POINT.

Chapter 8

Stating the main advantages gained by the student who adopts decimal notation in place of sexagesimal notation during triangle calculations.

It will be realized by now that the prompt $\angle = b/c$ INV TAN, is the most used prompt at the very beginning of a triangle calculation, particularly when sides b and c of the problem triangle are actually known, (or have been found by previous calculations).

This is also the first and easiest prompt to memorize, because of its design being based on the original capital letter T concept (as explained in the earlier text in chapter 3). It is also the most widely used during normal triangle calculations.

By using decimal degree notation throughout the whole of the triangle calculation process, we use the following method.

The author has found (through experience), that it is so much easier to calculate a triangle problem by using decimal degrees, (decimal notation) throughout, and to leave all calculated angle dimensions in this notation on all original drawings, notes, and rough sketches.

It is therefore considered preferable to leave this notation in place, unaltered, and the reader should not be tempted to convert these decimal figures back into degrees, minutes, and seconds (sexagesimal notation), too early in the calculation, for the following reasons:

1. If one needs to refer to previous calculations, they are much easier to use if retained in their original decimal notation. This avoids the extra work, and the necessity of converting angular dimensions of degrees, minutes, and seconds (sexagesimal notation), back into the required decimal degrees to enable immediate remedial calculations to take place.

2. It is only when the calculated work-piece is fully completed, and only then, if it is specifically required for a new drawing (or at a management request), that there is any need to convert the decimal calculation of angles back to degrees, minutes, and seconds (sexagesimal notation). The exception to this rule is when the degrees, minutes, and seconds notation is required by the drawing office for the general release of drawings to either the production machine shop, tool-room, or for dispatch to other companies.

In the event that your calculator cannot convert decimal degrees into degrees, minutes, and seconds with its existing keys, this can be achieved by using the following method;

Divide 1 degree by 60, giving us the figure of 0.016666666 (of a degree); we then put this into the calculator's memory by using the key X M.; now, (by using fig. 14 as a conversion, example), we enter 0.281642 divided by RM =, 16.89852 minutes. If we require seconds, we now subtract the 16 minutes from this figure (-16), = 0.89852; this figure is then divided by RM = giving us 53.9112 seconds, (we can finally round this up to 54 seconds if necessary). The full answer will then be 88 degrees, 16 minutes, and 54 seconds.

Now, with the retained figure 0.016666666 still in the calculator's memory, and it is required to convert 88 degrees, 16 minutes, 54 seconds into decimal degrees, we can enter the sequence 16 divided by RM = 960, enter +, 53.9 =, giving us 1013.9; we then divide this by 60, giving us 16.89833333. We then enter the sequence −, 16, =, giving us 0.898333333; this is then divided by RM, =, giving us 53.9 seconds. We can now put these figures together as follows: 88 degrees 16 minutes 53.9 seconds, thereby giving us the complete answer in so-called sexagesimal notation.

The correct method to use to find an unknown length of side or an unknown angle in a right-angled triangle

The following notes describe the method used to find the length of an unknown side (in this case a), and the unknown angle ($\angle$), found in the example right-angle triangle shown in fig. 15.

Referring to figures 15 and 16, the following is considered to be the best method to use in practice:

1. It is always good practice to draw a sketch first of the problem triangle on a separate piece of A4 paper, as shown in fig. 16 (this sketch need not be to scale).
2. We then offer the oriented probe to the problem.
3. We then ask ourselves the question;'What facts do we already know about the problem triangle'?
4. Answer: We already know that side $b = 28$ and that side $c = 53.065997$; we can now positively identify these two sides of the problem triangle, by giving them the individual notation letters b and c. (The other,or third, side of the triangle is of course identified with the letter a.)
5. Question: What facts do we wish to find out about the problem triangle?
6. Answer: We need to know The Angle (in decimal degrees) and the length of side a in the triangle (in this case and for practice purposes, these are shown in the sketch with question marks).
7. We now refer to the probe, and read off the appropriate length-of-side prompt (shown in the grouped off-center area of the probe), we now need to find the length of side a, and select the prompt which states; a = the square root of the sum of b squared, plus the sum of c squared. For this' we enter and calculate the sequence $28 \times 28 = 784.$, XM (storing this figure in the calculator's memory) we then enter the sequence, $53.065997 \times 53.065997 = 2816$; this figure is then added to the stored memory figure by pressing the key sequence, +, RM., =, giving us 3600.

We then press the INV key followed by the SQUARE ROOT key ($\sqrt{}$)' thereby obtaining the answer 60; confirming that the length of side a equals 60.

To find The Angle, we use the correct probe prompt sequence, $\angle = b/c$ INV TAN. (This, when spoken (under one's breath), will be 'The angle' equals b divided by c INV.TAN'; this will in fact be 28 ÷ 53.065997 followed by the sequence INV TAN.)

We therefore enter the sequence 28 ÷ 53.065997 =, giving us 0.5276449; we then press (in turn) the INV key followed by the TAN key, obtaining the answer 27.818139 degrees.

We have now reached the situation in the calculation where we now know The Angle, and we already know the length of side b; this situation now makes it possible to *recheck* the length of side a for practice purposes, by using theprobe prompt sequence $\angle b, a = b \div$ SIN $\angle$.

We do this by entering the sequence 27.818139, SIN., (display shows (0.466666662) X-M; we then enter the sequence 28 ÷ RM =, confirming that the answer is 60.

We can now move further on into the possibilities of these calculations, (again for practice purposes), to allow us to find the Other Angle (now marked with an X). To do this, we must of course remember that the total number of included angles contained within a right-angle triangle must equal 180 degrees, and that the square tag corner of the triangle uses up 90 degrees of this total, therefore the other two angles must share the remaining 90 degrees. We therefore enter into the calculator, 90 − 27.818139 = giving us 62.18186087 degrees, (the angle contained within the Other Angle now marked X).

It is now possible, having reached this stage in these practice calculations, to gradually increase our mental ability, by practicing the art of transposing (that is, to change the order of), the b and c lettering that is presently given to the triangle's sides, in order to check the angle of the Other Angle now marked X.

By transposing these individual letters mentally (in one's mind), it will be found there will no longer be the need (in this particular case), to draw another right-angle triangle for this calculation, but to go through the process of mentally imagining that the existing side c of the triangle presently being calculated, has now been changed to side *b*, and the existing side *b* of our triangle has now been changed to side *c* in our newly imagined triangle. The 90 degree tag (this being a short extension of side *b*) will also have been re-positioned into its new imagined place; this will then give us the opportunity of using the prompt 'Angle = b / *c* INV TAN'. We now therefore enter into the calculator the sequence 53.065997 ÷ 28 = (display shows 1.895214179), INV. TAN. giving us 62.18186087 degrees, thereby proving that our previously made calculation was correct.

An alternative and correct way of finding the angle X is to use the number 2 face of the probe, oriented so that it is pointing at angle X, followed by then using the appropriate prompt, 'Angle = *b* ÷ *c*, INV TAN'. We therefore enter the sequence (53.065997 ÷ by 28 = 1.895214179, followed by using the keys INV. TAN = to give us 62.18186087 degrees.

We have now proved that, having found The Angle and the length of one of the triangle's unknown sides, it is now possible to calculate the length of the triangle's other sides and its unknown angle X. This has been done by the process of mentally transposing the orientation and lettering of the probe previously being used to obtain the answer, then by using the probe and its prompts at its new correct orientation to confirm that the angle of X is correct.

[(Note that the word 'transpose' means to transfer; to change the order of; or to interchange.)]

Chapter 9

How one can use the assistance of the stepping stone approach method during trigonometry calculations

The stepping stone approach method is really a process of making a very methodical approach to the problem,; this is explained in detail by studying, firstly, fig. 17, followed by studying the fig. 17a example, which is used to obtain more dimensions and angles.

Fig. 17 shows a basic three-dimensional drawing of a square-based pyramid. (This drawing is based on a very much scaled-down version of the Great Pyramid found at Giza, in Cairo, Egypt). The approximate scale used, (in order to conform to the size of the available book's page), is 1:-2,414, (meaning that 1 mm of the sketch is equivalent to approximately 2,414 mm, (2.414 meters of the pyramid) on the ground.

These drawings show how, by being in possession of only the minimum of dimensions, it is still possible to obtain all the other required dimensions, by using the Probe and Prompt system combined with trigonometry calculations to complete a fully dimensioned sketch. (This explains, in this basic example, how dimensions not shown on a drawing can be found, and how problems can be solved on the shop floor by using this method.)

At the beginning of this series of calculations, the sketch shows just the basic known dimensional situation that exists prior to the stepping stone approach method being used.

It will be seen that it is now possible by using the probe to obtain all the other dimensions needed to complete the fully dimensioned sketch, as shown in fig.17a. This confirms how it is now possible to calculate all the pyramid's unknown dimensions by using the aid of the probe and its prompts, provided these three basic dimensions are actually known prior to starting the calculation.

This shows the completed dimensioned sketch, after complying with the appropriate displayed prompts, and by using the aid of the stepping stone approach method, a scientific calculator, and using practical triangulation calculations.

When one is working on a practical trigonometry problem, one often finds that, initially, there are insufficient dimensions with which to plan its calculation. The student is therefore forced to make every effort to search out (to glean) every scrap of information possible to be extracted from the (often-sparse) drawing dimensions provided.

These minimal (supplied) dimensions are then used to calculate many more helpful dimensions. These new accumulated dimensions are then in turn used to obtain the final sought-after critical angle or dimension required.

For the first stage in this calculation, it will be found necessary to calculate the diagonal length (across the corners) of the pyramid's square base. We do this by mentally dividing the base into four equal square areas; we then need to calculate the diagonal length across one of these squares.

The stated base dimensions of 95.4 × 95.4 confirm to us that this particular figure must in fact be square; we can therefore deduce from this that the diagonal line relative to the base's center line must have an inclination of 45 degrees. Having now concluded that The Angle is 45 degrees, and the length of side b must therefore be 95.4 divided by 2 = 47.7, we must now find the length of side a; we therefore orient the probe on the paper to point toward the base's center point, and select the prompt $\angle b$, $a =$, $b/\text{SIN} \angle$.

We therefore enter the sequence 45, SIN, (display shows 0.707106781) XM; we then enter the sequence, 47.7, ÷ by RM, =, giving us 67.45798693, the diagonal length of a. We now multiply this by 2, giving us 134.9159739, (this now gives us the total diagonal length, across corners, of the pyramid's base).

To find the pyramid's angle of slope relative to its base, we use half it's 'across the flat' dimension of 95.4, (this being 47.7), as line c of the probe, and its vertical height of 52.42161547 mm to form its line b.

It will be found much easier to sketch the problem triangle in isolation as shown in fig. 17a.

We now draw this isolated sketch, oriented at a similar attitude, with the probe oriented similarly (that is, with its tag and The Angle matching the sketch).

We then consult the probe and choose the appropriate prompt, 'Angle =, b/c, INV TAN'.

We enter the sequence 52.42161547 ÷ by 47.7 = (display shows 1.098985649) INV. TAN, giving us 47.69999997, this being finally rounded up to 47.7 degrees). This gives us the angle of slope (in degrees) relative to the pyramid's base when viewed in the 'across the flat' attitude.

The student is now advised to practice this new technique, by calculating all the angles and all the lengths of sides found to exist in the fig, 17a completed drawing, and by comparing the results obtained to those given in the drawing.

To explain fully how the Probe and Prompt system is used with the assistance of the stepping stone approach method, it will be found that only simple words and phrases are used to explain how a full sequence of calculations can be completed, eventually leading to the required (and much sought after), dimensions being obtained. The following practical example explains the basic method used in the process.

If one is attempting to cross a country stream by walking on a series of randomly spaced stepping stones to reach the other side, it is often the case that one is forced to deviate from a normal straight path, and to make several sideways steps in a zigzag fashion to the right or to the left of the intended path, before finally negotiating the stream. A similar situation does arise when attempting to solve (by calculation), those awkward and seemingly insurmountable trigonometry problems when using the the stepping stone approach method.

At first sight, this would seem to be an unnecessarily tortuous route to take in a calculation, but, by making these extra calculations, combined with the production of a large freehand sketch of the problem, it is possible to build up an accurate picture, not only on the paper, but also in one's mind's eye that gives the full extent of the problem being worked on. Vital experience can be gained during the course of making this sketch and making the extra effort required in calculating all of the new dimensions, but it now makes it possible, by careful attention to detail, to form a plan of action, that allows one to plan the extra moves or steps needed to resolve the situation.

It is very important that any sketches made should be drawn to a large enough scale; they must be of sufficient size to allow all the extra dimensions subsequently found, to be included where they occur and in their exact relative positions, (this accurate positioning of all the dimensions is very important).

When this dimensioned sketch is completed, it should now be carefully studied, and should now reveal, (as if by magic), the sought-after unknown angle or dimension originally required, or at least point out a calculation route needed to be followed to obtain it. By so doing, it puts one in a very good position to resolve the problem. One should not, at this stage, be tempted to shorten or round up the figures of any of the calculated dimensions or angles found, until the final usable figures have been obtained. It should also be noted that the full number of digits shown in the calculator's display should all be used. Failing to use them all will allow errors to creep into the calculation, (called an 'accumulated error)', as failure to carry out this practice could affect the accuracy of the final result.

It is quite possible, after making a careful study of the completed sketch, one will realize that more space should have been made available in the sketch to allow all of the calculated figures to be entered where they are found, without any need to overlap adjacent figures. It is therefore very important that the sketch is drawn to a sufficient size, so that it is capable of accommodating all those extra details in their correct situations. In practice, any cramping of the figures makes them very difficult to decipher accurately when referred to at a later stage.

Having obtained the final sought-after angle or dimension by using this method, we should now allow ourselves to experience a modicum of elation at the success of this achievement. After a long calculation, this experience of elation can be compared to the achievement of fitting in the final piece of a large and difficult-to-solve jigsaw.

Chapter 10

Problems likely to be found when studying issued engineering drawings and sketches

It is sometimes found (during the general engineering work-shop floor experience), that issued engineering drawings, when being studied by the shop floor personnel, are found to require either some extra clarification or the need for additional calculations to be carried out before they can be used confidently for the production of the component. These extra calculations tend to fall into the following categories:

1. To establish the exact length of material used up in the bend of a component, particularly in respect to the bends required in sheet metal, steel bar, or tube manipulation. This additional information is often desperately needed by the shop floor engineer to enable the material to be cut off to an exact cut-off length prior to bending. There is also the need to calculate the exact angle of bend, to provide the bending machine's setting-up datum points, to allow the work-piece to be produced on the bending machine with absolute accuracy, to comply with the drawing's strict dimensional requirements.

2. To obtain an easily usable and machine-compatible positioning of the drill, by converting the 'drawing given' angular spaced 'pitch circle diameter' dimensions (i.e. rotary pitch circle dimensions) into the more suitable co-ordinate dimensions required when the component is being accurately drilled, relying on the machine table's graduated scales as fitted to the

precision jig borer or the milling machine. These necessary calculations are carried out in order to convert the angular dimensions stated on the drawing, into linear dimensions that are more suitable for the index scales of precision machines, by using trigonometry calculations. These involve a modification to the given 'pitch circle' dimension (usually stated on issued drawings), into the more acceptable, linear, two-dimensional dimensions, these being more suitable for interpretation by the machine tool's longitudinal and transverse measuring scales that are normally fitted to the machine's transverse and longitudinal lead screws.

3. To convert the drawing given *taper angles* (for use mainly on center lathes and milling machines), from the traditionally given imperial-dimensioned 'taper per foot' measurement (normally shown on issued drawings), into the more usable (and desirable) linear dimensions of *x* and *y* co-ordinates calibrated into the preferred metric decimal notation. This then allows the machine operator to use the existing digital machine readout settings, to accurately set the angle to be machined on the work-piece, and also allows the use of the longitudinal and transverse index scale readouts (in conjunction with the actual known length of the work-piece) by utilizing its overall length dimension and calculations, to complete the required angle.

4. Converting the 'drawing stated' angular dimensions, into more accurate and user-friendly-sine bar' angular settings, (thereby allowing the use of a calculated and precise 'slip gauge'pile to allow the work setting of the angular magnitudes required, while using a sine bar set-up on the machine's tooling, or on the actual machined work-piece.

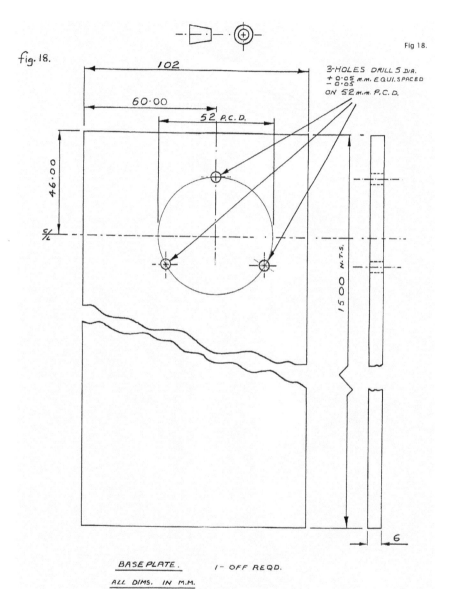

Fig 18.

fig. 18.

3-HOLES DRILL 5 DIA.
+0.05 m.m. EQUI. SPACED
-0.05
ON 52 m.m. P.C.D.

102

60.00

52 P.C.D.

46.00

C/L

1500 N.T.S.

6

BASE PLATE.   1- OFF REQD.

ALL DIMS. IN M.M.

MTL. 6mm. MILD STEEL PLATE.
TOL. ON HOLE CRS. TO BE +0.05
                        -0.05
OF POSITIONS SHOWN.

SCALE	1:1
DRN.	G.N. REED
DATE	10-1-14

A TYPICAL DRAWING SHOWING THE BARE MINIMUM OF INFORMATION SUPPLIED.
SEE figs. 18 a. b. c. d. e. AND f. FOR CALCULATIONS REQUIRED TO MAKE COMPONENT.

fig.18a.   THE EXTRA CALCULATIONS REQUIRED TO DRILL 3 – 5 m.m. HOLES, EQUI-SPACED
AND 18b.   ON A 52 m.m. P.C.D. WHEN A ROTARY TABLE IS NOT AVAILABLE OR THE WORKPIECE IS TOO
LARGE. IT IS CALLED 'ORDINATE' DRILLING, AND REQUIRES VERY ACCURATE MARKING OUT
FROM BOTH DATUM FACES.

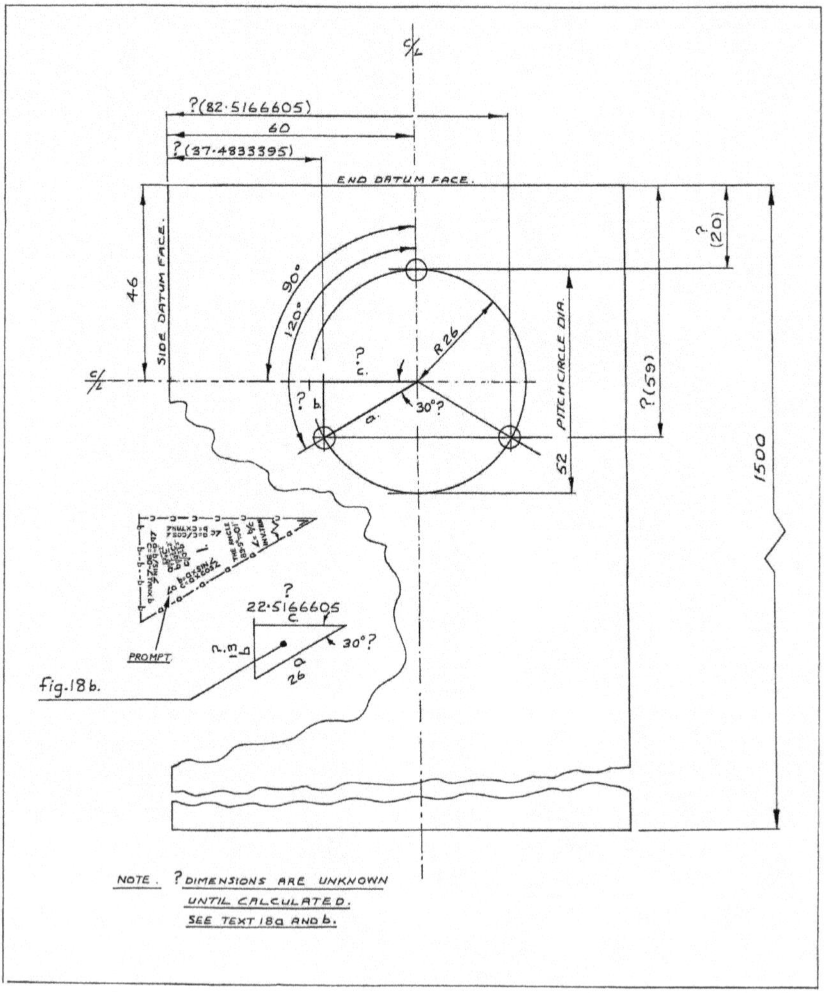

?(82·5166605)
60
?(37·4833395)

END DATUM FACE.

SIDE DATUM FACE.

46

C/L

90°
120°
?
c
R 26
? b
a
30°?

PITCH CIRCLE DIR.

52

?(59)

?(20)

1500

fig.18b.

PROMPT

?
22·5166605
c
30°?
b
a
26

NOTE.   ? DIMENSIONS ARE UNKNOWN
UNTIL CALCULATED.
SEE TEXT 18a AND b.

<u>fig 18c.</u> | TEN TIMES ENLARGEMENT OF fig. 18b, SHOWING HOW A SCALE METHOD CAN BE USED TO OBTAIN DIMENSIONS, WITHOUT USING TRIGONOMETRY.

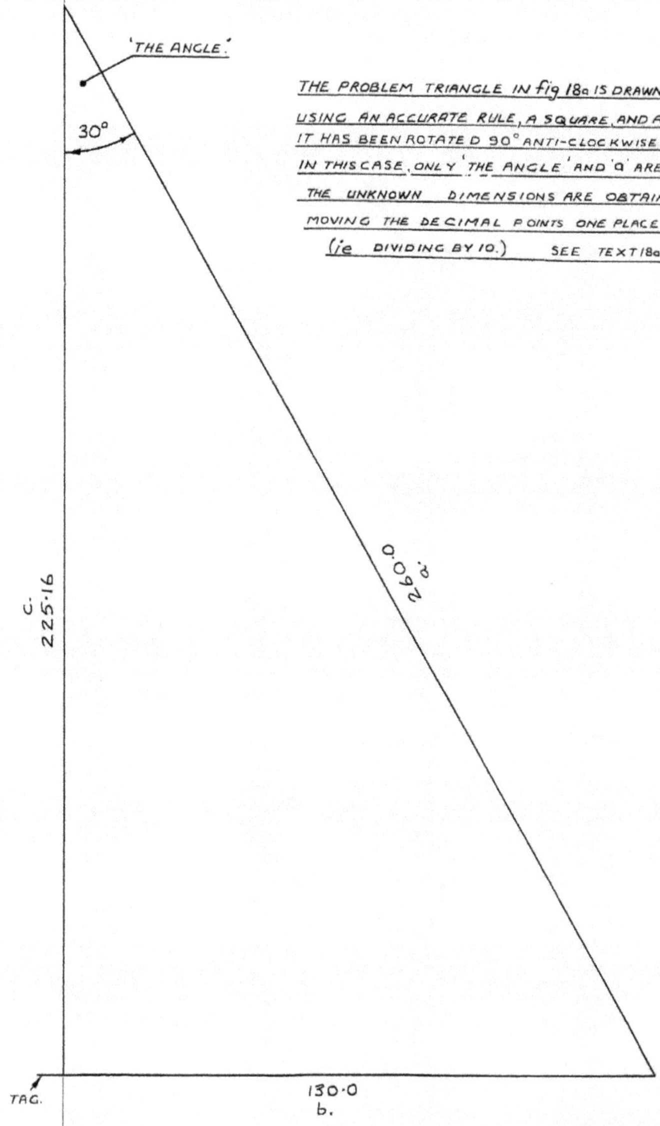

'THE ANGLE.'

30°

THE PROBLEM TRIANGLE IN fig 18a IS DRAWN 10X LARGER BY
USING AN ACCURATE RULE, A SQUARE, AND A PROTRACTOR.
IT HAS BEEN ROTATED 90° ANTI-CLOCKWISE (TO FIT A4 PAPER)
IN THIS CASE, ONLY 'THE ANGLE' AND 'a' ARE KNOWN.
THE UNKNOWN DIMENSIONS ARE OBTAINED BY
MOVING THE DECIMAL POINTS ONE PLACE TO THE LEFT.
    (ie DIVIDING BY 10.)   SEE TEXT 18a AND b. FOR DETAILS.

c.
225·16

a.
260·0

130·0
b.

TAG.

DIMENSIONED SKETCH FOR THE INSPECTION OF HOLES IN THE METAL PLATE fig.18, showing
THE METHOD USED TO OBTAIN KNIFE EDGED VERNIER CHECKING DIMENSIONS.
FOR INSPECTION DEPARTMENTS CHECKING PURPOSES.

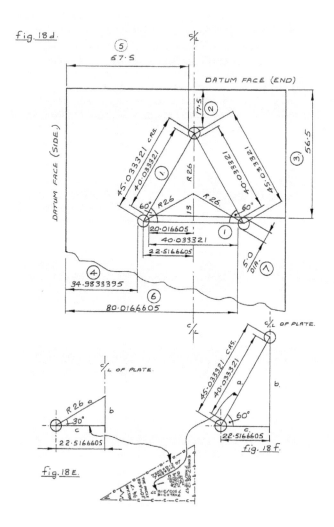

fig.18d.

fig.18E.

fig.18f.

Referring to the fig. 18a drawing of a metal base-plate, and fig. 18a and 18b, which contain the extra calculations required to convert the drawing's pitch circle angular dimensions into linear dimensions, when a rotary table is not available, also featuring fig.18c, which depicts a ten-times enlargement of the 18b problem previously solved by using practical trigonometry.

The practical scaling of engineering drawings

The fig. 18c drawing shows how one can scale the known dimensions to exactly10× full size, and use this method to obtain the required accurate dimension without the need for the student to either know or use trigonometry to obtain the answer.

Before going on to fully describe item 2, (co-ordinate drilling and its practical use on the shop floor), the author briefly explains that this is done by converting the drawing given 'pitch circle' dimensions into the more useable linear X and Y dimensions by using the Probe and Prompt triangle calculation system. It is also pointed out that it is possible to obtain the correct required dimension by the process of using a scaling method to obtain an unknown dimension on a drawing when it cannot be obtained by any other means. This operation can be performed without the student needing to have expert knowledge of trigonometry to obtain it. (An example of this problem is solved practically in fig. 18c), which shows how it is possible to find an elusive dimension in a triangular figure when all previous attempts at obtaining it have failed. We do this by drawing a ten-times-enlarged and very accurate sketch, by using our drawing skills to obtain the unknown dimension of a component by using just the supplied available dimensions.

Fig.18 a is a drawing describing three holes positioned on a pitch circle on a workpiece, which indicates the unknown dimensions being (marked with a question mark,) that need to be found.

Coordinate dimensioning

Fig.18-b shows the triangle calculation required to obtain the coordinate dimensions. The two drawings show a typical example of a mathematical conversion from *pitch circle diameter* to coordinate drilling dimensions (often found necessary on the shop floor). This calculation is required in order to change the drawing's given *pitch circle diameter* into the more usable linear coordinate dimensions, to enable the machine's *x* and *y* (longitudinal and transverse axis coordinates) to be used on the machine's measuring scales, or alternatively, to be used for marking out the work-piece for a manual drilling operation. These extra calculations are required to allow the accurate marking-out and the subsequent drilling of all three of the required holes accurately in the metal base-plate. This so-called conversion from *pitch circle diameter* to *coordinate drilling dimensions* often becomes necessary because of the unavailability of a suitable rotary table or dividing head normally used to perform this task. These pieces of tooling are normally required for obtaining the precise angular setting for the drilling and the secure holding of the work-piece during the drilling operation. Of course, when this tooling is not available, the work-piece must be marked out accurately manually followed by being drilled manually. This is done by positioning the drill over the intended hole's position solely by eye.

The dimensioned sketch shown in fig 18d is used for the accurate inspection of each hole's position in the fig. 18 plate. See also figures 18a, and 18c.

The 18c drawing depicts a ten-times enlargement of drawing 18b; this has been rotated (of necessity), anticlockwise by 90 degrees to fit into a vertical A4 page). This drawing describes how it is possible to scale a perfectly adequate dimension from an accurate drawing or sketch. This then gives us the option of using a scaled drawing method to obtain a required dimension when one finds oneself in dimensional difficulty. The use of this method allows the student to obtain an unknown dimension without the need to use trigonometry. The use of this method also has the advantage that it can be performed by a person who has no expert knowledge of trigonometry but who does possess the ability to draw to a very accurate scale.

We must first make an accurate scale drawing of the component's problem area (in this case, we will be using its now-known 30-degree angle to find the 'length of sides *b* and *c* in the problem triangle shown in figures 18-a and 18-b). For this, we use a sheet of A-4 paper, an accurate rule, and a plastic angle protractor. This drawing should be made as accurately as possible, and must include its lengths (×10) and the exact calculated angle of the problem; however, it is important that it is drawn to a scale that is *exactly* ten times full size (this exact scale measurement is very important) as shown in fig.18c).

Because this new drawing is now at ten times normal size, it will of course be very much larger than normal size; it will therefore be found to give us the distinct advantage that any dimension measured from it while using a scale rule will actually measure ten times larger than the existing problems dimensions being worked on (in figures 18-a and b).

Having completed this drawing, with its ten-times enlarged dimension now being 260 mm (instead of its original 26 mm), and with its angle of inclination being 30 degrees from its vertical base-line, a square is then used to complete a 90-degree line emanating from the 260 mm point relative to the vertical base line; this then allows us to accurately measure the newly found dimensions of b and c (which incidentally should measure 225.16 (rounded up to 225.2) mm and 130 mm), with a rule, (using decimal notation). These dimensions should be noted down on paper in preparation for the next stage in the process.

The foregoing moves now make it possible to obtain the unknown dimensions required in fig. 18-a, by the process of moving the decimal point of the now ten times larger) scaled dimensions, to a new position of one place of decimals to its left, thereby obtaining the actual dimension for the original life-size fig. 18-a drawing's required dimension (this being of course ten times smaller).

Having obtained the exact scaled dimension required, we are now able to answer the original problem questions, namely, that side b must be 13 mm long, and that side c must be 22.52 mm long.

To prepare this drawing, we must

1. Using A4 paper, in document mode, (positioned vertically), draw a light line vertically down, at the left-hand side of the paper positioned approximately a quarter of the horizontal distance across the paper's width, and this line to be to the full length of the paper. This now forms the vertical base-line of the required line c (shown in fig. 18-c).

2. From a position of approximately 40 mm. down from the top
   of the paper, a protractor is used to lay off, at a downward
   inclination to the right, a 30-degree diagonal line, to a length
   of exactly 260 mm. A square is placed on the first vertical
   line, and, using this line as its vertical datum, another line
   is laid off at 90 degrees to it, meeting the previously drawn
   30-degree line at its extended 260-mm point. (We can use this
   dimension because we already know that the pitch circle radius
   of the holes is 26 mm, and the angle (taken from the previous
   calculation) must be 30 degrees). We therefore use 360 divided
   by 3 = 120. This is then divided by 2 = 60; therefore 90 - 60
   = 30 degrees.

As previously stated, to obtain the sought-after dimensions
required, it will now be necessary to re-position the decimal point from
the larger-scale measurements so far obtained, (to reduce their value by
a factor of 10); we do this by moving the decimal point one place to its
left on all of the enlarged side length dimensions, followed by selecting
the two new dimensions required.

These sought-after modified dimensions are then re-introduced
into the earlier unresolved calculation in fig. 18-a. The figures can,
if necessary, be re-checked for accuracy by using the probe and its
prompt to confirm that the calculation is correct.

In the past, this method of discovering an unknown dimension
has been found very useful, but one should always remember that the
sketch that is used must be drawn very accurately in order to ensure
that the results will be completely accurate. The author also suggests
that this method of obtaining any unknown dimensions should only
be used as a last resort (i.e. when all else has failed). Its use, however,
can make a very time-consuming or difficult calculation, much easier,
quicker, and more satisfying to resolve, this being due in part to the
fact that one can now actually prove the problem practically, while
confirming it theoretically by using the Probe and Prompt system of
triangle calculation during the checking procedure.

The student is advised not to attempt to scale an officially issued
drawing to obtain an unknown dimension, because it may not be
exactly to scale. The originating draftsman would have probably

drawn the original drawing to its correct scale, but by the time it has been photo-copied by the firm's publishing department, the resulting copied drawing may well be untrue to its scale; therefore, any dimension *scaled* from this *issued* drawing should not be relied upon, because of the high probability that it is not true to scale.

*   *   *

There follows an exercise in the drawing of a 100-times enlarged drawing in order to obtain an unknown dimension. While we are considering ways of obtaining unknown dimensions by the use of greatly enlarged but accurate drawings, it should now be realized that it is possible to use this method of calculation to obtain the required unknown cutting radius on the internal screw-cutting tool shown in the fig. 43a screw-cutting drawing that depicts an internally screw-cut thread. We can discover the required and exact screw-cutting tool's tip radius by using an accurately drawn *100*-times enlargement of the internal screw-cutting tool's *calculated depth of thread*, by using calculations made using its tool's greatly enlarged cutting profile's shape. We can do this by using the result of extra calculations made to the *lower* fig. 43a internal screw-cut thread's *calculated depth of thread* formulas.

[At this point, the student should be made aware that to convert a millimeter dimension into an imperial (English inch) dimension, we must multiply the mm figure by 0.03937. Alternatively to convert an imperial (English inch) dimension into a metric dimension, we must divide the inch-dimensioned figure by 0.03937. To get some idea of scale, students should realize that a dimension of 0.001 inches (this being a thousandth part of an inch) will equal 0.02540005 mm and that a dimension of 0.001 inches will become 0.1 inches (2.54mm) when multiplied by 100. Therefore, when multiplying the fig. 43a drawing's 0.108 mm (0.00425196 inch) dimension by 100, it will now become 10.8 mm (0.425196 inches) on the new enlarged drawing. This enlargement to 10.8 mm makes the initial minute dimension of 0.108 mm much easier to recognize and understand, and will therefore allow the student to obtain extreme clarity regarding the actual size of the required tool's radius dimension by referring to this 100-times enlarged drawing.]

As a direct comparison, a plucked human hair from the head, will have an average diameter of 0.05 mm (0.0019685 inches when measured using a micrometer); the use of these very small dimensions should now make the student realize that he or she is now dealing with a tool radius that will actually be possessing minute dimensions, and, in doing so, indicates exactly how small the tool's radius needs to be accurately honed for it to accurately perform the screw-cutting operation.

The described formulas shown in the lower drawing of the *main* fig. 43a drawing indicate that dimension H multiplied by 0.625 mm will give us a dimension of 1.623797632 mm, but this only gives us the depth of thread without its tool's tip radius. The letter P, representing the pitch (of the 3mm thread), is then divided by 8 to give us the 0.375 mm flat. This dimension theoretically exists in the lower profile of the internally screw-cut thread.

H is then divided by 8 to give us 0.324759526 mm, the dimension from the theoretical 0.375 line to the lower vertex point of the V. These two numbers (1.623797632 and 0.324759526) are now added together to give us 1.948557158 mm the full calculated depth of thread, calculated from its lower vertex point. This dimension now enables us to draw the top horizontal line of our enlarged drawing. Also utilized is the known 60-degree included angle of the tool that makes up the profile shape of the internal screw-cutting tool.

To draw this figure accurately, we use a sheet of A4 paper positioned horizontally (i.e. in landscape mode). A base-line is first drawn horizontally across the paper at approximately 5 mm from the bottom of the page. A second line is then drawn horizontally at the height of 194.8 mm above the first baseline.

In the horizontal center of the page, we then draw a vertical line extending from the center of the baseline to form the object's vertical centerline, through to the top of the page. This basic drawn figure should now resemble a large capital letter H lying on its side.

The calculated linear dimensions obtained from the lower 43a drawing are now multiplied by 100. We next draw a 60-degree included angled V, with its vertex point starting at the baseline and its angled sides extending (equally side by side of its center-line) to reach the uppermost 194.8 mm drawn line.

Now, with a rule held horizontally inside the now-drawn V, and at a vertical height carefully selected to be sufficient to allow a 37.5-mm horizontal line to be established between the two sides of the V, we then confirm this position by drawing a pencil line horizontally linking the two angled sides. We follow this by drawing another horizontal line 10.8 mm below the 37.5 mm line fully across the V. The resulting narrow 10.8-mm lower area will now simulate the 0.108-mm gap that contains the tool's radius, and will include a minute area required for the necessary tool clearance.

If we now refer to the fig.43a drawing, we can see detailed in the lower area of the nuts' left-hand side, the total theoretical depth of thread indicated as 1.732 mm, and that the indicated 0.108 mm area includes the tool's tip radius generated from a theoretical 0.375 flat indicated in the P/8 dimension shown in the lower fig. 43a (ten times enlarged) drawing. Note that this tip radius is seen to occupy the major part of the 0.108-mm area.

At this point, the student should be made aware that any previously published details regarding the dimension of an internal screw-cutting tool's tip radius, should be considered as indecisive, as this radius is currently spoken of in engineering circles as being rounded and cleared beyond P/8, therefore up to now, the internal screw-cutting tool's tip radius has never been given any specific dimension.

The object of this 100-times drawing magnification exercise, is therefore to obtain the exact tip radius required on the tool so that it can be accurately honed onto the cutting tip of the fig. 43a internal screw-cutting tool that is to be used for this particular job.

We now select a sharp pencil compass, held with its sharp fulcrum point temporarily hovering over the drawn object's vertical centerline, the exact fulcrum point is then found by making adjustments to the compass's radius sufficiently to enable its pencil point to scribe an arc that exactly meets both the junctions of the 37.5-mm line and the 60-degree tool's angled V sides.

Having established this correct radius, the compass's fulcrum point is then finally pricked into the paper's vertical center line, followed by its pencil point being used to scribe an accurate arc and completing a full circle. (In practice, the diameter of the circle produced in this case will be approximately 50 mm; therefore, the scribed arc now produced below the 37.5 line, will possess a 25-mm radius, this being the so-called rounding of the tool's point. This arc's radius dimension is now divided by 100, to give us the actual 0.25-mm radius that is now required to be honed onto the tool's tip, as described in the fig. 43a drawing, and in section 19/8 of the script. When performed accurately, this 100-times enlargement exercise is found to be extremely useful when attempting to obtain an unknown tip's radius for a particular screw-cutting tool.

*   *   *

The coordinate method of marking out the workpiece

There follows an explanation of how one carries out the coordinate method of marking-out the workpiece, in order to redimension the actual position of the holes while using the drawing's 'pitch circle dimensions', to make the dimensions more compatible with the slides and indexing scales of the machine tool's x and y measuring scales.

This method of marking-out can be used for the accurate positioning and drilling of a metal plate when the machinist is using a precision machine tool to perform the task. This method of marking-out can also be used to obtain the positioning and final drilling of the plate by using just the hand and eye for the workpiece's final positioning prior to drilling the holes.

Fig.18 shows an example of a working engineering drawing, this being very similar to those normally issued to a research and development or production workshop, for the manufacture and drilling of a metal plate.

In practice, though, the machinist, on receiving this particular drawing for a job on the machine shop floor, would discover, because of the workpiece's extreme overall length, that it would be much too long (after being clamped into position) for it to be rotated to each of the required drilling positions while using a conventional rotary table indexing device. It would therefore become apparent because of this extreme length) that the workpiece would inevitably strike the drilling machine's vertical support column on being rotated between the required positions of its three holes. However, if the requirement had been for drilling a much shorter plate, that could be safely rotated, clamped, positioned, and drilled, through the necessary 360 degrees of rotation, without the workpiece's corners striking part of the support column of the drilling machine during its required rotation, then the rotary table could have been used.

The machine shop personnel will therefore be forced (in this instance) to perform a more involved work preparation process than would normally be required. This will include the calculation, marking-out, and center drilling of the workpiece to enable it to be drilled with sufficient accuracy using just manual positioning by eye.

Having now discovered it is not possible to use the assistance of a rotary table to accurately position the holes at their correct spacing to line up with the drill's centerline on this very long steel plate, then an alternative method must be found to complete the job satisfactorily.

The component will now require all the workpiece's holes being marked out individually and accurately, and each hole will need to be positioned and drilled separately by the workshop or tool-room fitter, while still retaining the positional accuracy of the holes as required by the drawing.

Because of the required accuracy stated on the drawing, this operation will now need to be performed by using the coordinate marking-out process, prior to drilling the three holes.

On completion, the plate, with its holes finish drilled in their correct positions, will need to possess sufficient accuracy to pass through the critical inspection department's checking system, in order to ensure that it complies with the dimensions shown on the issued drawing; therefore, it must be stressed that extreme care will need to be taken to ensure that the marking-out and drilling of the holes is completed accurately to enable the finished component to pass through the inspection department's unfailing scrutiny.

It is quite the normal practice in the engineering industry, for the draftsman, after completing the design process of a component (that usually involves a great deal of calculation), to leave out or deliberately not enter onto the issued drawing some parts of his originally calculated dimensions. These minute details (primarily needed during the design of the component) are often considered by the draftsman, on completion of his drawing, to be superfluous, and are not required to be passed on to the shop floor personnel, or included on the drawing for production purposes.

These important details not stated on the drawing are consequently deemed (by the drawing office staff) to be a fairly obvious area that can be dispensed with on the issued drawing as they will, if found necessary, be calculated by the receiving skilled workshop engineer immediately prior to his manufacture of the workpiece.

During the design stage of a product, the design draftsman will of course have all the necessary sketches and calculated dimensions readily available at his fingertips for the purpose of producing the required issued drawing, but for reasons of clarity, and because of the time-honored necessity for him to produce a neat, *uncluttered* drawing, many of the possibly useful design dimensions, and development details (which in practice would be found helpful to the workshop engineer), are purposely left out of the final issued drawing of the component. It is also worth mentioning at this point that it is generally considered (by the drawing office staff) to be bad practice for an engineering draftsman to duplicate or to put any additional dimensions on a drawing than is considered necessary for the manufacture of the component.

It is mainly for this reason that it is considered normal practice for the draftsman to exercise the utmost care in preventing himself from falling into the so-called trap of carelessly dimensioning a component twice on a drawing, for fear of falling foul of the drawing office rules which state that this is definitely not allowed. He will therefore be very cautious about entering any additional (or, for that matter, any possibly helpful) information on the drawing.

This practice can at times be taken just a little too literally by both the draftsman and the drawing office management. Experience has shown that the so-called office rules can cause a serious omission of an important or vital dimension from the issued production drawing, that tends to initially slow down the production of the component.

Because of this time-honored practice of omitting so-called unnecessary dimensions from issued production drawings, there is always the constant need for the shop-floor production staff to be on their guard when first consulting the issued drawing's details, and certainly before actually carrying out a so-called irreversible 'cut to length' stage regarding the components cutoff length of the material to be used.

Several cases have been recorded where a certain important dimension has been purposely left out of a drawing by the draftsman, so that on being approached for the missing dimension, weeks later, by the production shop-floor personnel, it will confirm that his designed component has now reached shop-floor level and is about to be produced; this so-called reminder then gives him the golden opportunity of visiting the machine shop floor, *adjusting* for any new *necessary changes to his design*, before production of the component is underway, thereby alleviating any serious repercussions should any change to his design have been required *after* production had been completed.

The shop-floor production staff must therefore be in possession of not only a reasonable knowledge of mathematics with which to deal with the everyday machining calculations found necessary for the manufacture of components, but also to possess a good working knowledge of trigonometry and in particular, a considerably greater knowledge of the most useful form of trigonometry calculation, called practical trigonometry.

The development of this particular mathematical expertise should be encouraged among the workforce, for it enables them to accurately carry out the 'extra' calculations, often found necessary in practice, to reproduce some of the dimensions or angles supposedly left out of the drawing by the draughtsman in the interests of neatness, in order to conform with the existing drawing office rules regarding non-cluttered drawings.

It must be said at this point that the author has yet to hear of any complaints from the shop floor personnel stating that too much information or too many dimensions were provided on an issued drawing.

Sadly, experience has shown that the necessary expertise in the use of practical trigonometry calculation is rarely to be found among the production staff currently working in the country's manufacturing establishments. The help and guidance contained within this book that deals with this subject will, it is hoped improve the workers' knowledge of the actual practical calculation problems regarding the triangles' dimensions and angles in the engineering shop-floor environment. There is also a distinct tendency in the engineering industry for the production management staff to turn a so-called blind eye to any 'unofficial' help given to the less experienced workers by their higher-skilled colleagues working on the shop floor.

There is often the need for the less-able workers to ask their shop-floor colleagues for mathematical assistance with their bending development lengths and their 'awkward' triangle calculations. There is a tendency in this case to ask their colleagues rather than the management staff for their advice to help solve their workpiece dimensional and trigonometry problems.

This 'unofficial' help given by colleagues is often freely given (when time is available) and does resolve the situation for the time being, by avoiding the use of valuable man-hours in the office when the problem can be resolved by his colleagues on the shop floor. The use of his colleagues' help can also avoid the embarrassment suffered by a worker who may be forced to admit to the management that he is unable to resolve this particular mathematical problem himself. This possible loss of self-esteem can occur when an employee is forced to seek the management's advice regarding a mathematical or trigonometry problem that needs to be resolved regarding a drawing. Taking this step can prove to be a rather degrading exercise when admitting to the boss that you are incapable of sorting out the problem yourself.

From the above situation, it will be realized that there exists an additional reason for writing this book on practical trigonometry, as it will enable its contents to be found informative and useful to all its readers, particularly those who currently receive this freely given help on trigonometry problems from their workmates. The current receivers of this help on the shop floor must now surely welcome the opportunity of developing their own ability to do it themselves and to learn the 'secrets' of workshop trigonometry calculations, at leisure in the privacy of their own home. It is unfortunate, but in this day and age, there seems to be very little spare time available in the industrial environment to allow this particular expertise to be acquired fully at work or on the job.

Should this mathematical help given by workmates be witnessed by the supervision, it does allow them (indirectly) to form opinions regarding the mathematical potential of each individual employee's capability in performing the intricacies of precision production work.

This particular help given to colleagues becomes abundantly clear to management when several workers are seen huddled together around a bench, scratching heads, clutching pencils, paper, and calculators, particularly when the orders of the day stipulate that there is to be no grouping.

The capability of workers to calculate their own work problems accurately at shop-floor level must surely be considered to be a distinct advantage, for, as previously stated, it avoids wasting time in lengthy office discussions, particularly when everyone is trying to deal with a very high workload.

If we now study fig.18, it will be seen that this particular drawing calls for three holes of 5-mm. plus or minus 0.05mm. diameter to be drilled in a metal plate; these holes are to be equally spaced on a 52-mm pitch circle diameter. As previously stated, on studying this drawing it will be seen that this particular job will be much too long to be drilled by using a rotary table, because of the high probability that it's extreme length will strike the drilling machine's upright support during the necessary rotational movements to reach the individual holes' positions. We will therefore be unable to use a rotary table to do this job; this leaves the fitter at a considerable disadvantage, by now having to mark out the workpiece fully for manual drilling.

The author will now explain the method normally adopted when a rotary table is available for drilling precision holes on a so-called pitch circle diameter, and where the workpiece's dimensions are deemed compatible with the available tooling.

The majority of machine shops possess a rotary table for use in the circular machining of plates or the circular drilling of engineering components that are normally dimensioned on a so-called pitch circle.

The rotary table is a very useful piece of tooling. It is used bolted down onto a pillar drill or a milling machine's worktable, by utilizing its inbuilt holding-down lug slots, and its holding-down 'T' bolts. It is most important to clean its underside areas thoroughly before use in order to remove any contaminants that could affect its accuracy with respect to flatness. Its cast-iron base is surmounted by a circular rotating table that incorporates machined T-bolt slots located in a cross pattern for use in clamping the workpiece securely down onto its top surface. This top surface is also provided with a number of circularly machined lines that can be used as an accurate aid to centralizing the workpiece relative to the table's centerline while using just the eye's perspective. The rotating table is also accurately engraved around its periphery with graduations marking out a full 360 degrees in one rotation; its base is also equipped with an adjustable zero index marking arrangement that incorporates a secure table locking device. The rotary table is also equipped with a geared operating hand wheel requiring exactly forty turns to complete one complete revolution of the main table. (One revolution of this handwheel will therefore provide exactly 9 degrees of rotation to the main table.) Most rotary tables also incorporate a dividing device that consists of a multi-drilled circular indexing plate. This plate is used together with a spring-loaded adjustable plunger that locates and secures the position of the indexing plunger relative to the holes in the plate. Therefore while 9 degrees can be achieved by one revolution of the handwheel, the indexed rotation can be further divided by using other plates with a differing series of holes designed to provide either separate degrees (or in some cases, parts of a degree), where accurate angular workpiece rotation is required. For example, if we choose an eighteen-hole series plate, then the circular distance moved between two of its adjacent holes (by use of its plunger) provides us with half a degree of rotation to the main table (other hole plates with differently spaced holes allow a multitude of spacings to be obtained for the rotation of the table and its workpiece). The hand-wheel mechanism can be disengaged when not required, to allow the table to be rotated freely by hand for workpiece concentricity setting-up purposes. This will be found necessary if one is using either a dial test indicator or a 'sticky pin' for workpiece concentric alignment, to ensure that its marked-out circular scribed lines are seen to be running concentrically relative to the table's centerline; also, one must plan the physical positioning of the workpiece clamps and its securing bolts to avoid contact being made with the dial test indicator's probe when being rotated.

The student may find difficulty while trying to centralize the workpiece accurately relative to the rotary table's top surface. There exists several methods we can use to achieve this.

Note that the rotary table possesses a centrally bored precision hole; this will usually be of 1 inch (25.4 mm) diameter. This precision-bored central hole can prove to be very useful should the workpiece also possess a central hole (even if this hole is of differing diameter, it can still be used to centralize the workpiece. To make use of this method of centralization' we need to turn on the lathe an accurately flanged plug from a piece of scrap mild steel or similar material, one end of which should be turned to exactly fit the central hole in the rotary table; its other end should be turned to exactly fit the workpiece's bored diameter. This so-called plug should also be machined to possess a thin flange (say, of 0.5 mm in thickness) positioned between the two turned diameters in order to prevent the plug actually falling through the table's bored hole. The use of this multi-diameter plug then provides a very accurate method for use in centralizing the workpiece relative to the rotary table's bored center line hole. (It is of course essential that the workpiece's pitch circle has been marked out and is concentric with centrally bored diameter, and that the workpiece is positioned (as described) to run concentrically with the center line of the rotary table.

As a matter of interest, if one is using a rotary table on a milling or jig-boring machine, this turned central plug's actual diameter can also be used to locate the exact center of the rotary table to enable it to be in line with the machine's center line. For this operation, we use a 'wobbler' center finder, to locate the diameter of the plug accurately, followed by indexing the machine's slides by use of its scales, to position the machine's center line over the center line of the rotary table. This is then followed by establishing a dimension that corresponds to the exact pitch circle radius of the job to be drilled; this then makes it unnecessary (when on batch work) to fully mark out each workpiece by just utilizing the angle graduations on the periphery of the rotary table to obtain the correct spacing for each hole.

Marking out a pitch circle on a flat component

However, if it is required to centralize a flat and round workpiece for the drilling of holes on a pitch circle diameter when it does not possess a central hole, it will be found the best course of action to first paint the area of the component to be drilled with marking blue, followed by marking out the pitch circle diameter very accurately by utilizing the flat surface of a marking-out table, a vernier height gauge, an angle plate, a center punch, and a pair of dividers suitably adjusted to the exact radius of the pitch circle required (while using a vernier caliper to check the exact dimension of the divider's final gap setting). The center punch is then used to indent the workpiece at the exactly marked-out center of the plate, followed by using this as a fulcrum point to permit the dividers to scribe the required pitch circle. We then use the vernier height gauge placed on the marking-out table (while the workpiece is being supported in a vertical position by an angle plate), in order to mark out the center line of the workpiece, followed by rotating the workpiece 90 degrees to mark out a second line to form a cross with its position now exactly on the workpiece's centerline. We now select the vertical line above the center-line to position the so-called master hole; this exact position is then marked out on the workpiece, at a point where it coincides with the line bisecting the scribed pitch circle, and so establishes the master hole's accurate position.

The workpiece is now placed on the rotary table with its clamps temporarily left loose sufficiently to enable the workpiece to be gently 'persuaded' into a central position. (Warning: do not forget to place a piece of medium density fiberboard or a suitable piece of flat plywood between the workpiece and the top surface of the rotary table in order to prevent the drill's point passing through and damaging the rotary table's surface). We then use a 'sticky pin' (made from a sharp sewing needle or a carefully sharpened panel pin) secured in a small piece of plasticine (or similar soft modeling clay) affixed to the sharp end of the center drill held in the chuck. The point of this pin is then used to aid the centralization of the workpiece. With the table out of gear and being rotated by hand, the scribed pitch circle diameter is then picked up by the point of the pin and any discrepancy in its alignment is adjusted by moving the workpiece (or alternatively the pin) laterally relative to the table until the scribed circular line is seen to be running perfectly concentric with the sticky pin's point, while the table is being rotated by hand. When true concentricity has been achieved, the workpiece is then finally clamped into position. The rotary table's handwheel is then re-engaged into mesh to use its worm drive gear, the table is then rotated by its handwheel until the centerdrill (or pointer) is aligned exactly over the master hole and its index scale is then set to read zero, while ensuring that its latent backlash has been taken up in a clockwise direction at this hole's position, in preparation for it being accurately rotated to its scale's new position to drill the second hole.

If the machinist is using a radial drill to do this job, it is considered to be more accurate to first use a pointer held in the drill chuck to ascertain the exact position of the master hole, prior to using the center drill to lightly indent its position. (A pointer is a useful tool made from a piece of 6 mm diameter tool steel ground at its end to possess a 60 degree included angle point). This pointer held in the chuck is then carefully positioned over the marked-out 'master hole's crossed lines, exactly at the point where all the lines coincide with the marked-out pitch circle radius line, using the eye and preferably by also using the assistance of a magnifying lens, in preparation for center-drilling this master hole. The rotary table's zero scale mark, (seen on its outer 360-degree scale), is then set to zero, and clamped. We follow this by rechecking that the handwheel's scale is also set to read zero. The machine is then switched on at a fairly high spindle speed (consistent with using a center drill at, say, 800 rpm) and the center drill's point is then used to momentarily spot the surface of the workpiece by it leaving a shallow indented 'witness' mark, at the exact position of the intended hole.

Should one be lucky enough to be producing a workpiece that possesses two of its holes diametrically opposite one another, (the workpiece in this case having an even number of holes and not the odd number of three holes as shown in our example drawing), then the hole's pitch circle diameter about to be drilled can be measured for accuracy by rotating the rotory tables handwheel twenty turns clockwise (equaling 180 degrees) then checking the spacing of these two opposing center-drilled indents with a knife-edged vernier caliper. To perform this check, and after spotting the first hole's position, and having turned the rotary table's handwheel clockwise by twenty turns (thus rotating the main table exactly 180 degrees), to position the opposing hole's position just under the center-drill's position. This new position is then spotted using the same centerdrill; the vernier caliper is then used again to check the distance between the two opposing drilled indents to check that they are correct. If the center distance is found to be correct, then all the holes required in the plate can be spotted and drilled by using the center drill, and then rotating the handwheel the exact distance required between each drilled hole to comply with the marked space positions previously set on its scale, (or by locating its plunger into its previously calculated hole position in its indexing plate). This is followed by removing the center drill and replacing it with the selected diameter drill, to finally drill all the holes in the component, using the identical previous scale or plunger settings for each hole.

It will be discovered that after drilling the first plate correctly (by using the settings that have been locked in to correspond to the correct pitch circle diameter), it will now only require the next plate to be centered, clamped, and the table rotated to the correct previously calibrated spacing for each hole, thereby allowing all of the holes to be drilled accurately on their correct pitch circle. With the rotary table now locked into this correct position, followed by aligning the next workpiece plate centrally and accurately clamped relative to the master hole's position, it will now be possible to drill this component and the following batch of similar components by just centralizing each one relative to the table's centerline, and relying on the previously set rotary table's marked scale (or plunger positions), to accurately position and drill all of the holes in all of the plates without the necessity of having to fully mark out all of the workpieces.

However, in order to assist the student in developing his own ability to cope with all eventualities, the author will now explain how it is possible to accurately drill the component featured in the fig. 18a drawing, by accurately drilling the required three holes equally spaced on a pitch circle without the assistance of the very useful rotary table.

There follows a description of how the workpiece can be accurately marked out by using trigonometry calculations together with the guidance of the probe and its prompts. The following notes describe how to mark out, position, and drill the holes in our example workpiece to the accuracy required on the drawing, without the need to use a precision rotary table or a precision dividing head to perform this operation accurately.

We know of course that a basic circle contains a total of 360 degrees. For the purpose of converting this drawing's given pitch circle diameter into the more workable linear dimensions, we must in this case divide this pitch circle into three equal parts, to give us 120 degrees angular spacing (on this pitch circle) between the three holes. We therefore subtract 90 degrees from the 120 degrees, giving us 30 degrees. This angle is now shown on the drawing as The Angle and is now used to begin the trigonometry calculations.

If we study figure 18a, it will now be realized that line a of this triangle is 26 mm long (i.e. half the diameter of the pitch circle).

We now orient the probe (using its side 1) in a sliding motion on the paper, and allow it to point in the direction of its 30-degree angle, confirming in our mind that by doing so we now know The Angle to be 30 degrees, and that the length of side a on both the workpiece and the probe's example sketch must be 26 mm.

We now need to find the actual length of the triangle's other two sides, namely side b and side c. We do this by reading off the appropriate prompt to use, while also realizing that we now know The Angle and the length of line a. The prompt to find b can be found on the now positioned probe. In this case we therefore use the prompt $\angle a, b = a \times \text{SIN} \angle$.

We now make a sketch of this helpful triangle on a separate piece of paper, and orient the probe and the sketch into its correct position with its tag as shown in fig. 18b.

Now, using a scientific calculator, we enter the key sequence 30, SIN, (display shows 0.5) x 26 = giving us the answer 13. This is now written in at side b of our sketched (18b) triangle. To find the length of c, we use the prompt $\angle a$, c =, a × cos $\angle$. We enter the sequence 30, cos. (display shows 0.8660254) × 26 = 22.5166605. We now write this figure in at side c of the 18b drawing. The unknown dimensions we wish to find for marking out the workpiece are shown identified with question marks (?) in the 18a drawing, and these must now be calculated using the drawings existing centerline dimension from its side datum face. We therefore enter the sequence 60, −, 22.5166605, =, 37.4833395. We now write this into the sketch as shown in the fig. 18a drawing.

We now enter 60, +, 22.5166605, =, giving us 82.5166605; this figure is now also written into the fig. 18a drawing. The dimension required for the positioning of the top hole from the end datum face is obtained by halving the pitch circle diameter and subtracting the result from the existing 46 mm drawing dimension, by entering the sequence 46 − 26 = 20.

We now need to calculate the exact positions of the bottom two holes in the plate, by calculating the dimensions relative to the end and side datum faces of the plate.

The drawing shows the dimension 46 mm to the center of the pitch circle; to this dimension, we must now add 13 mm, (calculated for the length of line *b* in fig. 18 b), giving us 59 mm.

At this stage the author wishes to add the following cautionary note before starting any marking-out operation on the workpiece. Firstly, when one is forced to use the coordinate drilling method to drill the holes (used in this case because of the absence of a precision rotary table), it is very important to check that the two datum faces of the workpiece have been correctly machined, and possess exactly square edges relative to one another, before they can be used as accurate datum faces from which to mark out the workpiece.

Any inaccuracies found, if not corrected, will affect not only of the outside shape of the finished product, but will also affect the exact positions of the three drilled holes relative to the workpiece's dimensions and the relative positions of one hole to the other.

Because we have now needed to use these outside datum faces of the plate as actual measuring surfaces, we are now required to check and verify that they are in fact exactly square, by using an engineer's square placed in close proximity to the workpiece's datum corner. Any misalignment found by eye must be rectified as necessary, by performing either a milling, or a filing operation, to ensure that the two chosen datum faces are perfectly square to one another before they can be used as measuring faces.

The datum edges of the plate, (after being corrected), should now possess an included angle of exactly 90 degrees. This can be checked (by holding the component and the engineer's square up to the light) to ensure that no daylight is visible between the two surfaces of the work when held in close contact with the inside surfaces, of the engineers square.

Having checked that these two datum faces are perfectly square, we can now mark out the position of the center of the pitch circle. The workpiece's surface area to be drilled should be painted with marking blue to aid the visibility of the marking-out. We must prepare for this marking-out procedure by securely holding the workpiece, firstly in a horizontal position in order to mark out its 60-mm center line, followed by holding it in a vertical position for marking out its 46mm center line, by using the flat surface of the marking-out table as a datum. With the workpiece held vertically, its rear face must be supported (for safety) against an angle plate, and securely clamped to the table to avoid any possibility of it accidentally falling over.

The required scribed lines are marked out on the workpiece by using a precision instrument called a vernier height gauge. This precision instrument is adjusted by hand to set its vernier scale to read the 60-mm dimension (its scribing beak will now be positioned at exactly 60 mm above the surface of the table); this setting will later be changed to the required 46 mm for the other dimension, and then for the other calculated 20 mm dimension required on the drawing.

The two scribed 60-and 46-mm lines are then accurately marked out at 90 degrees to one another, by using the process of rotating the workpiece, followed by re-clamping, then again using the vernier height gauge to scribe the calculated 20-mm line. These scribed measurements must be taken from the workpiece's chosen datum faces, which in turn must be held in close contact with the table's datum surface. As previously stated, it is important that the workpiece's back face is prevented from falling over by being clamped to an angle plate. The stability of this marking-out process can if necessary be assisted by the use of additional support in the form of a block of a suitably square-faced (and to advantage heavy), piece of mild steel material, and by using suitable clamps to safely locate this exceptionally long vertically held workpiece. The point where the two lines cross is then used to indicate the exact center of the intended scribed pitch circle'. This now allows a pair of dividers (that have been set to a radius of exactly 26 mm by using a vernier caliper to scrupulously check this dimension), to be used to complete the scribed pitch circle diameter,

while using its center-punched indent as a fulcrum point for the
dividers, previously marked out from the workpiece's two datum faces.

The position of this cross can be checked for accuracy by using
a magnifying lens and the vernier caliper gauge which is used to
measure and ensure that the crossed lines are in their exactly correct
positions. If previous use has shown that the height gauge in current
use has proved to be accurate, its measuring scale should be accepted
as being reliable,; this instrument will therefore be considered safe for
use in marking out the component. The author mentions this very
important point, because on rare occasions it can be discovered during
the working day, that the height gauge in current use has proved to
be inaccurate; therefore, one should always check the accuracy of
any height gauge before using it. We can do this by firstly carefully
cleaning under its base and then allowing its scribing beak to contact
the marking-out table's perfectly clean and flat surface, followed by
noting whether its vernier scale does actually read zero when this is
taking place. If the scale reads zero, then all is well, but if not, then
this inaccuracy should be reported to the management without any
delay, to allow it to be checked and corrected by the Inspection
Department because of its inaccurate scale.)

The 52-mm pitch circle is then accurately marked out on
the workpiece by using the dividers set accurately to the 26-mm
dimension. A quick check is then made of the scribed circle by using
the vernier caliper set to 52 mm to check its diameter. This accurate
diameter check will highlight any discrepancies found in the radius
that has been scribed.

The position of the three holes can now be marked out on the
workpiece; we do this by (once again), using the vernier height gauge
positioned with its base firmly in contact with the surface table, and
this time with its' vernier caliper scales reset (for the following two
operations), to the two new calculated dimensions, measured from
their respective datum face edges, namely 59 mm, and 82.5166605
mm, (note that the vernier caliper can only be visually set for accuracy
to one place of decimals on its vernier scale, so this will be set to 82.5
mm). We must ensure that the new marked-out horizontal and vertical
lines do actually coincide to form a cross at the exact point on the 52-
mm diameter previously scribed circle. The fact that they do coincide

will confirm that the marking-out is positioned correctly. These crossed points can now be carefully center punched, by using this tool to provide just a light indent in the workpiece, (by using just a light hammer blow), on a carefully positioned center punch.

Because we have not discussed in any detail the subject of drilling steel in all its various grades of toughness, and the necessity for the student to be able to grind his or her own twist drills, it is considered wise at this stage, to select the actual twist drill one intends to use for drilling the holes. It should be measured carefully to check that it is of the correct diameter and state of sharpness. We do this dimensional check by using a micrometer with its measuring surfaces placed across the drill's cutting edge lands, and not, (most importantly), across the drill's shank, (the drill's shank is often scored, or damaged, and may also be of a slightly smaller diameter than the drill's cutting edges).

The sharpening of twist drills

On collecting the requested twist drill from the tool stores, its state of sharpness should be scrutinized; if there is any doubt as to its condition, it should be resharpened to ensure that its cutting edges are perfectly sharp, and have been ground with adequate clearances. Its cutting edges should be of equal angle and length, (118 degrees included angle is the preferred grind angle for its cutting edges). As a precaution, the drill should also be rolled along a perfectly flat surface plate in order to check whether it is in fact straight (any wobble will indicate that it is bent). The spiral twist (or helix angle) given to the drill by the manufacturer, is designed to provide not only its means of removing the swarf generated by the drilling action, but also to provide the drill's actual cutting angle of approximately 25 degrees. The author personally prefers to thin the point of any drill about to be used; this modification permits a more accurate drilled hole, and allows the drill to centralize itself on the center-punched indent of the intended hole. It also provides less resistance to the drill's penetration into the workpiece. This thinning of the drill's point (the so-called web thinning') is performed by the careful use of the corner of the grinding wheel, to remove (by grinding), a small amount of the *central area of the inner cutting edge's point area*, thereby marginally increasing the length of the drill's effective cutting edges and reducing the area normally occupied by its chisel point. In consequence, this marginally extends the useable length of each of the drill's cutting edges and enables the drill to cut in the area much closer to its center line (this is considered by the author to be a distinct advantage). This deliberate reduction in the largely redundant area occupied by the chisel point's central core will be found to improve the performance of the drill's cutting ability. It is considered by the experts that in practice, the conventional relief grinding angle normally given to the chisel point of the twist drill fails to allow its central area to cut effectively, caused by allowing this area of the drill to merely rub its way through the hole and, in consequence, will cause less effective penetration of the drill into the workpiece and, in some cases, will also cause minute wobbling to occur during the process. This drill grinding fault is often due to the over-rotation of the drill in the machinist's fingers while he is in the process of hand grinding its cutting edge and its relief clearances. This over-rotation of the drill during its resharpening process is now known to produce a chisel point that possesses an unwanted and increased non-cutting area that is detrimental to the cutting performance of the drill.

Over the years, the author has developed his own method of regrinding the twist drill to enable it to drill very accurate diameter holes. When sharpening a drill at the grinding wheel's front face, the author has found it best practice to extend the elusive chisel point area of the drill's cutting edges sufficient for it to be slightly beyond and to the right of the grinding wheel's side edge so that this particular area is not actually receiving an identical clearance angle to that given to the rest of the drill's cutting edge. Having completed this handheld clearance grinding of one of the drill's cutting edges while being held at its correct angle and orientation, the drill is then rotated 180 degrees (away from the wheel), to allow its other cutting edges clearance to be positioned and similarly ground. The drill is then finally held at its correct relief grinding angle, (with its cutting edges (in turn) held in a perfectly horizontal plane, where the drill is then held to allow it to progress across the face of the wheel at this exact angle without any hand or drill rotation being given; this will then produce a very narrow flat along each of the drill's cutting edges that will extend all the way to the drill's point. This operation then allows the center part of the drill (the part that was originally largely occupied by the non-cutting area of its chisel edge), to receive a consistent cutting angle sufficient to allow the drill to cut along virtually the whole length of its two cutting edges that continues right up to its point, and its center-line.

This careful web thinning and making the previously described discreet modifications in the grinding of the drill's cutting edges will be found to improve the cutting action, which in turn will improve the diametrical accuracy of the drilled hole. A further check of observing whether the drill is cutting concentrically, and whether it is using both of its cutting edges equally, will be displayed to the student by the quality of the swarf produced during the drilling operation. A correctly ground drill (particularly when drilling steel or aluminum) will produce a curl of swarf that is ejected in the form of two distinctly unbroken equal spirals that tend not to immediately break up into a multitude of irregular-shaped chips.

One should, (after sharpening a drill) be very careful not to immediately plunge it into cold water to cool, as this can cause minute cracking at its cutting edges because it receives what is called a temperature shock. This can also cause a loss of its required hardness. Also, the drill's cutting edges should never be allowed to become 'blue' during the sharpening process, as this will cause the drill to overheat and reach at least 300 degrees centigrade, coupled with the possibility of losing its inherent temper, which will reduce its hardness and its 'length of serviceable use' between the need for it to be resharpened.

When drilling deep holes, one should avoid allowing the drill's flutes to become choked with swarf. This situation can be the cause of the drill jamming in the hole, thereby causing it to break while under power and extreme loading. We can overcome this problem by using what is called peck drilling; this is where the drill is only allowed to drill into the hole for a short depth and a relatively short period of time before being withdrawn for cleaning and applying cooling lubrication, before repeating the drilling operation several times to its required full depth, while still using plenty of coolant. Coolant must be used unless of course cast iron is the material being drilled. This particular material is best drilled without coolant because it has its own inherent drill lubrication qualities. This will avoid accidentally flushing the swarf produced into the precision moving slides of the machine, causing unnecessary wear to occur in its vital working parts.

[Some materials encountered during machining operations will of course be discovered to be much tougher than others; for example, when machining 18/8 grade stainless steel, it will be found that it can suffer from a serious problem called work hardening, while it is being turned or drilled. We counter this problem by ensuring that the drill or lathe tool in use is never allowed to rub or undergo a phase of not cutting, as this so-called rubbing action will have the effect of not only blunting the tool's cutting edges, but will also form a hard and virtually impenetrable skin on the surface of the workpiece being drilled or machined. Therefore, when engaged in making the last finishing cut on a lathe or drilling a hole in stainless steel, manganese steel, or mnemonic-alloy-materials (for example), one must (in the case of turning) always allow sufficient material to be left on the almost-finished workpiece's diameter immediately prior to making the last finishing cut, so that on the final cut, the tool can get under the work-hardened skin and finish its final cut. The amount of material that needs to be left on must be of sufficient depth to allow the point of the tool to actually get under the skin of the material being cut. Note that if a very light and shallow finishing cut is attempted, it will be found that the tool will just rub the surface of the workpiece, which results in a poor surface finish, an incorrect dimension being achieved, and a severe blunting of the tool's cutting edges. It will also be found that the tool in this case will not be removing the required amount of material to obtain the desired finished external or internal diameter. In this case, it is strongly recommended therefore that a dummy run should be made on the workpiece, to produce an initial diameter of exactly 0.1 mm (less in diameter for a boring operation, or more in diameter for an external turning operation) of the diameter required. To test the accuracy of the initial setup, we set the crossslide's scale to read exactly 0.1mm before its finished zero mark on its scale; the minimally oversized diameter turned must then be checked for accuracy by using a micrometer, before taking the final and critical finishing cut. To make this final cut, the crossslide must then be rotated to read exactly zero for its final finishing cut. The use of this method also applies if one is screw-cutting any of the other so-called work-hardening materials. It is therefore most important that the student or trainee engineer learns how to sharpen his own lathe tools and twist drills to allow them to cut accurate diameters.

[As a matter of interest, if the machinist finds himself in the position where he has the need to sharpen the twist drill he is using because it's not cutting effectively, and he possesses neither the ability to do this nor a suitable grinding gauge at his disposal, with which to obtain its correct grinding angle, he could give the following advice a try. The normally ground included angle recommended for sharpening a twist drill's cutting edges is an angle of 118 (included) degrees. I would point out that this particular angle of grind need not necessarily be strictly adhered to in practice. Other grind angles can be permitted and experimented with in order to get a particular drilling job completed, but there is of course the proviso that the drill's cutting edges must in all cases be of equal angle, of equal length, and its ground point must be exactly on its center-line. It is therefore quite possible to regrind the cutting edges of a typically ground 118-degree included angle drill so that it performs and works satisfactorily if it is reground to an included angle (of, (say for example, 120 degrees); this grind angle will still allow the drill to do its job perfectly well. In fact, by grinding this particular included angle on a drill it can be used with advantage on thinner materials where its full diameter can then be allowed to engage with the work-piece's existing outer surface '*before its point actually breaks through the material*'; this then maintains the drill's true concentric running ability throughout the whole drilling operation, (this can sometimes be found difficult to achieve when using a standard 118-degree included angle drill that has not been web thinned at its tip for drilling a particularly thin work-piece.

If we do not possess a correct 'drill grinding gauge' for the 118-degree standard drill, we can make our own gauge to check a *modified* 120-degree included angle ground drill, by using two hexagonal steel nuts. These must be in good condition and not possess any burrs, (and if possible, one should select new stock items); they should be of a chosen size that possesses a suitable length in their flats to match (and ideally exceed) the diameter requirements of the drill's cutting edges being sharpened. These nuts are placed so that two of their flats are securely held together in close contact, and positioned accurately in line'. These are clamped together by using a toolmaker's clamp on their outside flats. The resulting internal angle produced at the nut's junction is then used to check the newly ground drill's grinding angle. This gauge will now contain the required 120 degrees of included angle. Care should be taken to ensure that the nut's flats are being held correctly in place during their final assembly, by ensuring that both of the nut's flats are in true contact with the table's surface before being finally clamped together to establish a true relationship forming the gauge.

[(The drill can now be sharpened to its new 120-degree included angled cutting edges by using the usual off-hand grinding techniques and by then using this new gauge to check the accuracy of the drill's cutting angles, its cutting edges' comparative length, and the true centralization of the drill's point.)]

The text now continues by describing the actual 'drilling of the hole' procedure in detail.

A test hole, using the original selected drill, should now be drilled into a suitable piece of a scrap material of similar thickness. We do this to check that the experimental drilled hole is being produced to the exact required diameter. This diameter can be checked by using either a drill blank (of the required diameter, requested from the tool stores), or by a plug gauge (if this diameter of plug gauge is available), or by using a knife-edged vernier's internal measuring jaws to check the diameter of the drilled hole. This is a very wise procedure to adopt, for experience has shown that the general run-of-the-mill twist drills (especially the well-used ones that have suffered long use in service), cannot be relied upon to drill a hole of the exact diameter required; this applies even though its size is engraved on its shank and has been measured using a micrometer across its cutting edge's diameter (its outer land corners). A drill that has been re-sharpened many times will of course become shorter and will therefore suffer from web or core thickening as a result of being this shorter length. (This is a condition where the web of the drill gets progressively thicker as the drill gets shorter as it is being constantly reground.). In practice therefore, the central core thickness of the drill will be found (after long use) to become much thicker than required because its length is reduced by excessive grinding; this results in its chisel edge becoming progressively wider and thicker, resulting in a tendency for this area to be reluctant to cut. This condition therefore requires more extensive web thinning to be carried out in order to allow the drill to retain its original cutting ability; it should also be noted that the hardness and cutting ability of a shortened drill will also decrease as its re-ground cutting edges become nearer to its shank (thereby resulting in a much softer drill material being presented at its cutting edges). The drill in this case will therefore lose its?critical hardness as its cutting edges approach its shank; this 'softness' and its poor 'lasting cutting ability' will indicate to the machinist that this drill has now reached the end of its useful life and should now be scrapped and replaced.

Note that a used twist drill found to have been re-ground incorrectly (say by its previous user) may possess cutting angles and cutting-edge lengths at variance to the normally accepted drill-sharpening criteria, which state that an ideal drill cutting edge should be of equal length and of equal angle, and should possess an included angle of 118 degrees at its cutting edges. A faulty ground drill, in all probability will produce a hole that will be oversize relative to a drawing's stipulated limited diameter requirements.

An oversize drilling of a hole will usually be found to be the result of one of the drill's two cutting edges being ground to either an incorrect angle or to it being allowed to have a slightly longer cutting edge relative to its other cutting edge. In use, this longer cutting edge's surface will effectively displace the drill's center line (thereby increasing its effective cutting diameter); this causes the drill to make the hole much larger than required while it is being used in this off-centered position.

If one is attempting to drill a hole while using an off-centered drill, the imbalance of its cutting edges will allow the drill to be pushed slightly sideways by the amount of its eccentricity, thereby forcing its main cutting edge point to occupy an off-central position. The drill's longer cutting edge will then dominate the actual size of the drilled hole's diameter, while its true diameter will still be lying within the confines of the work-piece. The longer cutting edge will be traversing a larger arc than that of a correctly sharpened and centered drill; therefore, this allows the drill to produce a larger-diameter hole than is indicated by the diameter engraved on its shank, or the measurement made by a micrometer across its cutting edge lands.

If one analyzes this situation, it will be realized that one can actually vary the diameter of the hole the drill produces, by deliberately allowing the drill to be ground off-center to produce a required larger-diameter hole in the work-piece. This can be done by deliberately grinding the drill with a slightly off-centered point'. This practice can therefore be put to good use when the situation arises where a hole of a slightly larger diameter than a 'stock stock-sized drill is required.

This modification to the stock-sized drill will require the deliberate re-grinding of its cutting edges so that one of its cutting edges is ground with a slightly longer length than that of its partner; this then off-centers the drill's point (including its center-line) by an amount that is sufficient to allow an oversize hole to be produced. With practice, this modification will enable the drill to produce an accurate hole of considerably larger diameter than the original drill would have been able to produce if unmodified.

However, when inspecting and preparing a drill to produce an oversized precision hole, extreme care should be taken during the off-hand grinding process (including the amount of web thinning) of the drill's central chisel points edge. This grinding operation is undertaken to enable the full length of both its cutting edges to cut into the material without allowing the chisel edge to rub (often the case when web thinning has not been carried out at all or has been carried out inaccurately).

Should the machinist be in a situation where there is a definite need to off-center the drill in order to obtain a slightly larger hole than the stock-size drill will provide, then to drill this hole accurately in the intended workpiece, the modified drill must be supported (on the work-piece's underside) in the vice, by employing an adequate thickness of flat similar-grade scrap material under the work-piece so that the drill, after passing completely through the actual work-piece, enters the scrap material while still cutting at its maximum oversize diameter. This must be done in order for the drill to maintain the required off-center support, for its (larger) drilling arc. If one fails to use sufficient thickness of this underside-supporting material below the work-piece, then on breakthrough, the drill will (by its center now being unsupported due to the insufficient thickness of packing material), revert to using its 'normal' diameter of using the original chuck's center-line's support, which will cause the drill to drill the last part of the unsupported hole at its 'normal' diameter. It will then be found to leave a minute ridge of smaller diameter at the bottom of the work-piece's hole. Therefore, without the drill's point being guided by the thick packing's support, it will allow the drill to produce a normally centered hole of the basic drill's diameter for the last part of the unsupported portion of the hole in the work-piece. This practice of off-centering a drill to obtain a larger hole, should only be used when no other course of action is available, and this method must only be used to increase the required diameter of the hole by a small amount (say, by 0.1mm); otherwise, extended use of this offset mode of drilling will inevitably cause the drill to break (through metal fatigue) in the area just below its gripped shank area of the chuck's support.

Note that completing a precautionary drilling of a piece of similar scrap material and noting the size of hole the modified drill produces, will be a worthwhile exercise as it can prove that the diameter of hole it produces is correct. It is advisable to take this precautionary step before using a randomly chosen drill to drill an important drilled hole in the actual work-piece. This accurate check of the test hole's diameter will prevent the actual work-piece being scrapped because of its hole's diameter being incorrect. (Note that the author has only been forced to use this method of drilling an oversized hole when he has found himself in the awkward situation), where no other suitable drill is available.

If a specified critical diameter of the hole is stated on the drawing, (this will usually be indicated by its dimension being given (for example) as 5.00 mm, rather than just a plain 5 mm; in this case, a more accurate method must be used to ensure that the diametrical accuracy of the drawing-required hole size is observed.

We can of course ensure that the hole diameter will be *reasonably* correct if we initially drill all three of the holes using a smaller-diameter so-called reamer drill (therefore in this case we would use a drill of, say, 4.7-mm diameter), leaving approximately 0.3 mm of excess material in the hole. This would then be followed (without moving the machine relative to the work-piece) by using a straight-shank machine reamer of 5.00-mm diameter at a low spindle speed of, say, 125 rpm, and using either normal coolant or a thick reaming oil, (in this case, we must of course use the undisturbed drill chuck for drilling and holding the reamer's parallel shank), in order to finish the hole to the exact diameter required and to comply with the drawing's required tight-limit hole diameter.

By adopting the practice of re-grinding all of his own twist drills with a so-called thinned point', accompanied by grinding its angles equally and its cutting edges to possess exactly the same equal length), the author has found through experience and practice, that this allows the drill to actually cut and remove metal from the whole central area of the drilled hole that will extend up to the drill point's center-line. The drill will not now encounter any of the previously described rubbing action normally experienced in the drill's central area when using the traditional chisel-edged point drill design (as this is known to cause minute wobbling of the drill to occur). The added thinning of the drill's point will result in a drilling operation that requires much less effort on the part of the machinist to push the drill into and through the work-piece. Note that if one is engaged in the drilling of brass or bronze material, or if one is in the process of opening out an existing hole to a larger diameter (i.e. opening out a previously undersized drilled hole), then the drill must be backed off before attempting this operation. This backing off is done by reducing its spirally formed 25-degree normal cutting angle to effectively 0 degrees over the whole cutting surface of the drill's inner cutting area. This is done by using the corner of the grinding wheel to grind a very narrow flat (of, say, half a millimeter in width on the inside of its cutting edges to extend to virtually their full cutting length, in line with the drill's center-line. This backing-off operation to the drill reduces its normal cutting action into what is virtually a scraping action during its progression through the smaller hole. By performing this modification to the drill's inner cutting edges, the drill can then be safely used to open out the hole to its new and required diameter without fear of it *grabbing* the work-piece. This *grabbing* action will occur if a normally ground twist drill that possesses a normal 25-degree cutting angle is used to open out an existing hole. In this case, there will be the tendency for the work-piece to climb up the drill, and in so doing forcing the work-piece up and out of its holding vice (the student should now realize that a normal drill's cutting action, if called upon to open out a hole, will generate a marked similarity in its drilling action to that of a cork-screw being used to remove a cork from a bottle, but in this case, the cork will climb up the cork-screw (and in our case, the work-piece will tend to climb up the unmodified drill, flail around dangerously, and often cause the drill to break).

It may also be found necessary on certain occasions (when it is found that a 90-degree counter-sink is unavailable or is too short to reach the part that needs to be countersunk), for the machinist to similarly back off (by using off-hand grinding techniques) a standard drill of the required diameter, in order to provide its cutting edges with a similar zero-degree cutting angle, coupled with grinding both cutting edges to an included angle of 90 degrees, in order to be used as a replacement for the unavailable counter-sink. This modification can be achieved by regrinding the drill's cutting edges from its normal 118-degree included angle to its new required included angle of 90 degrees. The machinist must also allow sufficient clearance angles to be ground behind its newly modified 90-degree cutting edges to ensure that they are fully clearing the work-piece being cut, to enable it to accurately perform a counter-sinking operation without any fear of it grabbing or rubbing the work-piece bore due to insufficient clearance grinding being provided at its cutting edges.

After this rather long explanation regarding the grinding of twist drills and drilling practice, the script now returns to the previously unfinished job of ensuring that the marked-out crossed points on the work-piece plate have been center-punched correctly.

Having confirmed that the punched indents are in their correct positions, (it is good practice to use a magnifying glass to ascertain that they are correctly placed), they can now be given a final heavier hammer blow to provide the required larger and deeper indent necessary to allow the twist drill's point to self-center itself correctly, enabling it to drill the hole accurately. The use of the modified and thinned point on the drill will enable its self-centering action to take place more accurately than if one were using a basically unmodified chisel-edged pointed drill.

All three holes can now be drilled and reamed as necessary, by accurately positioning the twist drill's center-point, in turn, to coincide with the work-piece's three center-punched indents to complete the drilling of the work-piece.

Finally, on completion of the drilling operation, the holes must then be re-checked for diametrical accuracy to ensure they comply with the drawing's specified size limit of 5.0 mm plus or minus 0.05 mm on their diameters.

* * *

There follows a full description of the most accurate method to use when drilling precision holes in a work-piece when an accurate machine tool is available.

There exists in the engineering industry, a very precise method used for the precision drilling of holes in a work-piece. This operation is performed without the need to actually rely on the hand and eye to position the drill relative to the work-piece. The following method is used in the engineering industry when accurate diameters and an accurate positioning of the holes in a plate are considered very important.

The most accurate machines used to perform this task, are either the jig-borer, the jig-drill, or, (if a particular machine is equipped with an accurately calibrated lead screw), a precision vertical milling machine, that is preferably equipped with a digital readout facility.

The first two machines quoted are equipped with extremely accurate and calibrated measuring scales affixed to their lead screws. This built-in accuracy applies to the horizontal, vertical, and longitudinal axis of the machine, thereby making them capable of very precise positioning of the center-line of the machine's spindle and its drill holding chuck's center-line, relative to the center-line of the holes or bores required in the workpiece.

The procedure adopted by the operator to perform this task is as follows: The workpiece is first accurately placed squarely in position on the surface of the machine's work-table and temporarily clamped down, (we must not forget at this early stage that if the holes are to be drilled completely through the work-piece then we must interpose a piece of perfectly flat wood or MDF (medium density fiberboard) between the workpiece and the work-table to avoid any possibility of drilling (too deeply) into, and damaging the machine table's work surface (*this is an important precaution and must be observed*). A magnetically based dial test indicator (DTI) can then be used to check that the work-piece is set parallel to the machine's longitudinal work table Vs by allowing its probe to run along a convenient face edge in order to obtain a zero, zero, indication on its dial throughout a representative straight length of the workpiece. When set correctly, the workpiece is finally securely clamped down.

Alternatively, if two short metal tubes are cut to a length sufficient for them to bottom in the table's T slots while also being able to contact the datum edge of the workpiece, these will be found to be most useful pieces of tooling. They must of course have an exact diameter that fits with preferably a very slight interference fit in the table's T slots), these stops then allow the workpiece to be pushed up against them in order to obtain a true workpiece parallelism, thereby establishing that the work- piece piece will be in line with the longitudinal axis of the machine without the need to use a dial test indicator for this particular parallelism test. This setup will also be found to be particularly useful when dealing with batches of work-pieces, where these stop tubes are left in position to locate each component consecutively. The provision of an additional end stop located against the workpiece's side or end datum face being suitably clamped to the table, will also prove helpful for reproducing the exact longitudinal position of all the workpieces in the batch. It has been found that steel tubes are much easier to knock into the T slots rather than using solid blocks, as the tubes will deform very slightly and retain their fixed positions more readily.

The operator then uses a device called a center-finder (or edge finder), commonly called a wobbler (see fig. 49), to position, by means of using the machines' calibrated scales and lead screws, the datum edge of the workpiece relative to the center-line of the machine's (drill holding), Jacobs chuck. The diameter of the ball end of this edge finder must first be checked for diametrical accuracy by using a micrometer, (note that edge finders still exist with either metric or imperial diameter-locating balls). The metric diameter of this ball-ended location device is usually 5.00 mm. (The imperial alternative device is normally 0.200 inches in diameter; it is most important therefore that the particular ball's diameter being used is checked by using a micrometer to establish that the correct one is being used for this particular metrically dimensioned job.). With this edge finder secured in position in Jacobs chuck (at a point well away from the work-piece's now secure datum edge), the machine is switched on with a selected spindle speed set to a slow 250 rpm. The table's hand wheels are then used to move and position the work-table (which now includes the aligned work-piece), so that the tool's' rotating ball is brought into a position where it just touches the work-piece's side datum edge while carefully using both the longitudinal and traverse hand-wheels of the machine. Any further lateral machine movement after this point is reached will cause the rotating ball end to decrease its wobbling arc of motion.

As the wobbling ball progressively contacts the work-piece, its pronounced wobble will gradually reduce as the table adjustments are made, and will finally settle down to a situation where the ball will be rotating perfectly truly with no apparent wobble.

This is considered the crucial point of contact, and a mental note should be made at this time to check the machine table's scale's position, for without further longitudinal table movement, the longitudinal index scale should now be set to a (provisional) zero, using both the hand and eye. This is a critical point in the setting up process, for it is at this point where, with just the slightest extra movement of the table's hand wheel (and the table), the ball end will flick itself out of its true rotation in quite a violent manner, giving the impression that the ball is trying to climb up the side of the work-piece. It is therefore at this exact flicking-out point that its longitudinal scale should be reset to read exactly zero; this should then be re checked by repeating the same procedure and making any final adjustments to the scale as found necessary. This is the exact point where the longitudinal dimensions (and the later transverse dimensions) indicated by the wobbler are used first in the longitudinal plane of the machine, as the *preliminary* datum. Note that if the machine is equipped with a digital readout facility, this method of indexing can now be used in preference to using the machine's indexing scales for its setting up; therefore, if fitted, the digital readout should be set to read zero. (This particular method is considered an even more accurate way of setting the zero datum edge position on the work-piece). Having previously checked that the ball end of the center finder is 5.00 mm. in diameter, we now mentally halve this dimension to 2.500 mm (thereby giving us the ball's exact center-line). The machine's' table is now moved along longitudinally (or transversely in the case of later transverse operations), by using the appropriate hand-rotation Together with its scale, by exactly 2.500 mm, while checking the machine's calibrated scale (or its digital readout) for guidance. This move of the workpiece now positions the center line of the ball (and the machine and chuck's center-line) exactly over the work-piece's 'datum face' edge (the crucial point of the setting-up procedure); *its scale is then reset to read exactly zero.* The exact previously calculated dimension from the workpiece's datum face to the center-line of the hole is then used to move the table the exact amount required to enable the chuck's center-line to be in position now exactly over the intended hole's position, to comply with its newly calculated longitudinal dimension. This exact position can be checked practically by using a pointer (this being a 6-mm diameter HS steel tool bit, precision ground with a 60 -degree included angle sharp point) being held in the chuck to ensure the calculated move is correct. We now make either pencil or felt-tip datum marks on the machine's indexing scale (or a note of the digital readouts display to give the hole's exact center-line's position), to identify each of the three intended holes' locations; a note is also made to identify each hole's scale's position. (These notes become useful if we later need to identify these positions for drilling the other components when engaged in batch production).

This process of moving the center-line of the ball over the center-line of the intended hole is then repeated from the transverse work-piece datum face, followed by using the machine's transverse slide and scale (or its digital readout) to the exact calculated dimensions required from this particular datum face to the hole's center-line by using the transverse hand-wheel, its slide and its scale, to position the work-piece exactly over each of the intended hole positions; once again we must pencil in their exact locations on its indexing scale (or make an accurate note in the case of using a digital readout). It is now also advisable to make a sketch plan of the holes on a separate piece of paper with details of all the machine index scale readings so that they can be retained for each individual drilled hole's position. This will then allow the exact positions of the drilled holes to be repeated accurately during the following drilling operations required for each individual hole.

The recommended drilling sequence to use, when the ultimate precise position of the holes is required, firstly would be to spot drill all of the hole's locations with a shallow indent, using the machine's center drill revolving at 800 revs per minute, followed by checking the accuracy of the indents by using a pointer followed by using a vernier caliper to check their positions, then by drilling the hole using a previously selected smaller-*diameter* reamer drill' to drill the individual hole, followed by using a machine reamer (preferably of the parallel shank type that fits into the Jacobs chuck) without of course moving the drills or the reamer's position before drilling each position. It is very important that the machine is not moved between the drilling operation and the reaming operation for each hole. Note that if repeat settings are required for batch production purposes, then these must be retained by making an accurate reference sheet that will include the correct dimensions required for drilling the following production batch of workpieces. Note that when finally preparing the work-piece for drilling the holes, it is the usual engineering practice to go through the full procedure by first spotting all of the hole positions, using a switched-on and rotating (at 800 r.p.m), center drill, then inspecting the accuracy of the indented spots by using a vernier caliper, then fully center-drilling all of the holes, followed by individually reamer drilling and reaming all of the holes exactly at their longitudinal and transverse slide settings previously noted down on the reference sheet, while of course using a suitable reaming oil lubricant or coolant, (except in the case when using cast-iron material, which cut at their extreme end; therefore, these must be cleaned regularly, and all swarf removed before reaming each hole. For safety, a small brush should be used for this operation.

*   *   *

If, however, the issued drawing does not stipulate that extreme accuracy is required in the holes' positioning, or that the holes' diameter's are critical, then the drawing in this case will state just a plain 5-mm diameter is required for the drilled holes. This will of course allow a much wider tolerance of say plus or minus 0.5 mm on the position of each hole. In this case, the following simpler and quicker method can be used for this particular drilling operation.

The machinist may choose on this occasion to use a radial-type drill for this drilling operation. The drilling arm of this machine, (as its name suggests), can be rotated around its large-diameter and robust support column in a radial arc. This allows the drill holding chuck to be located and positioned by eye over the marked-out holes' position prior to its arm being locked to perform the drilling operation. This design of a drilling machine allows its complete drilling head assembly, complete with its chuck and drill, to be moved freely around and along its radial support arm as it is positioned (at will) to cover quite a large area of the drilling table, including the immediate work-piece area. This machine has the advantage that it possesses a selectable powered drive to the drill's feed mechanism, a feature found very useful for the repetitive drilling of large-diameter holes that require high downward pressure to be applied in order to complete the drilling operation while using the larger-diameter Morse-tapered fixation of the larger twist drill.

The method normally used for this drilling operation would be firstly to clamp the work-piece to the drilling table (not forgetting to interpose a suitable piece of perfectly flat wood or MDF underneath the workpiece to allow the drill to break through without damaging the table) and for the operator to use a hardened steel pointer (a piece of quarter-inch, (or 6-mm diameter) round tool steel, precision ground to possess a 60-degree included angle at its point), held in the Jacobs chuck and used to position the whole drilling machine's radial beam and chuck assembly by eye over the clamped work-piece into its correct position for drilling each individual hole. (In this case, the accuracy of the work-piece will be relying totally on the accuracy of the previous marking-out operation and the accuracy of the chuck and drill's positioning.). The radial arm assembly of the machine is then locked into this position, the center-drill is substituted for the pointer, coolant is applied, and the work -piece is center-drilled (by using preferably a high spindle speed of 800 r.p.m (this is the speed normally used for small-diameter center drills), to the depth of approximately half the depth of the center drill's 60-degree taper, positioned exactly over the center-punched indents previously made for manual drilling. The selected drill of the required diameter is then substituted for the center drill, coolant is applied, and a slower spindle speed is selected (say, 300 r.p.m, and more suitable for the drill in use. The hole is then drilled. This whole procedure is carried out independently for each of the three holes. However, this method of drill positioning is not considered as accurate as that performed by using a precision (lead-screw equipped) machine, which uses its calibrated lead screws scales, and a wobbler edge finder to obtain its accurate hole positioning that allows the accurate drilling of the holes to be carried out to their calculated coordinate dimensioned positions.

It will occasionally be found that the managements of engineering companies will be reluctant to allow their valuable precision machines to be used for just the run-of-the-mill, non-precision-type drilling work, if this particular work could be performed manually and by a bench fitter using a radial or pillar drill equipped with (preferably) a rotary table indexing device. In normal practice, the workshop's precision machines are often tied up on more important precision work, and this tends to prohibit their use for the less accurate run-of-the-mill type of drilling work.

At this point, it is prudent to bring to the notice of all engineering students when working on the shop floor, that there always seems to be a chronic shortage of clamps and 'holding-down T bolts' and their respective nuts and washers. These T bolts are used to clamp the work-piece securely down onto the machine's table by using their T-shaped head located into the T slots of the drilling and milling machine table, (or the rotary table, e.t.c.). The use of these T bolts enables the work-piece to be securely clamped down with the aid of suitable clamps. It is therefore often the case that these nuts, washers, and bolts, required for the job in hand, are not immediately available for use, these being difficult to find that possess the exact diameter, or the exact length required, particularly if the machinist needs to use pieces of threaded studding to securely clamp the work-piece onto the machine. This situation will often cause delay in starting the actual production of the component. To overcome this problem, the author has found it expedient (over a period of time), to make one's own complete set of T bolts, T nuts and clamps, in order to get the job done without unnecessary hindrance. It is advisable therefore for all students to use any spare time they may have available, to make their own T bolts, T nuts (and if possible, a few suitable drilled plates to be used as clamps), and to keep them in their own toolboxes to be instantly available for immediate use when required.

The basic hexagon head of a standard fixation bolt can (in emergencies) be modified to produce a clearance fit relative to the T slots in the milling machine's table, by machining two shallow flats *under* its hexagon head to allow it to locate into the underside of the slot by approximately one third of its head's thickness, (this is done in order to prevent the bolt rotating while being tightened or loosened). Standard nuts can also receive identical clearance machining to their hexagons, making them suitable for use in the machine's T slots when being used in combination with varying lengths of threaded studding.

When making one's own set of clamps from suitable scrap steel plate material, it is advisable to drill the necessary clearance hole (required for the holding-down bolt), in an offset longitudinal position to enable the clamp to be used in either it's near or its far area of clamping. Ideally, when setting up the clamp to tighten the work-piece down, the clamping bolt should always be positioned as close as possible to the work-piece; its packing at its other end should always be positioned considerably farther away from the clamp bolt as this then aids its clamping action, and ensures that the maximum pressure is being applied to the work-piece, and not to its packing piece.

Referring now to fig. 18-d, e, and f, this drawing shows a dimensioned sketch prepared to be used by the inspection department to accurately check the position of the holes in the metal plate as detailed in fig. 18.

Following the previous description of practical drilling and clamping methods used on the machine shop floor, we must now return to the preliminary calculations needing to be made before the drilling work can be started. We do this by using the calculations made previously in fig. 18-a and b that show us how the given pitch circle radius of 26-mm can be used in the fig. 18e calculation to allow us to find the length of side *c* in the triangle (the center line of the hole to the center-line of the plate).

For this we use the formula shown in the prompt $\angle$ *a*, *c* = *a* × COS $\angle$.

We enter the sequence 30, COS. (display shows 0.866025403), X-M; followed by the sequence 26,×, R-M, =, giving us 22.5166605, (this now gives us the dimension from the center line of the hole to the marked-out center line of the pitch circle. This is now multiplied by two, to obtain the center distance between each hole, to give us 45.033321 mm.

This dimension then enables a knife-edged Vernier caliper to be used for checking purposes; we now need to calculate the distance between the radii at the side edge of the holes relative to the workpiece's center-line. To do this, we therefore enter 22.5166605, and subtract half of the 5-mm hole diameter, (-2.5 mm.) to give us 20.0166605 mm; this is now multiplied by 2 to obtain 40.033321. This measurement is then used on the completed work-piece to check that it complies with the drawing, by using a knife-edged vernier caliper. This instrument is further used to check the dimension found between the radii of the side edges of each of the three 5.00-mm drilled holes.

On completion of the work-piece, this checking procedure would be carried out by the company's inspection department, who would perform this checking procedure by using a similar knife-edged vernier caliper.

It is possible that if the work-piece drawing calls for tight tolerances and a very accurately made component, then the inspection department may decide to use a rather more sophisticated inspection technique than was used for its manufacture. This would involve the use of a slip gauge pile (similar to that used in fig, 38), in conjunction with two 5.00-mm diameter drill blanks (used to obtain the 2.5 radii), or radius gauges, a clock gauge, and a precision angle plate mounted on a precision surface plate or table, to obtain the required accurate inspection of the component's dimensions.

This very thorough inspection of a component can and does cause a certain amount of dismay among many machine shop floor workers, particularly when it becomes known to everyone that the inspectors are using a more accurate checking procedure than was actually used in the manufacture of the component. This situation can cause a certain amount of discontent among the production workforce, particularly if the work-piece is found to fail this stringent check, and is found not to comply exactly with the drawing's stringent dimensional requirements.

Making an accuracy check of the holes in the drilled metal plate

To familiarize the student with the methods used by the inspection department, these fig. 18 d, e, and f drawings show the necessary dimensional check dimensions that will need to be verified on the component by the inspectors by using the following extra dimensions.

1. The dimension between the 'holes' edges should be 40.033321 mm. (shown in figures 18 d and in 18f).
2. We now add the dimensions 26 + 13 = 39. We then add 2.5, = 41.5; this is followed by the sequence 59 – 41.5 = 17.5 shown at 2 in 18d; (this is a dimensional check made by using a knife-edge vernier caliper between the side of the hole and the end datum face of the work-piece).
3. We now add 13 + 26, = 39. We also now add 20, giving us 59; we then subtract 2.5, = 56.5 shown at 3 in fig. 18d. This then establishes the vernier check dimension between the side of the hole and the end datum face.
4. We now calculate, 60 – 22.5166605 = 37.4833395; from this we subtract 2.5, = 34.9833395 shown at 4 in fig. 18d. This gives us the side of hole to side datum face dimension.
5. Now calculate 60 – 2.5, = 57.5 shown at 5 in fig. 18d; this gives us the vernier check of the side of hole to the side datum face dimension.
6. Now calculate 82.5166605 – 2.5, = 80.0166605 shown at 6 in fig. 18d, the side of hole to side datum face dimension.
7. It should be noted that if the drawing *had* called for the holes to be drilled to the more precise diameter of 5.00 mm (instead

of the less accurate requirement. of 5-mm diameter), this would then indicate the need for a more precise inspection to be carried out to comply with the diameter limit of + or – 0.05 mm, which calls for a much stricter check to be made by the inspector, in order to establish whether the work-piece would pass or fail his inspection process.

Note: The + or – 0.05 mm (0.0019685 inch) dimension would (in engineering terms), be called tight limits. Working to these limits requires extreme care to be taken during the manufacture of the component in order to comply with the drawing's critical so-called tight dimensions.

It should now be realized that the actual diameters of these drilled holes will be very important, since their 'side of hole' radii will be used to ascertain the actual location of the center lines of all three of the holes, relative to the datum face edges of the work-piece.

It would appear in this case, that it is the intention of the designer that the three drilled (or reamed) holes must be machined with extreme accuracy to enable them to line up precisely with an absent, but existing component, that has been retained elsewhere in the stock of the customer. This would explain why the holes must be drilled with extreme accuracy to allow both the components to be doweled together accurately by the customer, at a later date.

For this reason, we must always assume that extreme accuracy is required when drilling any hole unless informed otherwise; therefore, it is the best course of action always to work to the limits stated on the drawing and always to use extreme caution in the actual positioning of any drilled or reamed hole in a work-piece. Unfortunately, it is often the case, (throughout the whole engineering industry), that machinists are rarely told or made fully aware of the possible problems that could arise should a faultily drilled component fail to fit where it is required in the assembly, until being later informed by the inspection department that the component has been drilled wrongly and has failed their stringent inspection tests. Therefore, one should always bear in mind that repercussions are quite likely to occur if faulty work has been produced.

fig. 19.   BASIC 3-DIMENSIONAL CALCULATIONS TO FIND THE TOTAL C/L

LENGTH DIMENSION FROM 'Z' TO 'W'. INCLUDING TWO BENDS.

THE PROBES ARE PLACED IN A SIMILAR ATTITUDE TO THE
PROBLEM BEING WORKED ON.

THE DIMENSIONS OF THE FLOOR DIAGONAL ARE FOUND FIRST,
FOLLOWED BY THE BOX DIAGONAL 'Y' TO 'X', THEN A.O.B. AT Y. AND X.

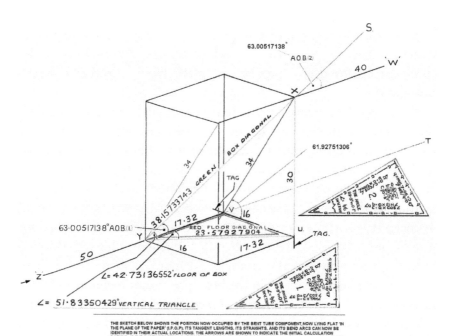

S.

63.00517138"

A.O.B.(2)

40   'W'

X

61.92751306"

T

34

BOX DIAGONAL

34

TAG

30

V   16

63·00517138° A.O.B.(1)

GREEN

38·15733743

17·32

Y   RED FLOOR DIAGONAL   U   TAG.
23·57927904

50   16   17·32

'Z'

∠ = 42·73136552° FLOOR OF BOX

∠ = 51·83350429° VERTICAL TRIANGLE

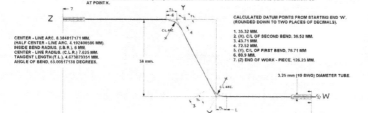

THE SKETCH BELOW SHOWS THE POSITION NOW OCCUPIED BY THE BENT TUBE COMPONENT, NOW LYING FLAT 'IN
THE PLANE OF THE PAPER' (I.P.O.P.). ITS TANGENT LENGTHS, ITS STRAIGHTS, AND ITS BEND ARCS CAN NOW BE
IDENTIFIED IN THEIR ACTUAL LOCATIONS. THE ARROWS ARE SHOWN TO INDICATE THE INITIAL CALCULATION
DATUMS. IF THE TUBES STARTING DATUM IS NOW TO BE HELD IN THE BENDING MACHINES CHUCK AT 'W'. THEN ITS
FIRST BEND WILL BE AT POINT Y, AND, AFTER BEING ROTATED 180 DEGREES, ITS SECOND BEND WILL NOW BE
AT POINT X.

Z

CENTER - LINE ARC. 8.384817171 MM.
(HALF CENTER - LINE ARC. 4.192408586 MM).
INSIDE BEND RADIUS. (I.B.R.). 6 MM.
CENTER - LINE RADIUS. (C.L.R.) 7.625 MM.
TANGENT LENGTH (T.L.). 4.673070351 MM.
ANGLE OF BEND. 63.00517138 DEGREES.

CALCULATED DATUM POINTS FROM STARTING END 'W'.
(ROUNDED DOWN TO TWO PLACES OF DECIMALS).

1. 35.32 MM.
2. (X). C/L OF SECOND BEND. 39.52 MM.
3. 43.71 MM.
4. 72.52 MM.
5. (Y). C/L OF FIRST BEND. 76.71 MM.
6. 80.9 MM.
7. (Z) END OF WORK - PIECE. 126.23 MM.

3.25 mm (10 SWG) DIAMETER TUBE.

W

34 mm.

107.32 mm.

Chapter 11

Dimensioning a tubular component to allow accurate bends to be produced, also how one calculates the dimensions of the three-dimensional triangles by using the box method of triangle calculation

The fig. 19 drawing shows firstly the method used to calculate the angles and dimensions of the three-dimensioned triangles that have been introduced into a three-dimensioned box in order to assist in their final accurate calculation. This drawing also includes a tube bending calculation exercise and a calculation check to explain the method used to obtain an accurate overall developed length of a small diameter tubular component, and the exact angles of bend required for it to be precision bent on a tube-bending machine. This is performed by using the results of the triangle calculations made using these inserted right-angle triangles in the box. The author therefore calls this method *the 'across the box' method of triangle calculation*; the following calculations also describe in detail the *stepping stone* approach method.

By successfully reaching this part of the book, the student will have now experienced and hopefully solved the majority of problems found while dealing with just two-dimensional right-angle triangle problems by using the Probe and Prompt system of triangle calculation. During this so-called learning process, he or she should have become very familiar with the use of the Probe and Prompt system by now as an aid to solving right-angle triangle calculations when they appear in the plane of the paper (called IPOP for short). We can now progress onward and upward by using a slightly more advanced method of triangle calculation that will often be needed when calculating three-dimensional objects. These three-dimensional triangle problems can be more easily explained and understood if we use the assistance of the fig. 19 drawing that depicts an angled, rectangular box containing all three dimensions of the worked-on problem, these being the object's depth, its width, and its height.

If we now refer to fig. 19, we are confronted with a drawing of the three-dimensioned angled box, complete with its three vital basic dimensions. This example of a boxed figure is shown with a depth of 16 mm, a width of 17.32 mm, and a height of 30 mm. A careful study of this sketch should reveal that it is also quite possible to introduce several right-angle triangles, at various attitudes within the figure's outline shape. The calculated right-angle triangles shown will be seen to occupy not only the plane of the floor of the box, but will also be positioned diagonally across the box from corner to corner, (top to bottom), in a so-called vertical plane, including one triangle that is inclined on a sloping plane to provide us with the correct angle of bend that will be needed for the accurate bending of a tubular component containing bends at Y and X, which also fits into the confines of the box's three dimensions and calculated angles.

By using the guidance of the prompts displayed on the probe and by also noting the positioning of the important 90-degree tag, we can now calculate the required (but as yet unknown) dimensions needed to solve the problems encountered; we do this by gradually adding to the drawing the newly discovered angles and dimensions relative to the various triangles placed within the sketch, as the individual calculations proceed. When we have completed the necessary calculation moves shown in the boxed drawing of fig. 19 and noted its additional 'Z to box' (and 'Z to Y') dimensions (these being the external leading-in dimension of 50 mm, and the external 'X to W' leading-out dimension of 40 mm), it will then be possible to obtain a *provisional* overall length of a tubular component that is seen to traverse the box diagonally between points Y and X by using the newly calculated vertex to vertex straight-line dimension (taken on the materials center line) of 38.15733743 mm from the vertex point Y through to the vertex point X. The *basic* overall length dimension of the component can now be completed by adding the two leading-in and-out dimensions of 50 mm and 40 mm enabling us to arrive at a provisional overall length (using in this case the work piece's straight center lines) to give us the figure of 128.1573374 mm. However, the student should now realize that this basic figure does not provide us with the exact length required for a work-piece that contains two bends along its length, because we have not included the *added* center-line radii of the two bends, and have not subtracted the four *tangent lengths* from its basic overall length. We will therefore need to re-calculate these dimensions to include the addition of the two bend arcs (located on the work-piece's center-line and the *subtraction* of the four tangent lengths from the basic overall length, to allow us to arrive at the exact overall length that includes its bends, to enable the accurate manufacture of the tubular component shown in the lower calculated drawing of fig. 19.

If the student has made an initial study of this box calculation method and has studied the trigonometry calculations required, he or she should note that the leading-in and leading-out' dimensions have been included in order to familiarize the student with the extra calculations required when obtaining the exact overall length of the work-piece that now possesses two bend radii at points Y and X while traversing across the box.

The length taken up by the two newly introduced bends (shown at points Y and X) will therefore need to be fully calculated in order to obtain their added center-line arc lengths and their four subtracted tangent lengths, while using an inside bend radius (IBR) of 6 mm.

Note that the name given to this 'three dimensional calculation' is the *across the box method* (as shown in fig.19).

The fig. 19 sketch also gives the student the opportunity of experiencing and using the stepping stone approach method, which guides one through using a methodical approach to the problem. A separate calculation method is also included that describes an alternative method that can be used to find the diagonal length of the straight center-line Y to X dimension when needed for checking purposes.

The text now returns to the initial calculations required to obtain the unknown dimensions and angles contained within the fig. 19 basic three dimensional box. To accomplish this, we will first need to find

The Angle of the floor of the box. We do this by studying the probe and reading off the appropriate prompt provided on it;, we therefore enter into the calculator the required sequence, $\angle$ = b/c INV TAN.

[(In some cases, the calculator used for some of the following calculations may be found to possess only eight figures in its display; therefore, should a ten-figure calculator be used to make these calculations, the resulting value will of course include two extra digits, and as a result, the total final number of digits produced will vary slightly. However, this difference will be found to be virtually insignificant in the final calculated result].)

Therefore, to find the corner angle of the floor of the box, we enter the sequence 16 ÷ by17.32 =, (display shows 0.923787528), followed by INV. TAN, giving us The Angle of 42.73136552 degrees, shown outlined in red.

To find the length of the diagonal line across the floor of the box, we use the given prompt $\angle$ c, a = $c/$COS $\angle$; we therefore enter the sequence 42.73136552, COS. (display shows 0.734543239) X-M; enter 17.32 ÷ R-M = 23.57927904, giving us the answer, (also shown in red).We now need to find The Angle of the vertical triangle (colored green), shown vertically and positioned diagonally across the floor of the box. For this we use the given prompt $\angle$ =, b/c INV TAN.

We therefore enter the sequence 30, ÷ 23.57927904, = (display shows 1.272303532), INV. TAN, giving us the answer, 51.83350429 degrees.

We now need to find the diagonal length of this vertical triangle. For this we use the displayed prompt $\angle$ b, a = $b/$SIN $\angle$.

We enter the sequence 51.833504, SIN, (display reads 0.786218379) XM, we now enter 30 ÷ RM = 38.15733743 mm, answer, giving us the Y to X' straight diagonal length across the boxed triangle.

The total length of the workpiece (by using just the basic straight center lines) from point Z through to point W can now be *provisionally* calculated as follows: 50 + 38.15733743 + 40 = 128.1573374 answer.

[However, there is an alternative method we can use to find the diagonal length from Y to X (should we need to find this quickly for checking purposes). We can do this by multiplying each of the box's so-called three dimensions (namely its width, its length, and its height) by themselves, as follows: 16 multiplied by 16 equals 256; 17.32 multiplied by 17.32 equals 299.9824; 30 multiplied by 30 equals 900. These totals are then added together to obtain 1455.9824; we then obtain the square root of this number by pressing the INV key followed by the SQUARE ROOT key, giving us 38.15733743, the answer.

[By using this additional method of calculation, we can now make an important check to confirm the accuracy of our previous calculation. We do this using the above method to produce (hopefully), an identical answer to that previously obtained for the Y to X' dimension obtained by using the Probe and Prompt method of triangle calculation.] This will be found to be a very worthwhile check of your previous calculation.

As a test of the student's mental agility, it is now important that we now find The Angle, marked V in the triangle X, V, U situated at the far end of the box. It will be seen that this triangle possesses a vertical height of 30 mm, which is currently being used as its side b, and the base length of 16 mm, which is currently being used as its line c in the triangle; therefore with this right-angled triangle now mentally placed in this new position, its The Angle can now be calculated by using the prompt $\angle$ =, b/c, INV TAN.

To obtain The Angle at V, we therefore enter 30 divided by 16 =, (calculator's display shows 1.875), followed by INV TAN, giving us the angle of 61.92751306 degrees.

We now need to find the length of side a in this triangle. We do this by using the prompt $\angle$ b a = b ÷ SIN $\angle$. We enter 61.92751306, SIN (display shows 0.882352941), XM; we then enter 30 divided by RM =, giving us 34 mm.

We now move on to consider an important (and now newly discovered) triangle in the box by now using this 34-mm dimension (found in the previous calculation) in this new triangle now being called its side b. Its The Angle position will now be at point Y. This new triangle will be seen to be leaning diagonally across the box to form a triangle X, Y, V. To find this triangle's The Angle, we therefore use this 34-mm length as its renamed side b, and its 17.32 mm side is used as its renamed side c. We then use the prompt sequence $\angle$ = b ÷ c, INV TAN. We enter the sequence 34 ÷ by 17.32 = (display shows

1.963048499) INV, TAN, giving us 63.00517138 degrees, which now gives us the important angle of bend required to bend the tubular work-piece of 3.25 mm diameter (10 gauge) tubular steel material from its basic start point at Z while bypassing its vertex points of Y and X (and following its soon-to-be-calculated bend center-line arcs) to finish at point W.

We now need to calculate the length of the two bend center -line arcs and the four tangent lengths of the two new bends now situated at Y'and X.

If we now view the box from the direction of the arrow shown positioned at point Z, it will be seen that our straight piece of tubular material (while still being in its straight and unbent mode), would initially pass through points Y and V and continue in a straight line to reach point T.

In our mind's eye, we must now visualize that we are now producing a bend in this tubular material at point Y, we do this by bending the tube bodily from its original lying flat position at point V and deflecting it by 34 mm to reach point X, thereby allowing the tube to traverse diagonally across the box to reach point S while also traversing the V to V center-line length of 38.15733743 mm) and obtaining an angle of bend of 63.00517138 degrees at point Y.

This is followed by making the second bend in the tube at point X from its temporary position at point S, (by using an identical angle of bend, at point X), where the tube material will now continue in the straight through to point W.

To calculate the length of each bend arc (this length will of course be identical for both bends), we can use the formula contained within the 'table of normal bend allowances, shown in fig. 32, and quoting in this case the figure that applies to 6-mm inside bend radius (IBR) for 10 gauge (3.25 mm diameter) material, by using the charts quoted figure of 0.133081412 in the following calculation, which is then multiplied by 63.00517138 (the angle of bend) to give us 8.384817171 mm, the length of the bend's full arc. We then divide this figure by two, to give us 4.192408586 mm, in order to obtain the 'half an arc' dimension, for use in our later calculations. This situation can now be identified in the lower drawing as being half of the C/L ARC.

[The length of the center-line arc can also be obtained by using the authors own quick alternative formula as follows: 0.0174533 × 7.625 (the tube's *center-line* radius) × 63.00517138 (the angle of bend) = 8.384817202, this being virtually identical to the previous calculation obtained by using the 'table of normal bend allowances shown in fig. 32.

To find the tangent lengths of the two bends (there will in this case be four, two for each bend), we use the author's own developed formula,*tangent length = CLR (center-line radius) multiplied by TAN of half the angle of bend*. We therefore enter the sequence 63.00517138, (the angle of bend) divided by 2 = 31.50258569, TAN, (display shows 0.612862865) × 7.625 (the *center-line* radius) = 4.673079351, thereby giving us the lengths of each of the four tangent lengths required in the following calculations. These are also referred to in the lower drawing using the abbreviated form 'T-L'.

It will be seen in the fig. 19 box drawing that the tubular component possesses three straight line center-line lengths, these being Z to vertex of Y, Y to vertex of X, and X to its starting end at W. Note that the center-line lengths at this stage in the calculation are comprised of straight lines that are situated between their vertex points, these dimensions being 50 mm from Z to Y, followed by the calculated length of 38.15733743 mm from Y to X, and 40 mm, from X to W. Note that the vertex points are also given the name 'tangent points' in our calculations.

It was originally intended by the author, that this box method of three-dimensional triangle calculation' would be used to explain basically to the student how one calculates the box's angles and the side lengths of the triangles it contains. It was then realized that if two bends were to be introduced into this box at points Y and X, then the extra calculations required for this would fully explain to the student how one can calculate the length of material required to produce two bends in a tubular component that traverses the box from point Z, across the box from point Y to point X and continued to point W.

[It should be clearly understood by the student, that this box calculation (which in this case is only using a very small-diameter tube) is being used purely to demonstrate, explain, and bring enlightenment to the student, of the extra calculations required should he or she be later engaged in the bending of accurate full-sized tubing. The calculations that would subsequently be made would therefore be very similar to those used when employing full-sized tubing, but everything will of course be to a very much larger scale.]

[A typical diameter of tubing used for normal bending in the engineering industry in the U.K would be from ½ inch (12.7 mm) diameter up to, say, 1 inch (25.4 mm) diameter, these being the most popular diameters used for general work-shop use. It should be noted that the student will also need to know the exact pitch circle diameter of the bending die that is used on the bending machine, and the exact diameter of the tube being used (in mm), in order to calculate the exact overall length dimension of tube required to finally establish the exact overall cutoff length required for the tubular component before bending.]

By introducing the two bends at pointsY and X, it will now be realized that there is the need to calculate the *center-line radius of each bend's arc*, minus the length *of each calculated tangent length*, in order to establish the exact overall length of material required for the manufacture of this component. The final calculated length of material required will therefore include the length of the two center-line bend arcs, minus the four tangent lengths, plus the lead-in straight portion, the straight portion between the two bends, and the finishing 'end straight' portion.

At this point in the calculations, it will be found very helpful if the student obtains a piece of small-diameter ductile steel (such as a piece of welding rod) or copper wire, with a diameter of, for example, 1.6 mm with a suitable length of approximately 130 mm. (The material chosen for the bends simulation could be either a small-diameter rod or a length of 3-mm diameter soft multi-cored solder, or possibly (if no other material is available at the time), a straightened-out metal paper clip for use as a bend test piece. The use of this test piece material will then allow the student, (after marking out the component's calculated bend points along its length using pencil), to physically bend the material at the calculated and marked-out points (by use of the thumb and fingers), to produce the approximate position and orientation of the bend angles, followed by checking the bend's angles with an angle protractor, to ensure they truly simulate the shape and orientation of the tubular component being calculated. The student, in our example's case, will need to rotate the 'test piece' by 180 degrees between its first and second bend to simulate the bending machine's sequence of operations being performed during the bending process. In this case, it will allow the material to be formed into the basic shape of a very lazy Z when placed in the flat in the plane of the paper (IPOP), on the surface of a flat, work-top, drawing board, or tabletop.

Performing this bending exercise practically, by using a short length of bent wire, will allow the student to convert his mental thoughts accurately and the theoretically calculated dimensions and angles into a practical object that simulates the shape of the finished component. This will then make it much easier for the student to visualize the shape of the object and to understand the bending machine's method of operation. (Note that the author still uses this bent-wire method to familiarize himself with the actions the bending machine needs to take when bending a component, and will apply particularly when the calculations being made are regarding a work-piece that contains a series of multiple bends that also possesses a series of multiple orientations.).

With the box now viewed from the direction of its arrowed end (now marked Z), it will be seen that point X in the box is 30 mm higher than its point Y.

If we now concentrate our attention on the 34-mm dimension shown running diagonally across the far end of the box (this dimension being identical to that shown at the left-hand near side of the box), we can now test our mental ability by visualizing the work-piece being rotated clock-wise, through the calculated angle of 61.92751306 (in practical terms, 62 degrees), to enable the work-piece to lie flat in the plane of the paper (or IPOP).

This new orientation of the tubular work-piece in its new flat situation 'in the plane of the paper' is shown in the lower part of the fig. 19 drawing.

By studying and viewing this lower drawing, the student should now realize that doing sonow makes it much easier for the draftsman to draw the object to its correct shape for his issued manufacturing drawing. It will also make it easier for the student to calculate the dimensions required to obtain the exact longitudinal points (or 'machine center back stops) required for the bend on the tube-bending machine. Each of the four tangent lengths and each of the two bends'center-line arcs (including each of the dies' clamping points,) can now be identified practically, while also identifying the correct positions that will be needed set the bending machine's center back stops for each bend prior to the machine operator completing the bending process. It will also be seen that the horizontal width of the component (while in its 'in the flat' IPOP form) will now be 34 mm, due to the work-piece having been rotated into this new position in the flat, in the plane of the paper (IPOP).

We have now reached the stage in the tube's calculations where the student (by using each *tangent length*, each *arc length*, and each *straight length* between the vertex points), can position all of the required bending dimensions in their correct places, similar to the lower fig. 19 drawing. We do this by using arrows and numbers in the series, 1 through to 7, with the two bends'*vertex points* being indicated by points X and Y (note also that these exact datum points will be referred to in a later paragraph).

By completing the foregoing calculations, the student should have by now improved his knowledge of the methods used in triangle calculation. He will also have completed several mental calculations with regard to the various positions adopted by the newly introduced triangles into the box required to solve the calculation problem, this includes the various positions adopted by the triangle's The Angle. In this case, we have made use of several right-angle triangles placed into various positions within the boxed problem, and while doing so we have used the 'Probe and Prompt's triangle calculations to obtain the correct angles and their correct triangular side lengths.

Completing this calculation exercise can now be considered a first in the art of placing right-angle triangles within a rectangular boxed problem, followed by using triangulation calculations to obtain the unknown but required angles and all the dimensions needed to obtain the fully calculated overall length of material required for the tubular component, complete with its 'angles of bend required to produce an accurately manufactured tubular component.

There follows the accumulated dimensions needed to set up a typical Pines hydraulic bending machine.

In this case with the work-piece's (gripped) end being held in the chuck at W, its datum points will be 35.32692065 mm at position 1, 39.51932924 mm at position 2 (also marked X to signify bend number 2), 43.71173782 mm at position 3 (the beginning of the diagonal straight), 72.52291655 mm at position 4 (the end of the diagonal straight), 76.71532514 mm at position 5 (marked Y to signify the position of bend number 1), 80.90773372 mm at position 6 (the beginning of its end straight), and 126.2346544 mm at position 7 (marked Z, being the outer end of the workpiece). Therefore for setting up the bending machine, it should be realized that the center-line of bend number one will be 76.71mm from its held end W, and the center-line of bend number 2 will be 39.52 mm from its end held at W.

Sheet metal and tube bending machines

There now follows a brief explanation of how a hydraulically operated tube bending machine is operated in practice.

The datum end (its working end W) is first positioned and secured in the bending machine's three-jaw holding chuck; this is now set to read zero on its 'angular rotation' scale (this chuck is mounted on the machine's main shaft and is located through the machine's longitudinal 'sliding carriage. The design of this sliding carriage assembly allows the whole chuck and its work-piece to be literally drawn along to its left *during* the bending operation while it is in the process of using up the total amount of the bend's arc *during* the bend's progression through the die of the machine. The stationary part of the machine's back slide is provided with *center back stops*; these provide the necessary longitudinal adjustment settings that are arranged for each individual *center of bend*' datum point. The bending die's clamp is then secured onto the work-piece in the area of position 6; this then allows the machine to produce its first bend at Y, also marked number 5 on the small drawing (this clamping arrangement ensures that the workpiece tube is drawn into and revolves in unison with the bending die during its bending of the tube).

During the making of its first bend at position Y, the center of bend (marked 5) of the bending die assembly (including the tubular workpiece), is rotated hydraulically in an anti clockwise direction to complete the full bend to a previously set angle of *bend stop*, of 63 degrees, this operation is taking place with the workpiece tube lying in a flat plane, which ends up in a position facing the direction of the operator. The workpiece is then released from its holding die and its calibrated holding chuck is rotated anti-clockwise by hand to the previously set rotational bend stop of 180 degrees (this now positions the work-piece in its correct attitude for completing its next bend). Its sliding chuck assembly is then advanced along its supporting slide into its second *center back stop* position (marked X number 2) by hand, in preparation for making the second bend), the workpiece is then again hydraulically clamped to the bending die in the area of position 3. The bending die assembly is then rotated by using hydraulic power in an anti-clockwise direction to provide the 63-degree *angle of bend* previously set on its *bend stop*.

The completed work-piece is then removed from its holding chuck to enable it to be checked for the accuracy of its bends, it's offset dimension, and it's overall length. This check will usually be made by using a precision checking jig that is normally provided for this purpose.

Chapter 12

Calculating a sheet metal component that contains one bend.

The fig. 20 drawing shows an 'as issued' drawing of a sheet metal component that contains one bend and one drilled hole along its developed length. The fig. 20 drawing is 'as supplied' to the sheet metal work'shop by the drawing office. It should be noted there is a noticeable lack of so-called full manufacturing dimensions on this drawing. This omission of vital dimensions will indicate to the skilled engineer, during his study of the drawing, that there will definitely be the need for several further calculations to be carried out before an accurate manufacture of this component can be made.

fig. 20.   SHOWING A BASIC   SHEET METAL DRAWING, AS ISSUED TO A MACHINE SHOP
FOR MANUFACTURE. NOTE THE MISSING DIMENSIONS ①, ANGLE OF BEND,
           ② LENGTH OF MATERIAL,③ FOLDING M/C BEND POINT,
           ④ AIR BEND M/C POINT, ⑤ MARKING OUT DIMENSION, FOR HOLE
           IN FLAT MATERIAL . ⑥ BEND VERTEX POINT DIMENSION.
           (THESE WILL NEED TO BE FOUND BY CALCULATION IN figs. 22.23.24.25.
           BEFORE THE WORK CAN PROCEED).
THE FULL SEQUENCE IS SHOWN IN figs. 3a. 22. 23. 24. 25. 26.27. TO COMPLETE THE COMPONENT.

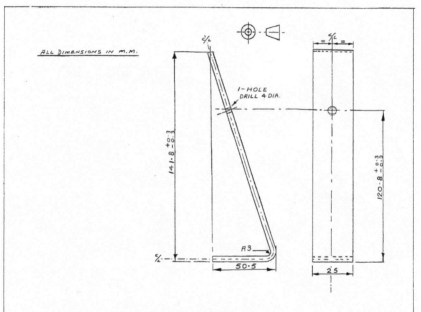

ALL DIMENSIONS IN M.M.

1 - HOLE
DRILL 4 DIA.

141·8 +0·3 −0·3

R 3

50·5

120·8 +0·3 −0·3

25

NUMBER   REQUIRED   20 OFF.

SHEET METAL BRACKET.		4 - 1250	
MTL.	1·6 mm. DURAL SHEET. L 72.	DATE	10 NOV. 13
SCALE.	N. T. S.	DRN.	G. N. R.
LIMITS.	+ or MINUS 0·3 m.m. ON ALL DIMENSIONS.		

Fig 22.

fig. 22.
A FREEHAND SKETCH OF DRAWING FIG. 20

SHOWING CALCULATIONS REQUIRED TO OBTAIN 6 FURTHER DIMENSIONS

WHICH ARE NOT SHOWN ON THE ISSUED DRAWING.

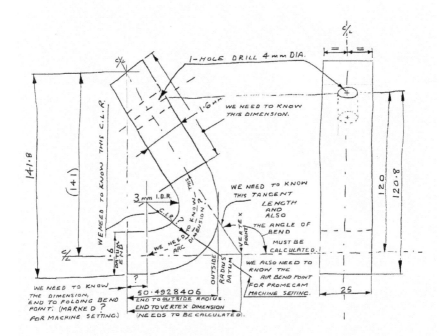

NOTE THAT A SKETCH NEED NOT BE NEAT AND TIDY
SO LONG AS IT IS EASILY AND ACCURATELY UNDERSTOOD.
THE DIMENSIONS IT CONTAINS ARE MORE IMPORTANT
THAN THE NEATNESS OF THE SKETCH.

ALL DIMENSIONS IN MM.

Fig 23.

fig. 23.   <u>SHEET METAL BRACKET. AS SHOWN IN fig 20,</u>
BUT INDICATING MISSING DIMENSIONS X, ANGLE OF BEND,
LENGTH OF MATERIAL, FOLDING M/C BEND POINT,
AIR BEND M/C POINT, MARKING OUT DIMENSION FOR HOLE
IN FLAT MATERIAL.

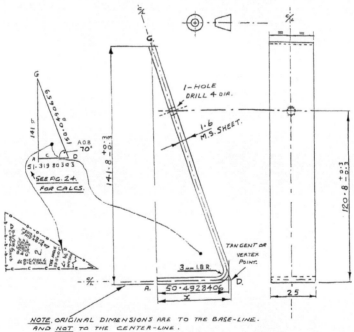

NOTE. ORIGINAL DIMENSIONS ARE TO THE BASE-LINE.
AND <u>NOT</u> TO THE CENTER-LINE.
THEY ARE ALSO TO THE OUTSIDE RADIUS OF THE BEND AND
NOT TO THE VERTEX POINT. (<u>WHICH IS REQUIRED</u>), FOR CALCULATIONS.

ALL DIMENSIONS IN MM.

Fig 24.

fig 24. **CALCULATIONS TO FIND ANGLE OF BEND AND TANGENT POINT (X DIMS),**
(ALSO SHOWN
EARLIER AS fig 3a **ON SHEET METAL BRACKET, fig 23, USING THE 'PROBE AND PROMPT' SYSTEM.**
FOR COMPARISON
WITH fig. 3.)

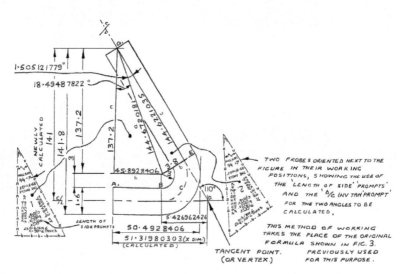

TWO PROBES ORIENTED NEXT TO THE
FIGURE IN THEIR WORKING
POSITIONS, SHOWING THE USE OF
THE 'LENGTH OF SIDE' PROMPTS'
AND THE 'b/c INV TAN PROMPT'
FOR THE TWO ANGLES TO BE
CALCULATED,

THIS METHOD OF WORKING
TAKES THE PLACE OF THE ORIGINAL
FORMULA SHOWN IN FIG. 3.
PREVIOUSLY USED
FOR THIS PURPOSE.

TANGENT POINT.
(OR VERTEX.)

ALL DIMENSIONS IN MM.

Fig 25.

FIG 25.   THE BEND CALCULATIONS FOR THE METAL BRACKET SHOWN
IN FIG 23, PROVIDE ALL THE DIMENSIONS REQUIRED FOR
ITS MANUFACTURE.

IN ORDER TO SHOW CLEARLY 'WHAT IS GOING ON' IN
THE BEND, THE SKETCH HAS BEEN ENLARGED
SUFFICIENTLY TO CONFIRM IN DETAIL ALL 6 OF THE
PREVIOUSLY UNKNOWN BUT REQUIRED DIMENSIONS
NEEDED FOR THE MANUFACTURE OF THIS COMPONENT.

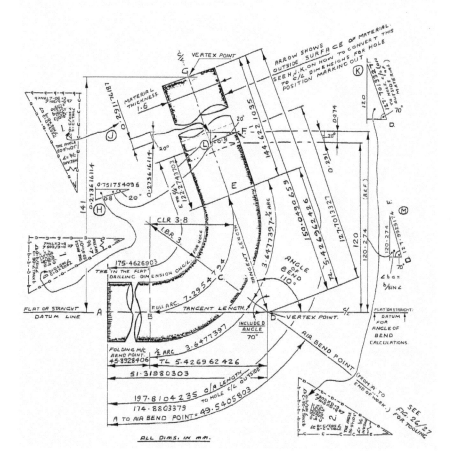

ALL DIMS. IN mm.

Fig 26.

fig. 26. THE PROMECAM HYDRAULIC 'AIR' BENDING MACHINE BEING USED
TO BEND THE METAL BRACKET SHOWN IN fig. 23.
THIS MACHINE IS USED WHEN EXTREME ACCURACY AND AN UNBLEMISHED
SURFACE FINISH IS REQUIRED ON THE FINAL PRODUCT.

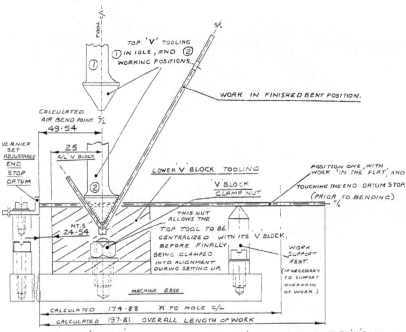

TOP 'V' TOOLING
① IN IDLE, AND ②
WORKING POSITIONS.

WORK IN FINISHED BENT POSITION.

CALCULATED
AIR BEND POINT. C/L
49.54

VERNIER
SET
ADJUSTABLE
END
STOP
DATUM

25
C/L 'V BLOCK

LOWER V BLOCK TOOLING

V BLOCK
CLAMP NUT

POSITION ONE, WITH
WORK 'IN THE FLAT', AND
TOUCHING THE END DATUM STOP.
(PRIOR TO BENDING.)

N.T.S
24.54

THIS NUT
ALLOWS THE
TOP TOOL TO BE
CENTRALIZED WITH ITS V BLOCK,
BEFORE FINALLY
BEING CLAMPED
INTO ALIGNMENT
DURING SETTING UP.

WORK
SUPPORT
REST.
(IF NECESSARY
TO SUPPORT
OVERHANG
OF WORK.)

MACHINE BASE

CALCULATED   174.88   'A' TO HOLE C/L
CALCULATED 197.81 OVERALL LENGTH OF WORK

NOTE THE 'AIR GAP', BETWEEN TOP TOOL AND WORK,/ ALSO UNDERSIDE OF WORK TO 'V' BLOCK.

ALL DIMENSIONS IN MM.

Fig 27.

fig 27. THE FOLDING MACHINE 'SET UP', FOR BENDING
THE METAL BRACKET fig. 23. THIS MACHINE IS USED
WHEN EXTREME ACCURACY IS NOT REQUIRED.

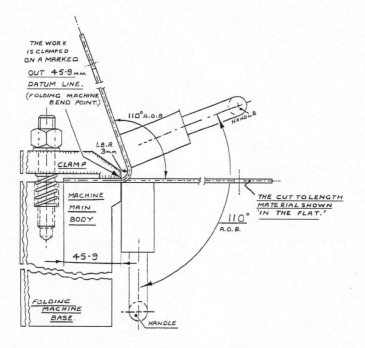

THE WORK
IS CLAMPED
ON A MARKED
OUT 45·9 M.M.
DATUM LINE.
(FOLDING MACHINE
BEND POINT.)

110° A.O.B

HANDLE

I.B.R
3 M.M

CLAMP

MACHINE
MAIN
BODY

THE CUT TO LENGTH
MATERIAL SHOWN
'IN THE FLAT.'

110°
A.O.B.

45·9

FOLDING
MACHINE
BASE.

HANDLE

ALL DIMENSIONS IN MM.

Fig 28.

fig. 28.   THE CALCULATION OF SPECIAL 'STRETCHED BEND,' ALLOWANCES.

SEE ALSO TABLE OF 'STRETCHED' BEND, ALLOWANCES fig.32a.

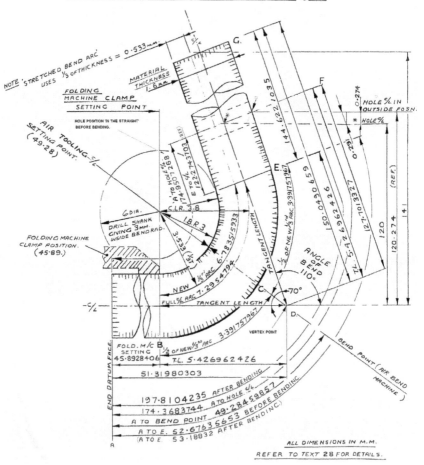

ALL DIMENSIONS IN M.M.

REFER TO TEXT 28 FOR DETAILS.

It must be said that this (fig. 20) drawing contains the minimum of displayed dimensional information, which is typical of an 'as issued' sheet metal drawing received from the drawing office by the workshop staff for the manufacture of a typical sheet metal component.

It will quickly be noticed that a large amount of manufacturing information is actually not contained within this drawing, indicating that only the absolute minimum of dimensions the sheet metal worker is required to know (by the drawing office), for the manufacture of this accurate component.

It is therefore considered quite ironic that the dimensions that have been supplied (while being totally adequate for the accurate inspection of the finished bent component,) will be found to be totally insufficient for the sheet metal worker to use to actually produce the component to the required accuracy.

In fact, this drawing, while being perused by the personnel of the sheet metal shop floor, will be found to require a total of *six extra* vitally important dimensions needing to be calculated before productive work on the component can proceed.

For this reason, the extra sketches and the calculations required to obtain these so-called *drawing office omissions* are shown very effectively in the drawings featured in figures 3a, 22, 23, 24, 25 and for the alternative *stretched version* of the component in fig. 28.

However, for the manufacture of this type of component, it must be explained that there are usually two quite different types of bending machine normally used for the manufacture of this component. They are shown in figures 26 and 27.

There exists in the industry one particular version of a precision sheet metal bending machine that is used where extreme accuracy and unblemished surface finish of the component is required. This particular machine is used for the bending of precision sheet metal parts used in the aircraft, and space industries; this machine is called the **Promecam** air bend machine.

The second type of machine used for bending the less accurate sheet metal components is called the folding machine or (the folder).

This folding machine is used for bending sheet metal components, where the components' accuracy is slightly less important, and the material's surface finish is not considered a critical factor, and it's therefore generally used for the run-of-the-mill production of sheet metal bending work, where the machine may be either hand operated or power driven.

The calculations required to find dimension *x*

When designing sheet metal and tube components, it is general practice in the engineering industry, for the draftsman to issue working drawings with dimensions given to the extreme 'outside edge' of the components' bend radius, (as is shown in the issued production drawing fig. 20). Adopting this practice makes the inspection of the finished component relatively easy to undertake *by the inspection department*, but the lack of certain vital so-called *missing* manufacturing dimensions, makes it more difficult for the sheet metal engineer to manufacture the component to the required accuracy, without the need for him or her to make a series of very important trigonometry bending calculations.

In the fig. 22 sketched example, and in order to prove this specific point, the author has shown the dimension of 50.4928406 mm being given to the outside radius of the finished component, (this dimension would of course be 'rounded up by the draftsman to 50.5 mm (as shown in fig. 20 drawing) before actually issuing the drawing). The author has left this figure in place in the fig. 22 sketch to assist in proving the point that this dimension is being given to the *outside radius*, and that what is required in a sheet metal bend calculation is the dimension *datum end to the vertex point*, which will need to be calculated.

The actual dimensions required by the shop floor engineer or sheet metal worker, to enable the manufacture of this component to the accuracy required, is for up to eight *other* necessary dimensions, which are unfortunately omitted from the drawing to be calculated. These vital but omitted dimensions would have given the sheet metal worker the exact position of the bend's vertex point (thereby allowing its angle of bend to be easily calculated), as shown in the fig. 23 drawing marked *x*.

This critical *x* dimension is given rarely (if ever) on a sheet metal drawing; it therefore becomes necessary for the sheet metal worker (or the sheet metal fitter) to perform several additional calculations, (with the help of the Probe and Prompt triangle calculation system), to establish this position accurately prior to carrying out all the other required calculations.

Using the Probe and Prompt system to aid this calculation will involve the use of two newly introduced additional right-angled triangles. These are carefully positioned in the fig. 24 calculation sketch, while being assisted by using practical trigonometry calculations', to link the two correctly oriented right angle triangles, in a particular triangle calculation method called *triangulation*; this is shown in fig. 3a and in fig. 24 (the modified replacement sketches and drawings).

In fig. 3a, we also see the *practical trigonometry triangulation method* being used to establish the unknown but required critical *vertex point*. Also shown is how we can find the vital *angle of bend* in the work-piece by adding the two triangle's calculated angles together followed by subtracting this answer from 90 degrees.

The author would also prefer that this particular angle is not given the name 'bend angle' (used frequently by the production staff throughout the whole industry), but is given the more descriptive name *angle of bend*, this being more appropriate for its use (explained in the later explanatory note regarding this widely used and named anomaly).

The old theoretical formula previously developed for the author's original earlier calculations (before the invention of the much simpler Probe and Prompt system now used for calculating triangles to find a component's vertex point), is shown in fig. 3: note that this old method of calculation has been totally superseded by the introduction of the new 'Probe and Prompt' *triangulation calculation method* shown in fig. 3a and in fig. 24.]

If one studies the old (*now superseded*) fig. 3 formula very carefully, it will be seen that it involves (what the author considers to be), a very complicated way of doing things, which can be considered (by the beginner) to be a rather mind-boggling experience, particularly to those students who are trying to grasp the subtle art of using practical trigonometry to solve their triangle problems.

It will now be found much easier to find the elusive and vital *vertex point*'s position, by using the aid and guidance of the Probe and Prompt system with triangulation, as shown in detail in fig. 24 and fig. 3a.

It is unfortunate, but due to the general engineering drawing office practice of producing research and development drawings that fail to include all the dimensions needed by the sheet metal engineer for the manufacture the component, he is forced to obtain these so-called missing but very important dimensions himself in order to get the job done.

It will therefore be found necessary (in practice), to use several additional trigonometry calculations to obtain all the missing dimensions that are currently left out of the issued drawing, as supplied to the shop floor. Because of this serious lack of helpful dimensions being provided on the issued drawing, the sheet metal engineer is therefore forced to use every scrap of information supplied on the issued drawing and combine this with practical trigonometry calculations sufficient to allow him or her to arrive at the much sought-after, so far unknown but required *dimension x*.

This so-called omission of vital dimensions will apply particularly to the issued research and development drawings that show a sheet metal, tube, and plate bend, without the necessary *vertex point*'s position being dimensioned, or including any other possibly helpful dimensions being shown on the issued drawing.

The experiences gained while in the process of solving these particular drawing office so-called omissions, when working on the shop floor, tends to indicate to the author that it is now an accepted practice for the drawing office staff to leave out some of the 'possibly helpful' dimensions with the expectation that the extra calculations required can be carried out by the shop floor engineers and sheet metal fitters to get the work-piece accurately completed. These elusive but required dimensions must therefore be obtained by using trigonometry calculation and by utilizing the meager information available on the issued drawing to obtain the required eight unknown dimensions, before work on the manufacture of the work-piece can actually begin.

In order to resolve this situation amicably and without causing unnecessary strife by making illegal visits to the drawing office, the author has found it best practice, (bearing in mind that we must not mark or alter the original issued drawing in any way, shape, or form),to make a large separate sketch on a sheet of A4 paper, containing the overall shape of the finished component, and including all the given known dimensions.

It is very important that this new sketch of the problem is made as large (in scale) as possible. This must be done so that sufficient space is left in the sketch to include all the newly found dimensions and angles to enable them eventually to be entered exactly where they occur. This sketch must be produced so that it is instantly readable and easily understood by the sheet metal worker for later use. (In other words, the use of the 'back of cigarette packet' drawing should be avoided if at all possible).

It is for this reason that the author's 'calculation' sketches are of necessity drawn to a scale much larger than true life size (as shown in the fig. 22, 24, 25, and 28 examples), and at first glance (to the so-called layman) they may seem to be completely out of proportion. However, there is a good reason for adopting this practice, as will be seen later.

One should also note that it is the dimensions contained within these important sketches that are the most critical; therefore (in this particular case), it may not necessarily be the scaled accuracy (or neatness) of the sketch that is required.

However, when one is trying desperately to interpret the unknown details of a small drawing, it will soon be realized that an accurately dimensioned large sketch will be found to be more helpful, particularly when it is made to a scale of at least ten times full size. This will then allow ample room for the inclusion of all the dimensions one intends to use for the final calculations used later on in the sketch.

Because of so-called drawing office omissions, the dimension marked x (in fig.23) will need to be ascertained on the shop floor by calculation. This will be found necessary because it is vitally important, that the *vertex point's position is established first*, before attempting to make any further bend calculations. As previously mentioned, it is most unfortunate (but quite usual in practice) for this particular dimension to be regularly omitted from the issued working drawing, as is shown in fig.20.

As previously stated, it is quite the usual practice for a dimension to be shown on the drawing given from its initial starting point at the beginning of the work-piece's left-hand side face, (the so-called left-hand side datum face), and this to extend to the right-hand edge of the work-piece's extreme outside edge of its bend radius. However, this particular dimension when given on the drawing will be found to be of little use to the manufacturing engineer who is attempting to produce an accurate component in the shop floor environment.

As previously stated, a dimension given to the work-pieces outside bend radius will only be found useful by the inspection department for checking the finished product's dimensional accuracy; however, this dimension will be found to be of little use to the workshop engineer or the sheet metal fitter who is actually making the component.

Having now completed the dimensioned sketch shown in fig. 22, (it must be emphasized that this is only a sketch), we can now make use of its contents to convert the given (so far) 'unusable for production' dimensions, (shown in the fig. 20 drawing), by transforming them into more easily usable, center-line dimensions, needed for calculating the component's bend angle and its overall length dimension.

This fig. 22 example sketch shows that the issued drawing dimension of 141.8 will now become 141.0. The dimension 120.8 will now become 120.0 etc. in order to produce the workable dimensions needed for the required *center-line*-based workpiece's bend calculations (which are essential).

[It is however quite the usual practice (on the shop floor), when dealing with a so-called simple bend of 90 degrees, that if the drawing dimensions have been given to the component's center-line, then of course these dimensions can be used without any change for the bends calculations].

The fig. 20 drawing is an 'as issued' drawing, produced by the drawing office for the manufacture of a sheet metal bracket. At first sight to the casual observer, and to the so-called man in the street, this drawing shows what would appear to be just a simple cut-and-fold bending job. In reality however, the dimensional content of this drawing, on reaching the workshop floor, will unfortunately prove to be decidedly unfit for purpose because of many important and helpful manufacturing dimensions not being shown on the drawing; these are as follows: (1) cutoff length, (2) angle of bend, (3) folding machine bend point, (4,) air machine bend point, (5) bend tangent length, (6) bend's vertex point, (7) bend center line arc dimension, (8) the important 'in the flat' dimension required for the drilled hole's position to enable an operator to drill the hole in its correct position prior to the bending operation taking place, (for this reason, the author has explained in detail all the necessary extra work that will now need to be undertaken by the shop floor personnel), to allow them to obtain the so-called missing drawing detail dimensions.

The drawing required to start making the initial calculations, is the fig. 20 issued drawing;' this is followed by the calculation drawings and sketches, shown in figures 3a, 20, 22, 23, 24, 25, and the stretched bend version shown in fig. 28.

The two typical bending machine drawings required to complete this bending job are shown in fig. 26, and fig. 27.

[Note that when the machinist is calculating bends with a center-line radius of either more or less than 90 degrees (examples of these details are shown in fig. 29 and fig. 31), the crucial vertex point's position will have to be determined by using trigonometry calculations. This need for extra calculation is due to the extreme rarity of the 'vertex point's position being given on the drawing office issued drawings, (this fact is shown missing in the fig. 20 drawing); this situation is rather disconcerting to the sheet metal worker on the shop floor, who is of course the person actually making the component.

The actual position of this vertex point will be shown (in later drawings and calculations) to be positioned either inside (in the case of a small angle of bend, as in fig. 29) or well outside the bend radii in the larger angles of bend shown in fig. 31. These variations will of course depend to a large extent on the ultimate angle of the component's angle of bend.

If we now refer to fig. 22 (the rough sketch of the problem), we see a very much enlarged freehand sketch detailing the problems found, when the sheet metal engineer is attempting to manufacture a product that fails to have the required information supplied on the issued production drawing, because of what is called a drawing omission.

It will be rapidly discovered that many of the critical required production dimensions are not included in the fig. 20 'issued for production' drawing; therefore, the use of the fig. 22 prepared rough sketch points out the dimensions that are *needed to be known* before the component can be produced to the required accuracy.

The fig. 23 drawing shows the inside bend radius of the component to be 3 mm; this dimension will of course need to be modified to provide the required bend *center-line radius* of 3.8 mm, by adding half thickness of sheet material, (the 3.8 mm dimension being the dimension the sheet metal engineer will actually require for his calculations). It should be noted that this dimension includes half the thickness of the material, this being measured to the *center line* of the material in its bend. Therefore, the total number of extra (so far unknown) dimensions that need to be correctly calculated in this *previously thought of as simple' sheet metal component* will be (1) the angle of bend, (2) the exact cutoff length of material to allow the component's length to be *cut off in the flat* (possibly by other production fitters, independent of the originating engineer), to its required dimension, (3) the folding machine bend point (indicated to the operator by his own calculated and scribed line on the surface of the sheet material), (4) the air bend point (the point calculated to be exactly half-way around the bend), and (5), if absolute precision is required, a stretched alternative cutoff overall length of material called the stretched length, to automatically take into account the recently introduced and vital factor now known as *stretch*, which is graphically shown in the *off center-line arc* of fig. 28 (note that the material must remain unmarked if the component is to be used in the aircraft, or space, industries, where any scratching or marking of the material's surface is not permissible because of the possibility of a crack emanating from the site of the scratch when in service), and (6) the marking-out dimension, required from the work-piece's 'starting point at end A (shown in sketch 22 as datum end' and in fig. 23 as end A, to the center-line of the 4-mm drilled hole; this dimension is needed to allow the hole to be drilled in the flat, possibly by other shop floor workers, independently, before the actual bending operation takes place. Note also that the dimension for this drilled hole will vary in length between the normal bends position and the 'stretched bends' position. This often-used batch drilling procedure must be mentioned here, in order to allow a production team to speedily complete a batch drilling of a whole quantity of pre-production 'in the flat' (un-bent) batches of blanks. This operation is often carried out by other machinists in complete isolation from the environment of the bending machine shop; its calculation must therefore be correct.

The fact that other personnel will be involved in the large batch production of the work- piece should make the student realize that all bend calculations undertaken must prove to be absolutely correct and faultless. If, in the case of large batches, the position of the drilled hole in the blanks has been wrongly calculated in the flat of the materials cut-off length', it will result in the hole's final position, after the bending operation has been carried out, being in a totally wrong place. The completed work-pieces, after bending, would therefore fail to be to the requirements of the drawing. This mistake could result in literally hundreds of components being wrongly drilled, rendering them to be scrap; this would then raise the serious problem of why this mistake occurred, and who was responsible for making this dimensional mistake. Therefore, every dimension shown on one's own sketch or drawing must be absolutely correct.

While working in the aircraft industry, it was found to be general practice that any markingout of a sheet metal work-piece (particularly for a drilled hole), should be very carefully performed by using a special marking instrument designed specifically to avoid any scratching or chemical degrading of the material in any way; therefore in this case, it must be used for reasons of safety on all aircraft parts.

Note that this particular marking-out instrument is to be used only within the confines of the intended hole to be drilled, (which is of necessity the very small area of material actually removed when the hole is drilled); special care should be taken when marking out this area to ensure that one does not impinge upon any material beyond the particular hole's surface area. This is a critically important point; therefore, to avoid any mistakes being made during the marking-out of a component, it is considered good practice to first mark out the component using just pencil, then checking that the results are correct, followed by very carefully marking out the component using a sharp vernier height gauge while ensuring that the scribed lines are kept within the area that is only occupied by the hole's final diameter.

Referring to the fig. 24 dimensioned sketch

The exact dimensions required (on the work-piece's center line) are now ascertained from the fig. 24 dimensioned sketch, starting from datum end A, to the vertex point D, then from the vertex point D to the top end of the workpiece at point G.

If we now study fig. 24 (this drawing being virtually identical to the earlier fig. 3a dimensioned sketch, the contents of which are now being re-introduced for instructional purposes), we see that in this enlarged sketch, (produced primarily to find the missing dimension x in fig. 23), that point A to vertex point D is found by using *practical triangulation calculations.*

In order to carry out these particular calculations' we must use the known (given) dimensions of, 3, 137.2, 141, and 1.6, and the accurate dimension of 50.4928406 mm, given to the work-piece's outside radius. This has been done to avoid using the fig. 20 drawing's 50.5 mm dimension in order to obtain a more accurate calculation.

How we find this elusive tangent point / vertex point D dimension, by the use of triangulation calculations, is now explained by referring to the fig. 24 drawn example.

Referring now to fig. 24, it will be seen that we use the 'probe's off-center length of side prompts to find the dimensions in both of the introduced sketched triangles, by entering the following sequence: 137.2 multiplied by itself gives us 18823.84, X-M; we now enter the sequence 45.8928406 (found by subtracting 4.6 from 50.4928406) multiplied by itself gives us 2106.152818, + RM (display shows 18823.84) = 20929.99282, followed by INV, SQUARE ROOT, giving us 144.6720181, thus giving us the diagonal length of this triangle. Now by using the outside edge prompts and by now actually knowing the length of two of the triangle's sides', and by also using the prompt Angle equals *b/c* INV TAN, we enter the sequence 45.8928406 divided by 137.2, = (display shows 0.334495922), INV TAN, giving us 18.49487822 degrees (answer). We now calculate the other triangle by entering 3.8 multiplied by it-self = 14.44, X-M. Enter 144.6720181 multiplied by itself = 20929.99282, - RM, (14.44) =, 20915.55282, INV. SQUARE ROOT, giving us 144.6221035, X-M, the answer.

Now by using the outside edge prompts, we enter 3.8 ÷ byRM = 0.026275374, followed by INV TAN, to give us 1.505121779 degrees (answer); we then enter the sequence + 18.49487822 degrees, = 20 degrees (answer) X-M; we then enter 90 degrees − RM =, 70 degrees (answer). Enter 180 degrees − 70 degrees = 110 degrees XM, (answer); this now gives us the required *angle of bend* in the fig. 24 problem. We now use this *angle of bend* to enable us to find the *tangent lengths* required.

To find the tangent length, we retain the previously displayed calculated number (110 degrees angle of bend) and use the sequence RM divided by 2 =, giving us 55 degrees (this being exactly 1/2 of the angle of bend), followed by TAN (display shows 1.428148007), followed by the sequence, × 3.8 (the center-line radius), =, 5.426962426 (thus giving us the required *tangent length* from B to D and from D to E). This is followed by the sequence +, 45.8928406 (the A to B dimension shown on figures. 24 and 25), =, 51.31980303 mm answer, (this now gives us the much soughtafter and required *dimension x* in the fig. 23 and 24 sketch, (and in the fig. 25 and 28 drawings).

To find the required length of arc (on the bends'center line), we now multiply the author's standard figure of 0.0174533 (the number the author uses universally to obtain 1 degree of arc in a calculation that deals with circumference), by 3.8 the (CLR (center-line radius) = 0.06632254; this is now followed by the sequence, ×, 110, (the angle of bend in degrees), =, giving us 7.2954794 mm, the answer, giving us now the total amount of material used up in the bend part of the workpiece, measured on the material's center line.

[Note: the author's method of establishing a standard figure for the calculation of one degree of arc in a circumference calculation is obtained by first using the arc of a radius of 1. This will give us a diameter of 2, which, when multiplied by π pi = 3.141592654), gives us 6.283185307; this figure is then divided by 360 (degrees), to give us the figure 0.017453292, (now rounded up by the author to 0.0174533, this now being the standard figure used for calculating 1 degree of arc), in calculations. It should be noted that in this particular case, to establish the actual center-line radius of a bend, we must use the addition of the drawing-stated inside bend radius (the dimension shown on most issued drawings); this must be added to half the thickness of material in use (in this case, this would be 3 mm. plus 0.8 mm), giving us 3.8 mm CLR (center-line radius). (This 'length of arc' information can also be obtained by using the bending chart's figures shown in fig. 32, for small-diameter tube, and sheet metal bend radii.).

To obtain the required 'in the straight' length from A to B in this component, we now subtract the newly found tangent length of 5.426962426 from the calculated straight length from A to D of 51.31980303, giving us the 'in the straight' length from A to B of 45.8928406 (note that this is also the folding machine bend point, as shown in fig. 25, and fig. 28).

To find the D to G dimension, we add the tangent length of 5.426962426 to the calculated E to G dimension of 144.6221035, to give us 150.0490659 from D to G.

For our convenience, we now add these newfound details to the previously drawn sketch.

To obtain the 'in the straight' dimension from E to G, we subtract the tangent length (5.426962426) from the D to G dimension of 150.0490659 to obtain 144.6221035; this being the 'in the straight' dimension from the end of the bend's center-line radius (at point E), to point G (the top vertex point), in figures 25 and 28.

To obtain the actual total length of material required to actually manufacture the component, we therefore add the calculated dimensions A to B, B to E (c/l arc), and E to G, thus; 45.8928406 + 7.2954794 + 144.6221035 = 197.8104235. This can now be called *the cutoff length in the straight for a component that is normally bent.* To find the *air bend point,* we use the A to D dimension, minus the B to D dimension, plus half of the center-line arc (shown as 3.6477397 in fig. 25); we therefore calculate 51.31980303, minus 5.426962426 + 3.6477397 = 49.5405803 mm answer.

By using the small sketch (k),top right-hand side of fig.25), we can find (by calculation), the exact dimension of D to L on the center-line of the 'straight' material to establish the exact point required to mark out and drill the 4 mm hole. We do this by using the 'prompt $\angle$ b, a = b/SIN $\angle$, by entering the sequence 70, SIN (display shows 0.93969262) X M, followed by the sequence, 120, divided by R M, =, 127.7013327, giving us the initial D to L dimension on the center-line of the material. We now subtract the D to E tangent length, 5.426962426, to give us 122.2743703; this is followed by adding the E to B center-line arc's dimension of 7.2954794, and the A to B dimension of 45.8928406 to give us 175.4626903, the dimension required for marking out the position of the drilled hole from point A through to point L, *enabling this drilling operation to be carried out while the material is still in the straight, in this normally bent component.*

In the unlikely event that a drilling jig will be required to drill the hole in the component *after bending*, then the jig's drill position (using fig. 25 as a reference) is obtained by utilizing the vertical height of position F relative to position L, by using the 20-degree angled slant height shown in the very small sketch at J; this shows the slanted dimension to be 0.291176187 mm, (rounded down to 0.291 mm). Its vertical height dimension will be 0.273616114 mm (rounded up to 0.274 mm); therefore, the vertical height of the drill's point (when entered into the jig on the work-piece's center line to contact the outside surface of the component) will be 120.274 mm, this dimension being taken vertically from the A–D work-piece's horizontal datum line, while using the 4-mm drill at a downward inclination of 20 degrees relative to A–D baseline. This calculated position is obtained by using the side *b* of the small sketch M.

The existence of a drilling jig would therefore make it unnecessary to mark out the work for the hole to be drilled in the flat. (Note that the marking-out of a component will only become necessary if the component is intended for use in general engineering, without a requirement to possess a perfectly scratch-free surface. However, the disadvantage in this case would be the extra expense of the jig's design and its manufacture. The management concerned would also need to consider the number of components needing to be produced in order to substantiate the extra cost involved in the design and manufacture of the drilling jig.

Sheet metal bending charts

We now refer to figures 32 and 32a (containing the two newly developed sheet metal bending charts), which show, at the top of the charts, details of all the current 'thickness gauges' of material presently being used, requiring either *normal* or *stretched* metric bend allowances for the manufacture of sheet metal components.

Shown down the left-hand side of these charts is the (drawing stated) inside bend radius the charts are being used for. The advantage of having the option of using either of the two new charts will allow the operator to choose whether he requires supreme accuracy, or just the basic accuracy in the finished component.

By using the charts' 'drawing stated' inside bend radii, the tabulated figures provide the exact *center-line arc* dimension that is taken up *by one degree of bend* in the sheet material being calculated; this figure is then multiplied by the angle of bend previously calculated for the component, thereby giving us the total amount of material that is being *used up* in the total bend's arc. There follows an explanation as to why only the *minimum useable* bending radii given below the zigzag line can be used practically when taken from the two bending charts featured in fig. 32 and fig. 32a.

It will be found that by using these charts will provide an accurate and much easier way of performing the important *length of bend arc* calculation, while the machinist is involved in rather complicated sheet metal bend calculations.

To reiterate, the two bending charts have been designed to be used to calculate the exact length of material used up in the bend when used for either *normal bends* featured in fig. 32, or the so-called *stretched bends* featured in fig. 32a.

The *inside bend radius* quoted in these charts ranges from the purely *theoretical* and *unusable* 0. 25 mm IBR that is situated above the zigzag line (this being the incredibly small IBR that is considered by sheet metal engineers to be virtually impossible to achieve without cracking the material, therefore the figures above the zigzag line have only been included in the charts purely for reference purposes) in order to provide a detailed *minimum* inside bend radii that may by chance be requested by a draftsman, manager, or designer, in order to comply with his or her *whim of fantasy* request requiring the absolute minimum of inside bend radius to be produced in the component. The tables shown below the zigzag line are to be used for the calculation of *normal bend arcs* indicated in the fig. 32 chart, and for the calculation of *stretched bend arcs* shown in the fig. 32a chart.

The student should note that the minute inside bend radii (as shown above the zigzag line) are included purely for theoretical purposes, and (where necessary), for making nonsense of an unrealistic request by management for a tighter bend radius than is possible to achieve. These figures can then be quoted to help the machinist prove a point (for example) during a works office production meeting.

The managements of engineering firms will often strive to obtain a perfectly square corner on the inside surface of a sheet metal component. It is in the author's experience that management may further insist that it is possible to manufacture a component that contains virtually no internal radius. This is an impossible task to achieve, but proof of this will need to be provided by the experienced shop floor engineers in order to convince the management that this is an impossible task; therefore, a full explanation of this impossibility can only be provided by their perusal of the chart's figures *above the zigzag line*, showing that this IBR is in fact impossible to achieve.

To re-iterate, this particular group of minute 'above the zigzag line' 'inside bend radii' should be considered by the practical sheet metal bending engineer to be purely theoretical; that is of course until a request is made for a component to be produced containing a 'critical minimum' inside bend radius'.

The *usable* range shown in the charts (i.e. those quoted below the zigzag line), are calculated to deal with material thicknesses of up to a maximum of 7.0 mm, with a stated inside bend radii ranging (for those *below* the line). to a maximum of 7 mm.

Students should note that sheet metal drawings supplied by the drawing office can refer to either the *inside bend radius of a bend* or the *outside bend radius of a bend*, and that these dimensions are used mainly for dimensional inspection purposes.

The two new charts have therefore been calibrated to comply with the *inside bend radius* only, making them applicable only to the inside bend radius dimensions stipulated on the majority of issued sheet metal drawings. The drawing office's preferred method of dimensioning drawings to the outside of the component's bend radius does, however, enable the component to be more easily measured for accuracy by the inspection department. But in the case of a component that is dimensioned *to the outermost edge of its bend radius*, will be of little immediate use to the shop floor sheet metal worker, without additional trigonometry calculations being performed (as shown in fig. 24), to cope with this. We must in this case first obtain the work-piece's vertex point (as shown in fig. 24 and 3a) to allow further calculations to be carried out in order to produce all the extra dimensions required to produce an accurate component.

The inspection department personnel will use either internal or external radius gauges, or the radius of a selected diameter of drill blank to establish that the correct radius dimension of the required IBR (inside bend radius) is present, during their checking process; in this case, the drill blank will be held in close proximity to the inner surface of the bend radius, and its closeness of fit will be monitored to confirm that the accuracy of the inside bend radius produced in the component is correct.

It is generally the case that sheet metal engineering drawings are dimensioned to the *outside bend radius* of the component; this dimension being only suitable for final inspection purposes. The drawing can also refer to the dimension of its inside bend radius for the manufacture of the sheet metal component but will only very rarely state its *center-line radius*.

In practice, however, it is the *center-line radius* dimension that is required by the sheet metal workshop engineer that will allow him to perform his bend calculations directly to enable the manufacture of the component to the necessary accuracy. He will therefore need to convert the drawing given inside bend radius (or its outside bend radius) into a more convenient *center-line radius* to allow him or her to calculate the other six or eight dimensions required before actually starting to make the component.

The fig. 32a chart, is considered the more advanced of the two charts, as it shows the calculated so-called stretched metric bend allowances for the bend, while using the most popular thickness gauges of the normally available sheet metal materials. This table need only be used when extreme dimensional accuracy is required in the finished product, or it is found that the basic calculations taken from the normal fig. 32 chart have produced a slightly overlong component, indicating that its material has stretched slightly beyond the normally accepted limits during the bending process.

*   *   *

If we now refer to fig. 22, showing a freehand sketch of drawing fig. 20, this sketch includes the calculations needed to obtain the six additional and required dimensions that are not shown on the issued fig. 20 drawing. It must be stated that these important dimensions are not normally shown on the originally issued drawings and will therefore need to be recalculated by using the Probe and Prompt triangle calculation system.

Referring now to fig. 23,this shows a sketched modification to fig. 20 and identifies the required position of the missing dimension *x* that will need to be used in place of using the issued fig. 20 drawing.

The fig. 23 drawing contains other so-called omissions that relate to the original fig. 20 issued drawing. These omissions are now revealed by reviewing the fig. 23 drawing in relation to the required fig. 22 dimensioned sketch, which now confirms the existence of the missing calculated dimensions required before any work on the manufacture of the component can proceed.

The required calculated dimensions are dimension *x*, the angle of bend, the cutoff length of material, the folding machine bend point, the air machine bend point, and the exact marking-out dimension required for the position of the 4-mm drilled hole *while the material is still in the flat and before the bending operation has been carried out.*

Note that the dimension given to the normal center-line arc length in the component (featured in the fig. 25 drawing) will differ in its 'in the flat' dimension, and in its hole drilling dimension when compared to the component that possesses a stretched arc length and the stretched overall length as featured in fig. 28. However, the overall dimensions of the finished stretched component and the finished 'normally bent' component should result in them being identical after bending.

There follows an example of a problem that can be encountered during the production phase of a batch of sheet metal components, where a request is made over the phone to quote some important dimension while the operator is still engaged in the early stages of his or her sheet metal calculations. This request is often made by the production management in order to plan the shop's future work-load. The student should be made aware that this situation can occur when the firm's night shift supervisors have just discovered the need for additional fill-in work to keep their night shift busy.

When faced with an instant request from a top management official for such vital information over the telephone, it is a very wise practice to refrain from instantly divulging any possibly unchecked dimensions in a hurry, without considering the possible consequences of one's rushed actions.

If crucial dimensions are requested, such as 'I only want the cutoff length', or 'I just want the position of the hole in the component', without the student first finding out the intended use of these critical dimensions, it is important to ensure that these instantly required figures just provided over the phone are perfectly accurate.

Failure to observe this precaution can result in the surprise production of literally thousands of faulty components completed overnight, possibly by semi-skilled workers, followed by the final bending operation being completed by the following day shift.

Should your instantly given dimensions, (given hurriedly over the telephone) eventually turn out to be incorrect, it will inevitably result in an enormous amount of scrap being produced, and the fact that you will be held responsible for this disaster must be seriously considered before allowing oneself to be rushed into giving any unchecked dimensions verbally over the telephone! It should be considered, therefore, to be of the utmost importance that students should heed this warning, or be prepared to suffer the consequences of their rushed actions.

Refer now to fig.24.

Fig. 24 is an enlarged sketch of the calculation problem found in fig. 20, showing the calculations required to find the angle of bend and the tangent point (x dimension), for our modified drawing of the sheet metal bracket shown in fig. 23.

This sketch shows the method used to obtain this vital vertex point's position, by using the Probe and Prompt aid in its calculation (this method will be found preferable to using the author's old, now superseded (and more difficult) theoretical method shown as an example in fig. 3.

Referring now to fig. 25, this informative sketch now shows, in the ultimate detail, the final bend calculations required for the sheet metal bracket previously shown in figures 20, 22, 23, and 24, and completes the calculations required to find all of the previously unknown dimensions needed for the manufacture of this metal bracket that contains a basic bend with one accurately located hole, (the calculations in this case have been made *without* the stretch in the material problem being considered).

If we now refer to the normal bending chart shown in fig. 32, it will be seen that it contains a newly compiled table of normal, run-of-the-mill bend allowances that establish the arc of one degree of bend (without the inclusion of any stretch factor in its figures). This chart provides the student with a golden opportunity for him or her to accurately calculate the center-line arc of a bend in a sheet metal component that possesses a 'drawing stated' inside bend radius ranging from 0 to 7 mm.

*   *   *

Fig. 26 Bend option 1

This sectioned sketch shows a typical example of the main working parts of a Promecam hydraulic so-called (air) bending machine being used to complete the bend in the metal bracket described in figures 20, 22, 23, 24, and 25.

This machine is used when extreme accuracy and unblemished work surface finish are vital in the finished product.

The name given to this popular air bend machine (as used in the aircraft industry), actually refers to the air gap left below the upper bend tooling relative to the work-piece, and to the clearance found below the work-piece and immediately above the bottom V block tooling, and does not refer to the motive force required to operate the machine.

This machine operates on hydraulics, and the force exerted by this particular machine on the component (during the bending process) is adjustable up to, in some cases, a full fifty tons when being used on

exceptionally long channels that require the absolute maximum motive
force to be applied to form the bend. The machine is normally used
set to a much lower pressure than this when used on the small bent
components described in figures 20, 22, 23, 24, 25, and 28.

However, when working on short sheet metal work-pieces (as used
for our metal bracket), the hydraulic pressure would be set to a much
lower setting for bending this 'short' easy-to-bend component.

Fig. 26 shows the setup for bend option 1 using the Promecam air
bend machine.

The final overall calculated dimension for setting up this work-
piece on the machine, is to allow the work-piece's bend to be accurately
aligned with the c/l of the machine's tooling, and is calculated to be
49.54 mm (as shown on the fig. 25 sketch/drawing). This dimension
is accurately set on the machine by the sheet metal engineer, and is
checked with the aid of a precision vernier caliper. This measurement
is actually taken from the machine's adjustable 'end datum stop', and
extends to the center-line of both the machine's top tool and to its
bottom V block. The horizontal center-line of the V block is measured
by simply halving the V' block's overall width dimension, followed by
using the vernier caliper to measure the distance from the V block's
vertical side face to the face of the end stop's adjustable datum.

The machine operator is protected from injury while performing
this bending operation by a series of light guards that prevent (by
switching off the machine's hydraulics), the operator's hands from
approaching the tooling during the bending operation (for safety
reasons).

Any interference made by the hands into this protective light beam
will instantly cut off the hydraulic power, and will stop any subsequent
movement of the machine's tooling.

The work must therefore be presented to the tooling carefully, and
at the correct square orientation relative to the machine's' end stop and
V block tooling; the operator is then required to stand back behind
the aforementioned guard while operating the machine from a remote
switch for reasons of safety.

This machine is ideal for the manufacture of precision sheet
duralumin angled parts for use in the aircraft and space industries.Its
advantage lies in its capability to bend parts with extreme accuracy,

while still maintaining the ability to produce bent work with a truly unblemished surface finish.

When the production detail fitter is producing the precision 'cut to length' material for this machine, he is not allowed to mark out, or place any scratch on the material's' work surface prior to bending. It is therefore essential that full and accurate calculations have been made to establish exactly the work-piece's cut-off length, and to obtain the machine's setting dimensions relative to the components' bend center-line radius, in order to obtain the exact bend datum points; this course of action avoids the need to mark out the work-piece prior to bending.

This 'no marking out' practice must be enforced on the workforce by the management in order to comply with the safety considerations necessary with regard to the resultant ultimate tensile strength of the actual component after bending; (any imperfection on the material's surface will ultimately reduce its tensile strength). It is therefore the normal practice in these industries, to avoid any scratches occurring on the surface of the actual component.

It has been discovered that any form of imperfection or scratch left on the surface of the material can be considered a dangerous condition and it is unacceptable for reasons of safety.

It has been proved through investigative tests (by the Air Accident Investigating Board), that a scratch can be a crucial starting point for a fatigue crack to develop; this in turn can be aggravated when the part is subjected to the prolonged stresses of an aircraft in flight.

A small scratch can, over time, extend into a large fracture ultimately causing failure of the component with disastrous consequences should the aircraft be in the air at the time.

*   *   *

Fig. 27 **Bend option 2: The folding machine**

This machine is shown set up for an alternative and slightly less accurate method that can be used for bending the sheet metal bracket.

This illustration shows the setup used when the ultimate precision is not required in the component. (The fully bent component is shown detailed in figures, 20, 22, 23, 24, 25, and 28).

This machine is normally used to produce the general run-of-the-mill sheet metal bends, where ease and speed of operation in this production environment is the requirement, and not necessarily the extreme accuracy of the finished component.

This machine's main advantage lies in the fact that work produced containing marked-out bend setting lines is fully acceptable provided the component is used only for general engineering work, where a so-called 'ultimate factor of safety requirement' is not considered critical.

In operation, the machine's bend mechanism is first adjusted to give the 3-mm inside bend radius to comply with the dimensions shown on the fig. 25 drawing. A piece of similar scrap material is first used for the preliminary inside bend radius check and for obtaining the correct 'angle of bend' setting.

The bend limit stop is then adjusted to give (in this case) the angle of bend of 110 degrees, plus approximately 2 degrees of extra bend to allow for the expected work-piece 'spring back' after the bend is completed. Having checked that the actual bend produced is correct while using the work-piece material, the machine is then locked into this position.

There is, however, a slight problem with this last statement and this applies particularly when one is working on precision aircraft parts on the previously described Promecam air bending machine.

While working on various contracts that included the manufacture of precision aircraft components, the author found that it was extremely difficult to obtain any suitable scrap duralumin sheet with which to actually set up the bending machines. This I feel is partly due to the usual practice carried out by the planners, who tend to allow only sufficient material with which to make the exact number in the batch specified on the drawing. The stored production material is bonded in the store-room and, as such, is required to go through a series of durability tests to check on its suitability for use on aircraft. The final piece of duralumin sheet issued for the manufacture of the components, will already have been cut to a rough overall sized sheet, and will be found to be of just sufficient dimensions to make the required number of items, and no more. This situation leaves the operator very little margin for error, and less possibility of leaving any piece of sizeable scrap left over on completion of one job to retain for future setting-up use on the next job. The saving of suitable scrap material for setting-up purposes is therefore very limited and results in just a small number highly valued offcuts being available.

It is imperative therefore that all bend calculations made, must be capable of being totally relied upon as being correct, for it is not possible to have a second chance with any spare material, (even for one extra component).

It should be noted that it is not possible to reclaim any wrongly bent components, by straightening them out and re-bending them correctly, because the material will have been stretched and weakened in the first 'wrongly bent' part of the operation.

Some of the sheet metal alloy materials will require heat treatment prior to being subjected to the bending operation. This is mainly due to the material being normally in a toughened state and liable to crack if subjected to bending without the usual hot brine bath's softening treatment. The issued material may not therefore be ductile enough for safe bending without this hour-long softening process prior to the bending operation being carried out.

The ductility of material so obtained may last for only one to two hours after the treatment, so all the bends in the batch must be completed before this time has elapsed.

From the foregoing, it will be gathered that it is necessary to plan in advance, both the availability of the softening process and the availability of the bending machine (at the right time); this will apply particularly when working in a busy machine shop with other production workers doing a similar type of work on their production sheet metal work-pieces.

fig. 29.

AN EXAMPLE OF A 30° ANGLE OF BEND. SHOWING CALCULATIONS
REQUIRED FOR MANUFACTURE.

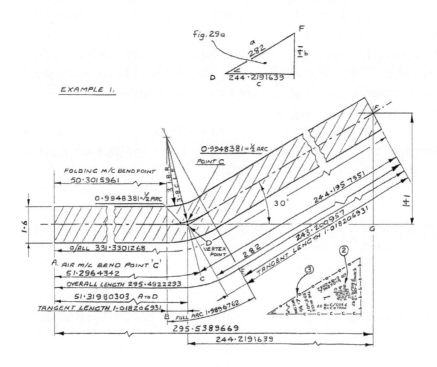

fig. 29a

EXAMPLE 1.

IT WILL BE SEEN IN THIS CASE THAT THE TANGENT LENGTHS ARE
VERY SHORT, AND COULD ALLOW THEM TO BE SHOWN ON ENGINEERING DRAWINGS.

ALL DIMENSIONS IN MM.

fig. 30.

AN EXAMPLE OF A 90° ANGLE OF BEND SHOWING CALCULATIONS
REQUIRED FOR MANUFACTURE.

Fig 30.

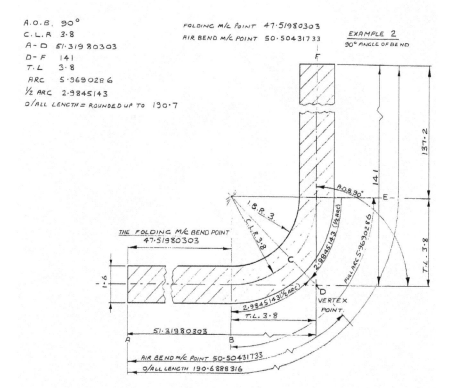

A.O.B. 90°
C.L.R 3·8
A - D 51·31980303
D - F 141
T.L 3·8
ARC 5·9690286
½ ARC 2·9845143
O/ALL LENGTH = ROUNDED UP TO 190·7

FOLDING M/C POINT 47·51980303
AIR BEND M/C POINT 50·50431733

EXAMPLE 2
90° ANGLE OF BEND

ALL DIMENSIONS IN MM.

Fig 31.

fig. 31.

AN EXAMPLE OF A 150° ANGLE OF BEND SHOWING CALCULATIONS
REQUIRED FOR MANUFACTURE.

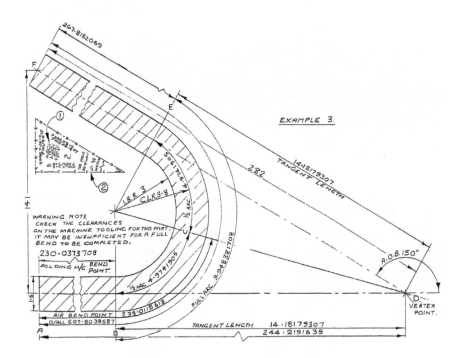

EXAMPLE 3.

 IT WILL BE SEEN IN THIS CASE THAT THE TANGENT LENGTHS ARE VERY LONG.
AND COULD EXPLAIN WHY THEY ARE NOT SHOWN ON ENGINEERING DRAWINGS,
DUE TO LACK OF AVAILABLE SPACE.

ALL DIMENSIONS IN MM.

Fig 32.

fig 32 TABLE OF 'NORMAL' METRIC BEND ALLOWANCES FOR 1° OF BEND WITH INSIDE BEND RADII FROM 0 TO 7 MM.

METAL GAUGE	24 S.W.G.	22 S.W.G.	20 S.W.G.	18 S.W.G.	16 S.W.G.	14 S.W.G.	12 S.W.G.	10 S.W.G.
IMPERIAL	0.022"	0.028"	0.036"	0.048"	0.064"	0.080"	0.104"	0.128"
METRIC	0.588 mm	0.711 mm	0.914 mm	1.21 mm	1.62 mm	2.03 mm	2.64 mm	3.25 mm
INSIDE BEND RADIUS M.M.								
0.0	0.0051327	0.0062204648	0.0079761658	0.0100559246	0.014137173	0.017715099	0.023038356	0.028361612
0.25	0.009494595	0.010567973	0.012339483	0.014922571	0.018500498	0.022078424	0.027401681	0.032724937
0.5	0.013857792	0.014931298	0.016702808	0.019285896	0.022863823	0.026441749	0.031765006	0.037088262
1.0	0.02258457	0.023657948	0.025429458	0.028012546	0.031590473	0.035168399	0.040491656	0.045814912
1.5	0.03131122	0.032384598	0.034156108	0.036739196	0.040317123	0.043895049	0.049218306	0.054541562
2.0	0.04003787	0.041111248	0.042882758	0.045465846	0.049043773	0.052621699	0.057944956	0.063268212
2.5	0.048764452	0.049837898	0.051609408	0.054192496	0.057770423	0.061348349	0.066671606	0.071994862
3.0	0.05749117	0.058564548	0.060336058	0.062919146	0.066497073	0.070074999	0.075398256	0.080721512
3.5	0.06621782	0.067291198	0.069062708	0.071645796	0.075223723	0.078801649	0.084124906	0.089448162
4.0	0.074944447	0.076017848	0.077789358	0.080372446	0.083950373	0.087528299	0.092851556	0.098174812
4.5	0.08367112	0.084744498	0.086516008	0.089099096	0.092677023	0.096254949	0.101578206	0.106901462
5.0	0.09239777	0.093471148	0.095242658	0.097825746	0.101403673	0.104981599	0.110304856	0.115628112
5.5	0.10112442	0.102197758	0.103969308	0.106552396	0.110130323	0.113708249	0.119031506	0.124354762
6.0	0.10985107	0.110924448	0.112695958	0.115279046	0.118856973	0.122434899	0.127758156	0.133081412
6.5	0.11857772	0.119651098	0.121422608	0.124005696	0.127583623	0.131161549	0.136484806	0.141808062
7.0	0.12730437	0.128377748	0.130149258	0.132732346	0.136310273	0.139888199	0.145211456	0.150534112

THIS TABLE IS DESIGNED FOR USE WITH THE DECIMAL NOTATION OF ANGLES, ie. 1.5 DEGREES AND NOT 1° 30' (SEXAGESIMAL.) IT WILL BE FOUND ADVISABLE TO USE ONLY THE ALLOWANCES GIVEN BELOW THE ZIG-ZAG LINE TO MINIMISE ANY CRACKING PROBLEMS,

IF THE MATERIAL HAS INSUFFICIENT DUCTILITY.

THERE EXISTS THE EVER PRESENT NEED TO PRODUCE 'TIGHT' BENDS, (WITH LITTLE OR NO INSIDE BEND RADIUS) ON THIN DUCTILE MATERIAL IS, IT IS DUE TO THE DEMANDS OF PRODUCTION MANAGEMENTS WORLD-WIDE,

FOR THIS REASON THE ALLOWANCES GIVEN ABOVE THE 'ZIG-ZAG' LINE HAVE BEEN SUPPLIED. THEY MUST BE USED WITH CARE TO AVOID CRACKING THE LESS DUCTILE MATERIALS DURING BENDING.

FOLLOW THE INSTRUCTIONS BELOW FOR USING THE ABOVE CHART.

IF FOR EXAMPLE A BEND OF 33° IS REQUIRED ON 20 S.W.G. (0.914 m.m) MATERIAL, WITH AN INSIDE BEND RADIUS OF 3 m.m.

REFER TO THE ABOVE COLUMN FOR 20 S.W.G.(0.914 m.m), FOLLOW DOWN TO 3.0 m.m. I.B.R. TO SELECT THE FIGURE 0.060336058.

MULTIPLY THIS BY 33, = 1.991089914 m.m.; THIS GIVES THE LENGTH OF MATERIAL USED UP IN THE BEND, ON NORMAL PRODUCTION

SHEET METAL COMPONENTS. NOTE. THESE METRIC BEND ALLOWANCES ARE UNIQUE TO THIS BOOK, PARTICULARLY

THOSE RELATING TO TIGHT BENDS THAT HAVE PREVIOUSLY BEEN UNAVAILABLE TO THE ENGINEER). METRIC BEND ALLOWANCES

HAVE ALSO BEEN UNAVAILABLE IN THE U.K. THERE BEING ONLY IMPERIAL BENDING ALLOWANCES AVAILABLE, REQUIRING CONVERSION TO METRIC DIMENSIONS.

THESE NEW ALLOWANCES GIVE AN EXACT DIMENSION TO THE SMALL AMOUNT OF MATERIAL USED UP IN TIGHT BENDS ON THIN

MATERIALS, BUT THEY ARE TO BE USED WITH CAUTION TO AVOID CRACKING.

IF WE TAKE FOR EXAMPLE AN EXTREME BEND, WITH AN INSIDE BEND RADIUS (I.B.R.) OF 0.0, ON THIN DUCTILE MATERIAL, IT WILL RESULT IN AN ALMOST SQUARE CORNER, BUT THE MATERIAL USED UP IN THE BEND WILL STILL BE MEASURABLE, AT 0.00513127 m.m. PER DEGREE OF BEND ON 24 GAUGE MATERIAL. (SEE ALSO STRETCHED BEND CHART FIG.32a, FOR PRECISION BEND ALLOWANCES.)

Fig 32a.   TABLE OF 'STRETCHED' METRIC BEND ALLOWANCES FOR 1° OF BEND, WITH INSIDE   BEND RADII   FROM   0 TO 7mm.

METAL GAUGE	24 S.W.G.	22 S.W.G.	20 S.W.G.	18 S.W.G.	16 S.W.G.	14 S.W.G.	12 S.W.G.	10 S.W.G.
IMPERIAL	0·022"	0·028"	0·036"	0·048"	0·064"	0·080"	0·104"	0·128"
METRIC	0·588 m.m.	0·711 m.m.	0·914 m.m.	1·21 m.m.	1·62 m.m.	2·03 m.m.	2·64 m.m.	3·25 m.m.
INSIDE BEND RADIUS m.m.								
0·0	0·003420846	0·004136432	0·005317438	0·007039497	0·009424782	0·011810066	0·015358904	0·018907741
0·25	0·007784171	0·008499757	0·009680763	0·011402822	0·013788107	0·016173391	0·019722229	0·023271066
0·5	0·012147496	0·012863082	0·014044088	0·015766147	0·018151432	0·020536716	0·024085554	0·027634391
1·0	0·020874146	0·021589732	0·022770738	0·024492797	0·026878082	0·029263366	0·032812204	0·036361041
1·5	0·029600796	0·030316382	0·031497388	0·033219447	0·035604732	0·037990016	0·041538854	0·045087691
2·0	0·038327446	0·039043032	0·040224038	0·041946097	0·044331382	0·046716666	0·050265504	0·053814341
2·5	0·047054096	0·047769682	0·048950688	0·050672747	0·053058032	0·055443316	0·058992154	0·062540991
3·0	0·055780746	0·056496332	0·057677338	0·059399397	0·061784682	0·064169966	0·067718804	0·071267641
3·5	0·064507396	0·065222982	0·066403988	0·068126047	0·070511332	0·072896616	0·076445454	0·079994291
4·0	0·073234046	0·073949632	0·075130638	0·076852697	0·079237982	0·081623266	0·085172104	0·088720941
4·5	0·081960696	0·082676282	0·083857288	0·085579347	0·087964632	0·090349916	0·093898754	0·097447591
5·0	0·090687346	0·091402932	0·092583938	0·094305997	0·096691282	0·099076566	0·102625404	0·106174241
5·5	0·099413996	0·100129582	0·101310588	0·103032647	0·105417932	0·107803216	0·111352054	0·114900891
6·0	0·108140646	0·108856232	0·110037238	0·111759297	0·114144582	0·116529866	0·120078704	0·123627541
6·5	0·116867296	0·117582882	0·118763888	0·120485947	0·122871232	0·125256516	0·128805354	0·132354191
7·0	0·125593946	0·126309532	0·127490538	0·129212597	0·131597882	0·133983166	0·137532004	0·141080841

THIS TABLE IS DESIGNED FOR USE WITH THE 'DECIMAL NOTATION' OF ANGLES. i.e. 1·5 DEGREES, AND NOT 1° 30' (SEXAGESIMAL.)

IT WILL BE FOUND ADVISABLE TO USE ONLY THE ALLOWANCES SHOWN BELOW THE ZIG-ZAG LINE, THIS WILL MINIMISE ANY CRACKING PROBLEMS LIABLE IN SHEET METALS OF INSUFFICIENT DUCTILITY. DUE TO THE DEMANDS OF PRODUCTION MANAGEMENTS WORLD-WIDE, THERE EXISTS THE EVER PRESENT NEED TO PRODUCE 'TIGHT' BENDS, WITH THE MINIMUM OF INSIDE BEND RADIUS ON VERY THIN MATERIALS. IT IS FOR THIS REASON THE ALLOWANCES GIVEN ABOVE THE 'ZIG-ZAG' LINE HAVE BEEN SUPPLIED. THEY MUST BE USED WITH GREAT CARE, TO AVOID

FOLLOW THE INSTRUCTIONS SHOWN BELOW BEFORE USING THE ABOVE 'SPECIAL' STRETCHED ALLOWANCES, CRACKING THE LESS DUCTILE MATERIALS. DURING BENDING.

MULTIPLY THE INDICATED BEND ALLOWANCE BY THE ANGLE OF BEND REQUIRED, THIS WILL GIVE THE TOTAL LENGTH OF 'STRETCHED' MATERIAL USED UP IN FORMING THE BEND

IF FOR EXAMPLE, A BEND OF 33° IS REQUIRED, ON 20 S.W.G. (0·914 m.m.) THICKNESS MATERIAL, WITH AN INSIDE BEND RADIUS OF 3 m.m., WE REFER TO THE ABOVE COLUMN FOR 20 S.W.G. (0·914 m.m.) THICKNESS MATERIAL, FOLLOW THIS DOWN TO 3 m.m., INSIDE BEND RADIUS, THEN SELECT THE FIGURE 0·057677338; MULTIPLY THIS BY 33 = 1·903552178 m.m. THIS IS THE TOTAL LENGTH OF MATERIAL USED UP IN A STRETCHED BEND CONTAINING 33° ON 20 S.W.G. (0·914 m.m) MATERIAL. THE SO CALLED 'STRETCHED BEND' IS WHERE AN ALLOWANCE HAS BEEN MADE IN THE CALCULATIONS, TO TAKE INTO ACCOUNT THE EXPECTED 'ELONGATION' OCCURRING DURING THE BENDING PROCESS.

'STRETCHED' BEND ALLOWANCES ARE TO BE USED WHEN EXTREME ACCURACY IS REQUIRED IN THE FINISHED PRODUCT. THEY ARE TO BE USED WHEN FINISHED WORK, (USING 'NORMAL' ALLOWANCES,) RESULTS IN THE PRODUCT BECOMING UNACCOUNTABLY OVERLONG DURING THE PRODUCTION PROCESS, DUE TO THE OVERSTRETCHING OF MATERIAL DURING BENDING.

Fig. 32 shows a calculation table containing 'normal arc' bend allowances, used for normal sheet metal components, in metric notation.

Fig. 32a shows a calculation table containing 'stretched arc' bend allowances used for stretched sheet metal components in metric notation. (See fig. 28 for the stretched version of a similar sheet metal bracket.)

Either of the two tables can be used to calculate the amount of material used up in the arc of a bend when needed in the following two cases,:

1. If a component has the specific requirement to possess the ultimate in precision, this can be obtained by using the stretched arc bend allowances as shown in fig. 32a. This allows for the stretch in the material's bend being taken into account by calculation prior to the bending process being carried out.

If the requirement is for just an accurate component containing so-called normal production accuracy during its manufacture, we can use the 'basic bend arc' allowance chart shown in fig. 32.

The actual allowances given in the two bending tables can be independently compared to reveal the minute differences in the amount of material that is consumed in a normal basic bend's arc, compared to the material that is consumed in the 'stretched precision' bend arc (requiring of course slightly less material in its arc, resulting in a slightly shorter cutoff length' of material being required.).

*　　*　　*

Refer now to fig. 25.

The fig. 25 sketch has been prepared to show clearly what is going on in the actual bend of the work-piece material. It also shows the full calculations required for the manufacture of a precision sheet metal work-piece containing one basic bend and one drilled hole (without in this case the stretch of the material being considered).

Refer now to fig. 28.

The fig. 28 drawing shows how it is possible, by using just a small calculated adjustment to the length of the component's *center-line arc* (in this case with it being calculated at 1/3 of the thickness of the material). It will be found that we can actually program out the amount of stretch (found by experience and exhaustive tests), to occur in the material during the bending process.

By using the fig. 32a newly calculated stretched bend tables, it will be found that it will provide the ability to absorb the expected stretch during the bend calculations, thereby obtaining an extremely accurate and exact overall length of material required for the 'in the flat' stretched material's cutoff length,; this extreme accuracy requirement will be found very useful on components used in the aircraft industry, where this amount of precision is always required.

This small modification to the calculated figures allows the operator to manufacture the component with the ultimate precision.

Note that the basic original bend calculations (shown in fig. 25) show the total theoretical calculated length of material required for the manufacture of this component.

This given calculated dimension will be found to be perfectly adequate and acceptable when used for the normal run-of-the-mill type of sheet metal bending work; however, it should be noted that this calculation fails to take into account any stretching of the material that occurs in the bend during the bending operation.

For this reason, the author has used the new stretched classification, in the fig, 32a chart, to describe this phenomenon. It explains perfectly the inevitable increase in the length of material that occurs during the bending process, and takes this *stretch factor* into account in the calculated figures.

This modification to the overall length of material required, is made by simply substituting and using a new '1/3 of thickness arc' (as shown in the fig.28 stretched version), in place of the original calculated 'center-line arc' dimension generally used for the normal bending arc calculations as shown in fig. 25.

While in the process of obtaining the new stretched overall length of material required, we will also need to recalculate new positions for the 'air bend point', and the position of the 4 mm diameter hole on the center-line of the work-piece (particularly if the component is to be 'drilled in the flat").

The dimension given on the drawing for the folding machine's bend point however, willremain unaffected, for up to this point, the material will not have been subjected to any stretch.

In practice, when one is consulting the issued working drawing of a component, one should always make a mental note of the accuracy required by the design engineer or the design draftsman, in the finished product.

This is usually stated on the drawing, by the component being given a definite size limit on its critical dimensions; this is done by indicating a 'plus' or 'minus' 'limit' given to its critical dimensions.

To comply with the drawing, therefore, the component must be kept within these so-called limits to enable the component to pass through the critical inspection test that will inevitably follow, after its manufacture.

There follows a general explanation of what actually happens to the internal core of the material during the bending process.

When sheet metal is subjected to the forces of bending, there is the tendency for the material to be deformed around its inner radius areas, and to be stretched around its core and its outer radius areas. This results in a small (and quite often unexpected) gain in the overall length of the material in the component, (if one consults and compares the dimensions with the original basic theoretical calculations shown in fig.25 to the 'stretched version shown in fig. 28), the extent of this gain in length between the two is quite small but is still significant; and this actual increase in length should be noted down for future reference purposes so that it can be used on similar materials on later work-pieces.

The overall length of the example component shown in fig. 25 will have increased by approximately 0.5 mm due to stretch in the material over and above the original normal basic calculated length. One should note that components that contain multiple bends will be subjected to an accumulation of stretch that will be occurring at each bend, and this must of course be taken into account (by the deliberate shortening of the *cutoff length* of the material by these calculated stretched amounts) before finalizing the overall cutoff length. Also, the position of the hole when being drilled in the flat on the stretched work-piece material, will need to be positioned to a slightly foreshortened dimension to allow for the stretch that will be occurring in the bend.

To reiterate, for precision work-pieces, we must take into account this gain in overall length that occurs during bending, (this may differ by a small amount from the original calculated length shown in fig.25). We do this by consulting fig. 28, which contains the new 'stretched bend' calculation for this particular workpiece. This modification to the initial calculations will be found necessary if one wishes to achieve an extremely accurate precision component.

The overall length of material originally calculated for the component (in fig. 25) will therefore need to be reduced (for precision work) by using a smaller calculated amount being used up in the 'third of thickness arc' as shown in fig.28.

The thickness (or gauge) of material used, the magnitude of the required inside bend radius, coupled with its 'workability, can also marginally affect the amount of stretch encountered during the bending process.

When the material is soft and ductile, it requires only a minimum of effort to be applied to form the bend. The material tends to flow around the bend with little resistance to deformation. However, when the material is tough and less ductile, it requires considerably more effort to be put into the bending process,; the material is reluctant to flow freely around the bend, and develops the tendency not only to stretch, but also to build up stresses in the material during the process.

All materials suffer from a condition known as spring-back. This is a condition, found in the bending process, where the material needs to be 'over-bent' to a greater angle than is finally required, so that when released from the machine, it will spring back to its correct and intended angle of bend.

If we take our bracket as an example, it will therefore need to be over-bent by at least 2 extra degrees than the 110 degrees required, so that its spring-back will allow it to return to the required angle when released from the machine.

This could explain why the angles of the V block used on the air bend tooling shown in fig. 26 have been designed to allow the work to be over-bent by a small amount, in order to allow the natural spring-back of the material to produce the correct bend angle when released from the confines of the machine.

Note that the words 'over-bent' and 'spring-back' are shop-floor engineering terms and are not (as yet) incorporated in the normal dictionary.

The methods used to overcome this stretch problem have been developed over many years of experience in the sheet metal and tube bending industry.

This method was developed as a way of standardizing a repeatable allowance that could eventually be used for the calculation of stretched bends, while still using the tooling and the common sheet metal materials available on the work-place shop floor.

This revelation has solved the ageold problem that has occasionally been found in practice where the workpiece has appeared to have grown unaccountably longer than the anticipated calculated length *after bending*.

It is now possible with confidence, to subtract the anticipated amount of stretch from the originally calculated theoretical center- line length of material, by using the new 'stretched bend arc'calculations shown in the fig. 32a table to solve the problem.

Until now, there has always been a small but unexplained growth in the length of the work-piece when being finally subjected to the scrutiny of the inspection department's close study of the drawing's dimensional requirements.

This age-old problem has always been explained away in the past by workers (and occasionally by the management) as tooling error or operator error, and not as we now know it to be, just plain stretch in the material.

In fig. 25, for a normal bend we have calculated the bend center-line arc to be 7.2954794 mm; now, in order to reduce this slightly to allow for the stretch in the material for a stretched bend, we must re-consider the length of the arc at a radius of one third of the thickness through the material, (instead of using the radius to the center-line of the material).

This particular 1/3 radius dimension was decided upon after carrying out a series of exhaustive tests to ascertain the most accurate radius to use in calculations, for obtaining an average stretch that is likely to be found in most of the known sheet metal and tube materials in current use.

The method we use to do this in our example component, and in the calculation of the 'stretched tables' is as follows:

The component's material thickness is 1.6 mm. We divide this thickness of material by 3, to obtain 0.533333 mm. We now add this to the drawing that has a given inside bend radius of 3 mm, which now gives us the new *1/3 of thickness bend radius* of 3.5333 mm.

We now multiply this by 0.0174533 (the author's standard mathematical shortcut for obtaining 1 degree of arc in a calculation);, followed by multiplying the answer by 110 (the angle of bend), to obtain the new 1/3 of thickness arc length of 6.783515933. (A study of this figure will show that this new *stretched 1/3 arc length* is slightly shorter than the original center-line arc length. This can be checked by the calculation 7.2954794 minus 6.783515933 equals 0.511963467) (roughly a half millimeter). We now substitute the new stretched 1/3 arc (6.783515933) in place of the original center-line arc of 7.2954794; we then re-calculate all the new dimensions as follows: taking the A to B dimension of 45.8928406, plus half of the stretched 1/3 arc (3.391757967);, to give us 49.28459857, to be used for the stretched air bend point; we now add the remaining half of the stretched 1/3 arc (3.391757967), giving us 52.67635653. We now add the 122.2743703 (E to the 4-mm hole center line) to obtain 174.9507268, (the A to center line of the hole position for drilling the hole *in the straight* before bending).

For the overall length of material required in the flat before bending has taken place, when using our now stretched material, we take the A to B dimension of 45.8928406 plus the new third arc dimension of 6.783515933 plus 144.6221035 = 197.29846 mm, we must therefore take into account that this dimension will grow by 0.511965467 mm in order to achieve the 197.8104235 mm overall dimension. This will make this length identical to the *un*stretched fig. 25 version to suit the required drawing dimensions, *after the bending process has taken place*.

*   *   *

An explanation of the stretched arc in the bend material

Explaining what is going on relative to the bend's arc center line

To allow the reader to fully appreciate 'what is going on *inside* the material of a normal sheet metal bend, the following drawings (29, 30, and 31) are shown in a very much enlarged form, (approx. ten times full size), in order to fully illustrate this point.

A study of these drawings will give a valuable insight into the relative positions of the required air bend point, the folding machine bend point, and the vertex point on all three of the various bends shown, including details of how to obtain the vital overall length of material required. To establish the actual bend points and the overall length of material required, the author has standardized his main bend calculation formula as follows. It is therefore considered a very wise decision for the student to commit this important formula to memory as follows: *Tangent length equals the center-line radius, multiplied by the tangent of half the angle of bend.* (Note that in this case we are referring to the *angle of bend* and *not* the *bend angle*, which of course will be a completely different angle, (except of course if we are referring to the angle of 45 degrees, where it is identical.)

It is considered a wise practice to get used to using this particular formula on all bend calculations, even if the bend in question is the most commonly used 90-degree *angle of bend* design (if the student follows this universal method of bend calculation, it will help to maintain total uniformity in all single bend calculations).

Fig. 29 Example 1 (The full development calculation of a 30-degree bend while using 1.6mm. thickness of sheet-metal material).

The calculation method used in fig. 29 is as shown; we first make a sketch and note the length after we have calculated the A to D length).

(1) In fig. 29 a, we therefore sketch a right-angle triangle and enter the known dimensions 244.2191639 and 141.

(2) We will need to find the length of *a*, so for this, we will need to study the probe and use the 'length of side' prompts to indicate to us that the length of side *a* will equal the length of side '*b* squared (that is, multiplied by itself), *plus* the length of

side *c* squared (multiplied by itself), followed by then finding the square-root of the result. We carry this out by entering the sequence; 141 multiplied by 141 = 19881, X M; we then enter the sequence 244.2191639 multiplied by 244.2191639 = 59643.00002, +, R M, =, 79524.00002, INV, SQUARE ROOT, to give us 282, the length of side *a* (the D to F dimension).

(3) To find the angle of bend, we use the most popular prompt ∠ = *b*/*c* INV TAN; we therefore enter the sequence 141, divided by 244.2191639, =, (display shows 0.577350269), INV TAN, thus giving us the important *angle of bend* of 30 degrees.

(4) To find the tangent length, we again use the author's (popular) formula: *Tangent length equals the center-line radius multiplied by the TAN of half the angle of bend*; in this case, we therefore enter the sequence 15, TAN (display shows 0.267949192) × 3.8 = 1.018206931, giving us the answer.

(5) To find the bend's *center-line arc*, we use the author's constant, 0.0174533, multiplied by the *bend center-line radius*, multiplied by the *angle of bend*, as shown; we enter 0.0174533, × 3.8, × 30, = 1.9896762 answer.

(6) To find the overall length of material required we add together the lengths A to B and B to E (*the c/l arc*) and the straight length E to F, as follows: 50.3015961 + 1.989672 + 243.200957, to give us the total cutoff length of 295.4922251.

(7) To find the air bend point in fig. 29, we take the A to D dimension, minus a *tangent length*, plus half the *center-line arc*, as shown: 51.31980303 − 1.018206931 + 0.9948381 = 51.2964342, answer.

Fig. 30 Example 2, (the development of a 90-degree bend while using 1.6-mm material)

To keep uniformity in our bend calculation methods, the student may initially find it best to go through the normal procedure of finding the tangent length, which is, '*Tangent length equals the center-line radius multiplied by the TAN of half the angle of bend.*' In this case we would therefore enter the sequence, 90, divided by 2, =, 45, TAN, (giving us 1), × 3.8 = 3.8. Therefore giving us proof that for a

90-degree angle of bend, the *tangent length* is in fact identical to its *center-line radius*.

Of course, once learned, this process used for obtaining the *tangent length* (for this particular angle), need not be repeated, for we now know that for a bend of 90 degrees, the *center-line radius* is used in place of the *tangent length*.

We know that a circle contains 360 degrees; we therefore need to know the exact amount of circumference that is occupied by *1 degree of arc* on the bend's center line of the material. We can do this by using the author's standard of 0.0174533 in the calculation; this is then multiplied by the known radius, then multiplied by the known angle of bend.

The long calculation that *could* be used to obtain this figure would be to use the arc of radius 1, giving us the diameter of 2; this is then multiplied by pi (π) (3.141592654), to give us 6.283185307. We then divide this by 360 to give us 0.017453292, which is exact, (but in this case this can be rounded up to 0.0174533 to give us the standard (shortened form) that is used in calculations, to obtain *one degree of arc*, in all situations).

To find the length of the center-line arc, we now use this new constant figure, 0.0174533; this is first multiplied by 90, then by the c/l radius, (3.8), to give us 5.9690286. As we also need to know the *half arc* dimension, we divide this *full arc* by 2 to obtain 2.9845143 (the *half arc*).

To find the folding machine bend point, we take the A to D dimension of 51.31980303; we then subtract the *tangent length* of 3.8, giving us 47.51980303 (answer).

To find the *air bend point,* we again take the A to D dimension of 51.31980303; we subtract the *tangent length* of 3.8, giving us 47.51980303; we then add *half the arc* (2.9845143);, giving us 50.50431733, (answer).

To find the overall length of material required we add together the A to B dimension, the B to E (*full arc*) dimension, and the straight E to F dimension as shown: 47.51980303 + 5.9690286 + 137.2 = 190.6888316 (rounded up to 190.7 for use on the shop floor).

Fig. 31 150 degree AOB Example.3 (The development of a 150-degree bend using 1.6 mm sheet material)

It should be noted that the *vertex point* in this particular case will, of necessity, be positioned a considerable distance off to the right of the drawing, (and of the actual component,) this being due to the *tangent lengths* in this case being extremely long; (this could of course be one of the reasons why engineering draughtsman tend not to include *tangent lengths* in their issued drawings), due to this lack of space.

(1) To find the length of D to F, we use the prompt *a* equals the square of *b* plus the square of *c* added together, then finding the square root of the answer;,

(2) We enter the sequence 141 ×141 equals 19881, X M; we enter the sequence 244.2191639 × 244.2191639 equals 59643.00002, + R M, = 79524.00002, INV SQUARE ROOT ($\sqrt{}$), giving us 282, the answer.

(3) To find the *angle of bend*, we first use the prompt $\angle = b/c$ INV TAN, followed by the sequence 180 - $\angle$;

We enter 141 ÷ 244.2191639, = 0.577350269, INV TAN, (30 degrees) X-M; we enter 180 − RM = 150 degrees (the angle of bend).

We then use my popular formula to find the *tangent length*: 'Tangent length equals the center-line radius multiplied by the TAN of half the angle of bend.

With 150 still displayed, we ÷ this by 2 to give us 75; we then enter the sequence TAN (3.732050808) × 3.8 = 14.18179307 giving us the required *tangent length*.

(4) To find the total center-line arc of the workpiece, we enter (the new) constant formula in the following sequence: 0.0174533 multiplied by 3.8 multiplied by 150 =, giving us 9.948381. (answer).

(5) To obtain half the arc, we now ÷ this figure by 2 to give us 4.9741905 (half the arc);

(6) With this figure still displayed, we add 230.0373708 (A to B dimension), giving us 235.0115613 (the air machine bend point).

(7) To obtain the overall length of material required, we add 4.9741905 (the other half of the arc), plus 267.8182069, (the E to F dimension), giving us 507.8039587 (the overall cutoff length); this figure is now rounded down to 507.8 for use on production.

(8) To obtain the folding machine bend point (which is also the A to B dimension), we enter 244.2191639 (the A to D dimension) minus 14.18179307 (the tangent length), giving us 230.0373708.

There is a definite confusion in the industry regarding the actual meaning of the name 'angle of bend' when used in calculation explanations. The author now gives his reason for wishing to clarify this anomaly.

It should be noted that the angle of bend shown in fig. 31 is 150 degrees.

The author must point out that there is often confusion among many office and shop floor workers regarding the difference between the term *angle of bend* and the term *bend angle*, a term commonly used in the industry.

Most engineers, (the author included), prefer to use the term *angle of bend*, because it describes precisely the angle through which the material needs to be actually bent from its original straight-line shape into its finished shape after the bending process has been carried out; (the outdated term *bend angle*, on the other hand, is still often used in the engineering industry, (quite wrongly in the author's opinion, to refer to the included angle of the bend, which is of course a completely different angle, unless of course the workpiece in question possesses a bend of exactly 45 degrees, (where in fact it is identical).

[This situation in the work-place does lead to a certain amount of uncertainty, ambiguity, and confusion, when deciding which angle is being referred to by others, or by the management regarding an angle in the work-piece.]

A study of figures. 29, 30, and 31, will help to confirm in the student's mind the importance of knowing the method used to fully calculate a bend, and how to establish the actual position of the *bend's vertex point*, for it is from knowing the exact position of this critical point, that further bend calculations can be accurately carried out.

The absence on an *issued* drawing of this critical dimension will necessitate a considerable amount of extra calculations being made by the sheet metal worker on the shop floor to obtain it. The three previous examples show how one can obtain the *tangent length* by using the above calculation method as an alternative to finding the position of the *vertex point* described in the fig. 3a and fig. 24 sketched triangulation calculations to obtain it.

Note that it is now possible to use this completely new triangulation method in calculations to find the exact position of the critical so-called missing vertex point.

By using this new 'practical and easier to calculate' method as shown in fig. 3a and in fig. 24), we can now establish this critical *vertex point*'s position and so allow the necessary further calculations to be accurately carried out.

fig. 33.

Fig 33.

SHEET METAL CONE.

THE CALCULATIONS REQUIRED FOR ITS MANUFACTURE ARE SHOWN
IN figs. 34, 35, AND 36.

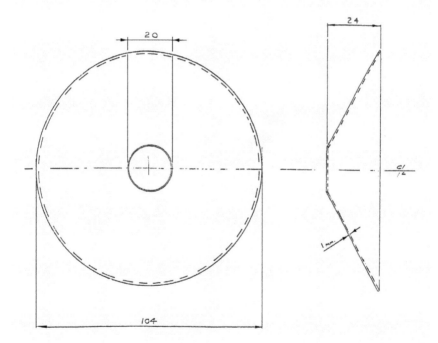

ALL DIMENSIONS IN M.M.
MATERIAL 1 M.M. MILD STEEL SHEET.
NUMBER REQUIRED 5.
LIMITS :— $^{+0.5}_{-0.5}$ ON DIMENSIONS SHOWN.
SCALE :— NOT TO SCALE.

Fig 34.

fig.34.

A THREE DIMENSIONAL VIEW OF CONE SHOWN IN *fig.33.*

THIS VIEW IS USED AS AN INTRODUCTION INTO THE CALCULATIONS

REQUIRED TO COMPLETE THE JOB (SHOWN IN FIGS. 35 AND 36.)

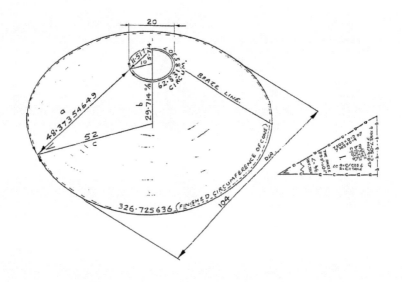

ON COMPLETION, THE TWO ROLLED ENDS ARE BRAZED TOGETHER.

THE JOIN IS THEN HANDWORKED SMOOTH USING A FILE.

ALL DIMENSIONS IN MM.

fig. 35.

CALCULATIONS REQUIRED TO MAKE THE DEVELOPMENT 'BLANK'
FOR THE CONE SHOWN IN figs. 33, 34, AND 36

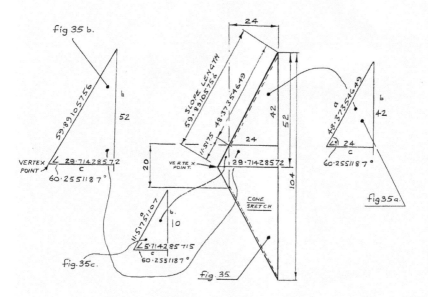

fig 35 b.

fig.35c.

fig.35a.

fig. 35.

FOLLOW THE CALCULATION SEQUENCE AS SHOWN ABOVE IN 35a,b,AND c

ALL DIMENSIONS IN MM.

fig. 36, a and b.

THE DEVELOPMENT OF THE "IN THE FLAT" BLANK, FOR THE CONE SHOWN IN figs. 33, 34, AND 35.

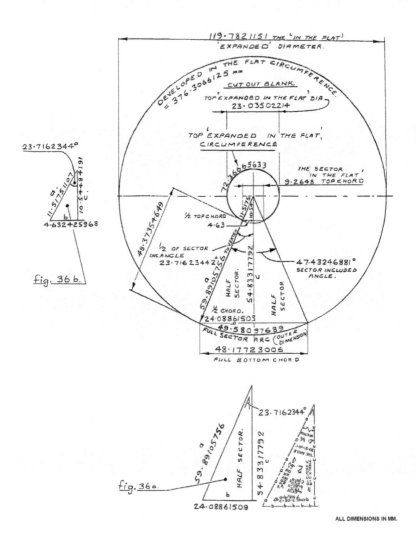

fig. 36 b.

fig. 36 a.

ALL DIMENSIONS IN MM.

Chapter 13

The spinning and development calculations required for the manufacture of the sheet-metal cone

Fig. 33. The drawing office supplied drawing for the cone's manufacture

In order to provide adequate dimensional details for the cone's manufacture, the fig. 34 drawing is the author's helpful three-dimensional sketched view of the cone depicted in the originally issued fig. 33 drawing for the component's manufacture. Note that the fig. 34 drawing is being used as an introduction into the calculations that will become necessary before one can actually manufacture the component. The extra trigonometry calculations required are shown in detail in figures 35 and 36.

The fig. 33 issued drawing contains the absolute minimum of basic dimensions required for the cone's manufacture. The student should note that these minimal 'drawing office supplied' dimensions are very basic and will be considered by the sheet metal engineer to be insufficient for the manufacture of this component without further calculations being made.

For this reason the figures 34, 35, 35a, 35b, and 35c sketches and drawings should be carefully studied as they contain a more practical and easy-to-follow series of development sketches that contain all of the calculated dimensions needed to manufacture this coned sheet-metal component. A thorough in-depth study of these extra drawings will be found extremely helpful to the student, as they point the way through the various sequences of calculations required (in fig. 36a and 36b), that culminate in the necessary *in the flat' dimensioned blank*, shown in fig. 36.

(Also note that the production fig. 33 issued drawing, will only be found useful by the inspection department for use in checking the dimensions of the finished product.)

On reflection, it will be realized, that the fig. 33 drawing fails to contain sufficient dimensions with which to produce the work-piece with any certainty or accuracy, without further calculations being carried out.

The fig. 33 drawing will be seen to contain conventional drawing dimensions typical of those issued by the drawing office to the shop floor. These dimensions are consideredsufficient for the manufacture of the component.

As previously stated, the dimensions supplied on the issued drawing are indeed sufficient for the inspection of the finished component by the inspection department, but any helpful manufacturing dimensions have not been shown on the drawing.

When a student is issued with this particular drawing for the manufacture of this component, it will soon be discovered that it lacks sufficient manufacturing dimensions. There will therefore be the need for extra calculations to be made, at shop floor level, to obtain them. He or she may find difficulty in deciding how to actually start making this cone-shaped component from the dimensions stated on the fig. 33 drawing.

Spinning the coned work-piece

There are actually two main methods used for the manufacture of this design of sheet-metal cone when it is made from steel or aluminum sheet material.

The two possible methods used for its manufacture can be performed as follows: -

1. The usual method used for manufacturing this work-piece is by calculating the unknown dimensions required, followed by marking out and producing a blank from the sheet-metal material, then rolling it into shape, followed by welding or brazing the joint, as described in the following notes, or

2. By using a production process called spinning, where a disc of (usually), aluminum (the preferred material due to its ductility), is formed over the contours of a previously machined hard-wood conically shaped former, securely attached to the faceplate of a center lathe. A circular work-piece disc of material is then held between the flattened machined end of the former, and a purpose-made (suitably modified) hardwood circular pressure plate is secured to the tailstock's rotating center's bearing housing, which secures the work-piece disk under pressure against the machined former by sandwiching the flat disc between these two pieces of tooling.

With the lathe switched on at a moderate speed of, say, 300 rpm, the 'spinning into shape' process' is then performed by a hand operation using tallow as a lubricant and using a process of gradually inducing the rotating disk of material to take up the shape of the former, by using a long wooden handled and hardened steel ball-ended tool. This tool is made from a piece of one-inch-diameter silver steel with its end suitably turned into a ball shape (by the use of a female radius lathe tool), of approximately 25-mm diameter (or in some cases slightly smaller, depending on the size of the cone being made). This tool is necked down to approximately 15-mm diameter to allow its shank when suitably heated to be bent to 30 degrees in order to produce an offset ball that induces a 'castor' action during its use. This tool is finally hardened by heating to a cherry red followed by being plunged into oil prior to it being firmly fitted into a long hardwood stock. In use, the ball of this tool is lubricated with tallow (to avoid the cone's material picking up and adhering to the ball during use). The tool is supported between two vertical prongs secured into an inverted rake-shaped fixed tool bar, secured in the lathe's tool post. The action used in this operation is similar to the motion of an oar in a row-lock (in boating terms). The smooth action aimed for is of a stroking motion against the face of the material in order to carefully 'persuade' the material to form the correct shape over the former. The handheld tool must be allowed to 'castor' in the hand as necessary so that a small rotation of its shaft will gradually induce the material to be formed over the former into its desired shape. This is a rather noisy and clattery operation, and must of necessity be carried out in several stages due to the operation being considered a skilled process that requires several separate annealing operations to be carried out to re-soften the work-piece sheet several times during the forming operation. This softening treatment eliminates the inevitable work-hardening that will occur in the material. This so-called softening operation is necessary because of the continuous work hardening endured by the material during the forming process. When nearing completion, the large accumulated area of excess material that has built up at the face plate end of the setup is removed by using tin snips, while being most careful to leave sufficient material on the component to allow an accurate parting-off operation to take place at the workpiece's longitudinal dimension required. One does have to decide at this stage whether it is best to part off the excess material at the component's major diameter of 104 mm in the first operation, by using a conventional parting tool (while relying on the frictional close fit of the component relative to its former to transmit sufficient drive for the parting-off operation to be achieved, or alternatively, to firstly part off the component at its smaller-diameter minor end, relying on the 'holding pressure of the rotating 'hardwood block / support center combination' to hold the work-piece sufficiently firmly to allow the parting-off operation to take place, relying on the pressure exerted from the lathe's tailstock's support to hold the component in place.

*   *   *

Producing an 'in the flat' blank by using trigonometry calculations

If the machinist has no choice in the method of production, and finds that the manufacture of the fig, 33 component must be made by using the 'mark out', cut out, and roll' process, then additional calculations using the detailed dimensions and sketched triangular figures contained within figures 33, 34, 35, 36, 36a, and 36b sketches and drawings, will need to be used. It is considered advisable therefore for the student to be made fully aware of the engineering methods and the number of extra calculations that will be needed to enable the production of the *'in the flat' blank* shown in figures 36, 36a, and 36b.

The following calculations are rather lengthy and tend to become rather involved. They must therefore be studied with care in order for the student to appreciate fully that they are indeed necessary, because of the insufficient *manufacturing* dimensions given on the issued fig. 33 drawing.

It will be found helpful and informative to the student, if a paper tracing (or a cutout photocopy of the drawing) is first made of the 'expanded in the flat' image of the work-piece drawing, shown in its fully dimensioned 'in the flat' blank stage, in fig. 36.

The outline of this paper tracing should be carefully cut out, and the excess material around its outside diameter, its inner sector, and its expanded inner bore removed.

The raw ends of this tracing can then be brought together to form the finished conical shape. This then shows in a very practical way, the changes in diameter and overall shape of the object as they occur in the finished product, after the full manufacturing process has been completed.

It will be realized during the handworking and shaping of this tracing that although its shape may not be perfectly accurate because of the accrued errors in the drawing and in the later photo-copying process, the diameters of the blank and its bore will actually be larger when in its original blank form, than when it has been reduced to its finished coned shape. (As a matter of interest, the appearance of its actual finished shape will be found to be quite similar (though smaller) to that of a traditional pre-war household lampshade).

The student will now realize that the dimensions supplied in the fig. 33 drawing are totally insufficient to allow the immediate manufacture of this cone-shaped component, without considerably more calculations being made to obtain the so-called missing dimensions required for its manufacture.

The following notes and sketches explain in detail how one performs the extra calculations required to produce the vital 'in the flat' dimensions that allow the final production of the development blank as shown in (fig. 36, 36a, and b,).

(1) To calculate the cone's 'in the flat' manufacturing dimensions, we will first need to consult the sketch fig. 35, followed by

(2) producing the calculated and supporting dimensioned sketches 35a, b, and c as shown on the same drawing.

(2) For undertaking the necessary calculations, and finally drawing the sketch fig. 36, (the development blank), we will need to perform the supporting calculations and sketches shown in figures, 36a and b, on the same drawing.

(3) If we now refer back to fig. 35, this shows the necessary calculations required. We first find The Angle and the length of the cone's slope; this information is duplicated in the informative fig. 34 drawing, which will be found helpful to the student, as it shows a dimensioned three-dimensional sketch of the finished component.

In fig, 35 (which shows the side view calculations), we first use the stated diameter of the cone (104mm) diameter minus its bore (20mm) to obtain 84mm as shown in the sketch 35a.

To obtain The Angle we make a sketch as in (fig. 35a), and carefully select the prompt required to obtain The Angle. We therefore enter the sequence $\angle$ =, b/c INV TAN, i.e. 42 ÷ 24 = 1.75, followed by INV TAN, giving us the angle of 60.2551187 degrees.

Now, by retaining this displayed figure of 60.2551187 in the calculator's memory by using the XM key, and referring to sketch fig.35 b, we now use the prompt $\angle$ b, a =, b / SIN $\angle$, to find the total slope length to the triangle's vertex point.

We therefore enter the sequence RM, SIN (0.868243142), X-M; we now enter the sequence 52 ÷ RM (0.868243142) = 59.89105756, giving us the total slope length from the outside diameter to its theoretical vertex point.

We now return our attention to fig. 35a, in order to calculate the length of the shorter (work-piece) slope length, and use the prompt $\angle$ b, a =, b / SIN $\angle$.

We enter the sequence 60.2551187 SIN (0.868243142) XM; we then enter 42 ÷ RM = 48.37354649, X M, shown in fig. 35a.

We now enter 59.89105756 – R M (48.37354649) = 11.51751107. This gives us the slope length of the 'in the flat' bore (shown in fig. 35c); we then × this by 2, to obtain (what will eventually be) the blank's bore diameter of 23.03502214 mm, shown in fig.36.

We now need to find the 'in the flat' dimensions of the sector and the bore, to enable them to be marked out and later removed (as excess material), together with removing the other excess material from the 'in the flat' outside diameter of the blank.

Fig.36, shows in detail all the necessary working dimensions of the blank, using the guiding triangular sketches, fig. 36a and 36b to assist with this marking-out.

To obtain the expanded diameter of the blank (while the material is still in the flat), we use the calculated vertex to outside diameter slope length, (shown calculated in fig. 35b) of 59.89105756; we therefore enter the sequence 59.89105756 × 2 = 119.7821151 XM, establishing the new 'in the flat' diameter of the blank.

We now enter the sequence RM, × INV π. (3.141592654) = 376.3066128, XM (the 'in the flat' circumference of the blank' in mm); We now enter the sequence RM ÷ by 360 = 1.045296147 mm giving us the amount of material actually taken up by 1 degree of the blank's circumference per degree of arc at the radius of 59.89105756 mm; (this figure should be noted down for future reference).

If we now refer to figures 33, 34, and 35, we see that the diameter of the finished cone is 104 mm.

To obtain this finished diameter's circumference, we multiply 104 mm by p. We enter the sequence 104 × INV ϖ (3.141592654), = 326.725636; we now store this into the calculator's memory by pressing the XM key. We now enter the calculated circumference of the expanded blank (376.3066125), followed by the sequence, minus RM, (326.725636) = 49.5809765 mm; now giving us the outer arc dimension of the sector (which is to be removed). We now enter the expanded blanks circumference of 376.3066125 mm ÷ 360, =, giving us 1.045296146 mm per degree of arc from the full circle. We now divide the cut out sector arc of 49.5809765 mm. by 1.045296146 to give us the included angle of the sector of 47.43246881 degrees; we now divide this by 2 to obtain 23.7162344 degrees, (by doing this, we now obtain half the angle of the sector).

If we now refer to fig. 36, with the wish to find the length of c (the vertical height of the half sector triangle, (base to vertex point), and knowing the angle to be 23.716234 degrees, we use the prompt $\angle$a, c = a × COS $\angle$; we enter the sequence, using the previously calculated angle 23.7162344, COS (0.915548667) × 59.89105756 = 54.83317793, as shown in the calculation in fig. 36a;, noting that this gives us the vertical height of both the full and half sector of the blank.

We now have the need to find the length of side b of the triangle (which is half of the full sector chord) by using the prompt shown in 36a, $\angle\ a$, $b = a \times$ SIN $\angle$; we enter the sequence 23.7162344 SIN (display shows 0.404207207) × 59.89105756 = 24.08861503, giving us half of the bottom chord.

We now × this figure by 2 to obtain the full bottom chord of the sector, namely 48.17723006; this is now retained in the calculator's memory by pressing the X M key. (As a matter of interest, should we wish to find the difference in length between the sector chord and the sector arc, we could now enter the sequence 49.5809768 − R M =, giving us 1.403746826 mm, the difference in mm.

To obtain the dimensions for marking out the top chord of the sector, we first refer to fig. 35 c (where we obtained the short slope length from the vertex point to the 20 mm diameter hole of 11.51751107; we now enter the sequence 11.51751107 × 2 = to obtain the 'in the flat' expanded diameter of the central bore to 23.03502214, this figure is now multiplied by (ϖ) using the key sequence × INV (ϖ) (3.141592654) =,to obtain the hole 'in the flat' circumference of 72.36665633 mm as shown in fig. 36. We enter this into the calculator's memory by pressing the X M key.

We now enter into the calculator the finished diameter of the cone's hole (shown in fig. 33, and 35, as 20 mm) and use the sequence 20 × INV, π (3.141592654) =, to obtain (what will be) the finished cone's hole circumference of 62.83185307 mm in fig 34.

We now use the key sequence INV, X M (72.36665633) - R M =, 9.534803258 (this gives us the full arc of the sector's top chord). We now refer to fig.36-b for the calculations required to obtain the top chord for the marking-out operation.

Referring to fig. 36 b

To find the angle of the half sector at its vertex point we use the prompt $\angle$ =, b/c, INV TAN; we enter 4.632425968 ÷ 10.5448419, =, (0.439307294) INV TAN, giving us 23.71623442 degrees.

We transfer this into the calculator's memory by pressing the X-M key.

We now find the length of side c in fig.36b using the prompt $\angle\ a$, $c = a \times$ COS $\angle$; we press RM, (23.71623442), COS (0.915548667), × 11.51751107, = 10.54484191.

To find side b in the half sector (which is actually half of the top chord) we use the prompt $\angle\ a$, $b = a \times$ SIN $\angle$; enter the sequence 23.7162344 SIN (0.402207207) × 11.51751107 = 4.632425968.

This is now × by 2 to obtain the full 'in the flat' top chord of the small sector's dimension, giving us 9.264851936 mm, for marking this out on the 'in the flat' blank.

Finally, we carefully mark out the blank using a center punch, a scriber, an accurate steel rule, and dividers; after the marking-out operation, we carefully remove the excess material (from the sector, the bore, and the outside diameter), using tin snips or similar precision hand tools.

Following the cutting operation used to remove the excess material, we follow this by physically setting up a hand-operated rolling machine to roll-form the sheet material blank into its final coned shape, to produce its correct diameters and dimensions in order to comply with the fig. 33 drawing. Note that the sector's cut edges should be offered into the rollers of the machine while they are being oriented exactly square to the rollers' surface. (Note that if one fails to adopt this 'exactly square' offering of the material into the machine, its cut ends will not eventually come together into perfect alignment on completion of the rolling operation, and will therefore produce a twisted and largely unusable shaped object that will undoubtedly render the component to be scrap.

This forming operation into its fully coned shape is carried out until its ends are seen to actually butt up at their intended joint faces to form the complete cone. The join is then either brazed or welded (as specified on the drawing) to complete the work-piece.

The outside circumference of the developed blank (originally 376.3066125 mm.) will now have been reduced to 326.71563 mm on reaching its final conical shape and now gives it the required finished outside diameter of 104 mm.

The original 'expanded in the flat' bore circumference of 72.36665633 (with its expanded bore diameter of 23.03502214 mm), will now have become 20 mm in diameter in its finished coned shape.

fig. 37.

STEPS SHOWING THE CALCULATION OF 'UNEQUAL,' (OR NON-RIGHT ANGLE) TRIANGLES,
USING THE 'WONKY GABLED HOUSE' METHOD
OF CALCULATION. (SHOWN DETAILED IN fig 6b, 6c, AND 37f.)
THE KNOWN LENGTH OF SIDES ARE. P. S. AND B.
THE UNKNOWN ANGLES TO FIND ARE 'G'. 'H'. AND 'W'.

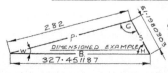

DIMENSIONED EXAMPLE

---

fig. 37a. STEP ONE.

WE NEED FIRST TO OBTAIN ANGLE 'G'.

USING THE FORMULA $\cos G = \dfrac{(s^2 - B^2) + p^2}{2 \times p \times s}$ INV. COS

THE METHOD USED TO OBTAIN THIS IS
EXPLAINED IN THE TEXT.

THE ANGLE (AFTER CALCULATION) FOUND TO BE
150·2868189°

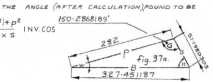

fig. 37a.

---

fig. 37b. STEP TWO

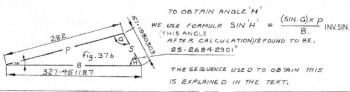

fig. 37b

TO OBTAIN ANGLE 'H'

WE USE FORMULA $\sin H = \dfrac{(\sin G) \times P}{B}$ INV. SIN.

(THIS ANGLE
AFTER CALCULATION) IS FOUND TO BE,
25·26842901°

THE SEQUENCE USED TO OBTAIN THIS
IS EXPLAINED IN THE TEXT.

---

fig. 37c. STEP THREE.

TO OBTAIN THE 'WONKY' ANGLE 'W'
WE USE THE CALCULATION 180 − (G + H) AS SHOWN
150·2868189 + 25·26842901 = 175·5552479
180 − 175·555248 = 4·444752088°

THIS ANGLE AFTER CALCULATION FOUND TO BE
4·444752088° AND CALLED
'WONKY ANGLE 'W'

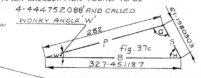

fig. 37c

---

fig. 37d. STEP FOUR.

TO OBTAIN THE VERTICAL HEIGHT OF
THE TRIANGLE WE USE THE 'PROBE' AND ITS PROMPT
$\angle a\,b = a \times \sin \angle$, USING THE 'NOW'
KNOWN 'ANGLE 'W' 4·444752088°
IN THE CALCULATION.

WE OBTAIN THE BASE LENGTH OF THE 'LARGE' TRIANGLE
BY USING THE 'PROMPT' $\angle a, c = a \times \cos \angle$ WHICH
(AFTER CALCS) GIVES US 281·1518921.

fig. 37d.
LARGE TRIANGLE

---

fig 37e. STEP FIVE.

TO FIND THE BASE LENGTH OF THE 'SMALL' TRIANGLE
WE SUBTRACT 281·1518921 (BASE LENGTH OF 'LARGE' TRIANGLE)
FROM 327·451187 GIVING US 46·2992949

THE SMALL TRIANGLE

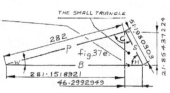

fig. 37e.

ALL DIMENSIONS IN MM.

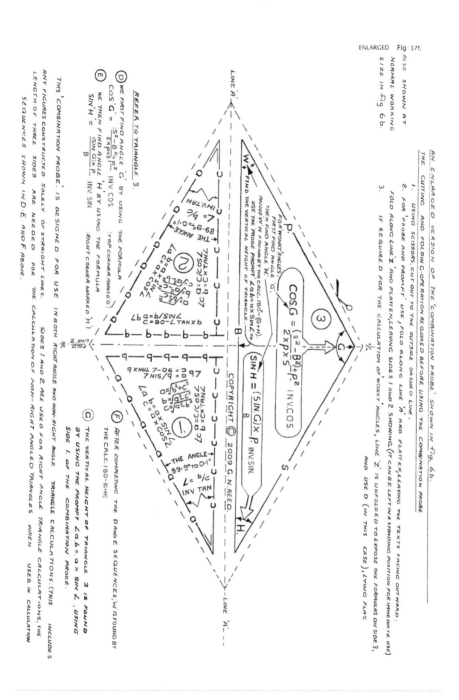

AN ENLARGED VERSION OF THE 'COMBINATION PROBE' SHOWN IN fig. 6b.

THE CUTTING AND FOLDING OPERATION REQUIRED BEFORE USING THE COMBINATION PROBE

1. USING SCISSORS, CUT OUT TO THE OUTSIDE DASHED LINE.

2. FOR 'PROBE AND PROMPT' USE, FOLD ALONG LINE 'A' AND FLATTEN, LEAVING THE TEXTS FACING OUTWARD. FOLD ALONG LINE 'Z' AND FLATTEN, LEAVING SIDES 1 AND 2 SHOWING. (IT CAN BE LEFT IN A STANDING POSITION FOR IMMEDIATE USE)

3. IF REQUIRED D FOR THE CALCULATION OF WONKY ANGLES, LINE 'Z' IS UNFOLDED TO EXPOSE THE FORMULAS ON SIDE 3, AND USE D (IN THIS CASE), LYING FLAT.

ALSO SHOWN AT NORMAL WORKING SIZE IN fig 6b.

REFER TO TRIANGLE 3.

(D) WE FIRST FIND ANGLE 'G' BY USING THE FORMULA
$$COS\ G = \frac{(S^2 - B^2 + P^2)}{2 \times P \times S}\ INV.COS$$

(E) WE THEN FIND ANGLE 'H' BY USING THE FORMULA
$$SIN\ H = \frac{(SIN\ G) \times P}{B}\ INV.SIN$$

LINE 'A'

P ← FOR 'WONKY' ANGLES:
FIRST FIND ANGLE 'G'.
THEN FIND ANGLE 'H'.
ANGLE 'W' IS FOUND BY THE CALC. 180°-(G+H).
USE THE PROBE PROMPT $\angle a.b.= a \times SIN \angle$ TO
FIND THE VERTICAL HEIGHT OF TRIANGLE.

W → THE ANGLE 89·9°TO 0·1°
$\angle z = b/c$
INV.TAN

③
$$COS\ G = \frac{(S^2-B^2+P^2)}{2 \times P \times S}\ INV.COS$$
$$SIN\ H = \frac{(SIN.G) \times P}{B}\ INV.SIN$$

②
$\angle b\ a = b/SIN \angle$
$c = 90·7\ TAN \times b$
$a = b/c$
$SIN\angle$
COPYRIGHT © 2009 G.N. REED.

B

H

S ↓

① ← THE ANGLE 89·9° TO 0·1°
$\angle z = b/c$
INV.TAN
$\angle b\ a = b/SIN \angle$
$c = 90·7\ TAN \times b$

(F) AFTER COMPLETING THE D AND E SEQUENCES, 'W' IS FOUND BY THE CALC. 180°-(G+H)

(C) THE VERTICAL HEIGHT OF TRIANGLE 3 IS FOUND BY USING THE PROMPT $\angle a.b.= a \times SIN \angle$, USING SIDE 1, OF THE COMBINATION PROBE.

LINE 'A' ---

THIS 'COMBINATION PROBE', IS DESIGNED FOR USE IN BOTH RIGHT ANGLE AND NON-RIGHT ANGLE TRIANGLE CALCULATIONS. (THIS INCLUDES ANY FIGURES CONSTRUCTED SOLELY OF STRAIGHT LINES. SIDES 1 AND 2 ARE USED FOR RIGHT ANGLE TRIANGLE CALCULATIONS, THE LENGTH OF THREE SIDES ARE NEEDED FOR THE CALCULATION OF NON-RIGHT ANGLED TRIANGLES WHEN USED IN CALCULATION SEQUENCES SHOWN IN D.E. AND F ABOVE.

USING THE WONKY GABLE HOUSE METHOD OF TRIANGLE CALCULATION.
NAMING THE PARTS OF THE GABLE END.
DIMENSIONING THE INDIVIDUAL SIDES OF THE UN-EQUAL ANGLE TRIANGLE.

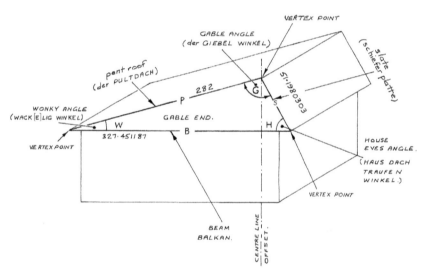

VERTEX POINT

GABLE ANGLE
(der GIEBEL WINKEL)

51·1980303

pent roof
(der PULTDACH)

P 282

G

S

slate
(schiefer platte)

WONKY ANGLE
(WACK[E]LIG WINKEL)

W

327·451187

GABLE END.

B

H

HOUSE
EVES ANGLE.

(HAUS DACH
TRAUFEN
WINKEL.)

VERTEX POINT

VERTEX POINT

BEAM
BALKAN.

CENTRE LINE
OFFSET.

TO FIND ANGLE 'G' USE FORMULA $\dfrac{(s^2 - B^2) + P^2}{2 \times P \times s}$ INV.COS

TO FIND ANGLE 'H' USE FORMULA $\dfrac{(SIN\,G) \times P}{B}$ INV.SIN

TO FIND ANGLE 'W' USE THE CALCULATION $180° - (G + H)$

TO FIND VERTICAL HEIGHT USE 'PROBE AND PROMPT', UTILISING KNOWN LENGTH OF
SIDE 'P' AND THE 'NOW KNOWN' ANGLE 'W'. USING 'PROMPT' $\angle a$ b = a × SIN $\angle$.
OR USING THE 'NOW KNOWN' ANGLE 'H' AND SIDE S. USING 'PROMPT' $\angle a$ b = a × COS $\angle$.

Chapter 14

There now follows a *preliminary* explanation of the Wonky Gabled House method of calculating the non-right angle (or the so-called unequal-angled) triangle.

This is a preliminary explanation of how one-uses the Wonky Gabled House method of triangle calculation to obtain all three of the included angles contained within the non-right angle (or the unequal-angled) triangle).

We should first make a study of the fig. 37g Wonky Gabled House drawing, which depicts a simulation of the gabled end of the roof of a terraced house. We should then study the fig. 37f drawing that shows the 'cut and fold' Combination Probe' used as a guide for the calculations made in the examples shown in fig. 37 through to fig. 37e to complete the whole triangle's calculation. During a careful study of the fig. 37 calculation examples, it will be seen that these separate sequences explain the full series of calculations required to enable us to obtain all three of the unknown angles contained within the Wonky Gabled House non right-angled triangle. It will be realized after this study (and by using a little imagination) while using the later process of introducing an additional vertical line into each of the triangular figures shown at fig. 37d, and e, which makes it now possible for them to transform the original triangle's unequal angled shape, into a new so-called back-to-back (side *b* to side *b*) orientation within the original figure.

The ultimate aim of using the Wonky Gabled House method for solving the un-equal-angled triangle, is for the student, on reaching the latter stages of the calculation (as shown in fig. 37d, step 4 and five), that we use this inclusion of the two right-angled triangles into the original un-equal triangle in order to easily obtain its vertical height, and to obtain the exact length of its 'new' sides *c*, in both of the two newly introduced right-angled triangles, as shown in figures 37d and 37e.

To reiterate, to complete the final stages of this calculation, we actually introduce the two new right-angle triangles into the original problem triangle; this introduction then simplifies the final calculation of finding its vertical height and the length of the two new sides *c* of the introduced triangles by then using the original Probe and Prompt calculation system (resulting in the calculated vertical height of all three of the triangles being virtually identical).

The following additional notes describe the design features built into, and developed for using the Wonky Gabled House method of calculating the non-right-angled triangle.

The principal aim of showing the fig. 37 calculation examples, is for the student to obtain and retain (by using its easy to follow steps, One, Two, Three, Four, and Five,) the method used to obtain the magnitude of each individual (so far unknown) included angle of each of the three included angles contained within the Wonky Gabled House triangular figure.

The sequence of the formulas that need to be followed to complete this full calculation process, are described on side 3 of the Combination Probe (and duplicated in figures 6b, 6c, 37 f, and 37g).

In order to encompass the whole range of triangle problem shapes the student may encounter in the school, college or workplace, and to aid in the simplicity and the easy understanding of the sequences required, we should now refer to the figures 37, 37a, b, c, d, e, f and 37 g, where the correct formulas for use in obtaining the required included angles are shown in the six fig, 37 sequenced examples.

Fig. 37g, shows the dimensioned sketched example of the problem triangle we are about to work on in order to calculate all its included angles. It is hoped that these calculated examples will convey to the reader that it is only necessary for the student to possess the ability to calculate two main shapes (or configurations) of triangle, from the infinite number of shapes that could possibly be encountered.

It must, however, be mentioned at this juncture, and before we finally undertake the calculations necessary to complete those required for calculating the non-right-angle triangle, there also exist two major and vitally important geometric shapes that are found to be absolutely vital in the world of mechanical engineering. These being

1. The six-sided hexagon, closely followed (in importance) by
2. The eight sided octagon.

If we first consider the hexagon, its design goes back in world time to the insect period of evolution, where historians consider that the wild honey-bee developed (purely by trial and error, over millenniums of time), the six-sided honeycomb structure for the storage of their nectar and brood larvae. It is therefore no accident that this structure's shape was found capable of providing the maximum storage area for their honey or brood without entailing any wasted space that would normally occur if a circular tube design had been used to form the cell structure's shape.

In the field of engineering, the hexagon shape has been found to be ideal for use in the design of the common nut and bolt-head; this shape is a perfect match for the design of the open-ended, socket, ring spanner, or the hexagon socket spanner, that is universally used in the engineering industry and the motor trade for the tightening and loosening of bolts and nuts e.t.c.

The (eight-ided) octagon is often used for the larger diameter nut /bolt applications on pipes, particularly in the plumbing field, where, for example, the securing flange nut of a U-Ks domestic central heating system's immersion heater allows this design to provide sufficient space for the internal electric contacts within the confines of its larger inside cavity. The necessity of leaving sufficient inside space for its internal electrics has in consequence led to this shape comprising of eight relatively short length flats, which necessitates the use of a large ring-type spanner in place of using the normal open-ended spanner normally used for secure fitting and tightening purposes.

However, by using a process of elimination, (based on the current trend to develop our mathematical expertise on a need-to-know basis), considering the essential number of geometric shapes that are actually required to be known by the student for the purposes of understanding the contents of this book, and enabling the student to deal with all triangle calculations, the extended list of existing triangular shapes has been systematically reduced so that it now includes only three of the most important shapes.

These are considered to be,

(1) The right-angle triangle, (as shown in fig. 1, and 2,)
(2) The non-right-angle triangle, also referred to in the text as the unequal-angled triangle, (for convenience purposes), (as shown in fig. 37, 37g and 37f).
(3) The equal-angled triangle, (where all the triangle's sides are of equal length), which can be calculated in a similar way to (2), the non-right-angle triangle, by using the same Wonky Gabled House triangle calculation method, i.e. dividing the figure vertically into two parts in order to obtain two right-angle triangles (back to back), followed by using the basic Probe and its Prompts for the last of the required calculations.

By adhering to this relatively small number of geometric shapes found in triangular figures that require calculation, we have avoided allowing ourselves to enter into the quite formidable task of naming all the geometric shapes known to exist in the field of geometry.

The other geometric shapes referred to, (the lesser-known ones), are of course beloved by those who set the so-called stumbling-block questions in examination papers. These particular questions usually have the requirement to name the most obscure geometric shapes (rarely encountered in true life), which results in the student failing to answer them correctly, thereby restricting the total number of valuable examination points possible to obtain in an examination.

The author has now realized, (as a result of many years' practical experience in the engineering industry), that even if one actually knows them all, and can actually recite the names of the multitude of differing shapes, this particular knowledge is of little use when confronted with a complicated practical triangle problem that needs to be solved straightaway on the shop floor, or in the workshop design office.

As a result of this realization awareness, the author has resisted the temptation to name each individual geometric shape known to man, as there is no advantage in the student knowing them all for the purposes of understanding the calculations undertaken in this book.

This section therefore deals with the practical simplification of the age-old problem of calculating all the angles in the unequal-angled triangle.

Suffice it to say that the student can get by in virtually any triangle calculation situation by using either the Probe and Prompt aid to resolve the problems found in the right-angle triangle, Combination Probe to calculate all the problems found in both the right-angle and the unequal-angle triangle.

The use of these calculation aids totally frees the student from the need or the ability to recite, the specific names of the whole multitude of possible shapes that actually exist in the universe.

In the earlier chapters, we dealt with the calculation and solving of problems occurring in right-angle triangles, by use of the probe and the basic Probe and Prompt system.

Its disadvantage is that this basic system is only capable of assisting in the triangle calculations needed to solve just right-angle triangle problems.

It is of course hoped that by now, through practice and acquired experience, the reader will have reached a reasonably expert level in using the basic Probe and Prompt system to calculate all his or hers right-angle triangle problems.

This recently gained expertise should by now be sufficient to allow him or her to actually move up a notch and to be in the position where he or she can now gain the additional ability of understanding and coping with problems that occur in the more advanced non-right-angled or unequal-angled-triangle situation featured in the Wonky Gabled House method of triangle calculation as shown in fig. 37g.

This more advanced calculation system does involve a slightly more complicated series of calculations to be undertaken by the student before the required answers can be obtained.

It follows that a calculation situation can occur (though in the author's experience, this only occurs occasionally), when the student will need to deal with and calculate the angles of the non-right-angled triangle.

While performing this so-called Wonky Gabled House method of calculation, it will be considered an advantage if one uses a methodical and painstaking approach to the problem.

It is inevitable, therefore, that from time to time while performing right angle triangle calculations, the reader will reach a stage in a calculation where the basic Probe and Prompt aid will have reached the limit of its ability to assist, and will therefore be unable to assist the student any further, because the problem triangle being worked on is found not to contain a right angle from which to continue the calculations, (the complete calculation of the non-right-angle triangle is now assisted (in the latter stages as shown in fig, 37d step 4, by the introduction of two right-angle triangles into the main outline shape of the worked on figure, in order to obtain its vertical height. This move can only be undertaken after two of the non-right-angle triangle's included angles have become known through calculation, and this can only occur after the first three stages of the figures. 37a, 37b, and 37c, have been completed in the calculation.

This stumbling-block situation can occur even though the actual dimensions of all three side lengths of the problem non-right-angle triangle have been discovered, and are fully known.

In order to solve this elusive 'find the angles' problem, we can now use the newly designed Wonky Gabled House method of non-right-angle or unequal-angle triangle calculation shown in figures, 6b, 6c, 37f, and 37g. The sequence of calculations required when using this method is fully explained in meticulous detail in the calculation examples shown in fig. 37, 37a, 37b, 37c, 37d, and 37e.

When using this calculation method, it will be realized that its main advantage lies in its ability to break down the whole triangle calculation problem into five individual, easy-to-solve steps. These steps are then worked through in stages until the final complete answer is obtained, (hence a methodical approach to the problem is advisable).

The student must however always follow the strict calculation sequence as indicated in the instructions for each of the required steps; these are lettered D, E, F, and G, on side 3, of the fig. 37f Combination Probe, shown in fig.37f and 37g.

Fig. 37g

In order to make the Wonky Gabled House method of triangle calculation more easily understood, the fig. 37g drawing is designed to depict the shape of the gable end of the roof of a terraced house. It contains all of the formulas and information required to calculate the three unknown angles in the non-right-angle triangle.

The student will find it a distinct advantage at this early stage in the calculation process, if he acquires and commits to memory the correct calculation sequences needing to be adopted. This triangle's side and angle identifying letters, should be carefully studied with the aim of imprinting them into the memory, in order to assist in their practical use.

The student, while using this method of triangle calculation, should be fully prepared to accept that this particular-shaped triangle does actually contain the gable end of the roof of a terraced house; this name was chosen because the roof of a house contains a very familiar shape and is recognizable by everyone.

The student is advised to pay particular attention to the pitch angles (or slopes) of its roof relative to its two angled sides.

In the example roof shown in fig.37g, it will be seen that the left side of the roof slopes downward at a shallower angle than that of its right-hand side, thereby providing a basic similarity to the old pent roof design used on houses and buildings of a past era.

For identification purposes, the left side of this roof is given the initial letter p (for pent roof'), thereby giving this side of the roof a positive identification letter for use in the following calculations.

The outer surface of the right side of this imagined roof is slated, and is therefore given the initial letter s (for slate), again for identification purposes.

The point of the roof, (normally in housing parlance given the name gable point, and in mathematical terms, its vertex point), is therefore given the initial letter G for 'Gable angle' as its identification.

It will be seen in this case that point G is offset to the right-hand side of the main figure; this then establishes (for our calculation purposes) its lateral position, to aid in its easy identification. It should be noted that this lateral offset position is being used only as a calculation guide, and is not intended to simulate the exact shape of the problem triangle being worked on (which will in all probability be at variance to this). This particular shape is being used to give only an indication of its relative position in our fig. 37g example of a worked-on non-right-angle triangle.

The sequence of calculations now needed to be performed by the reader, will require the positive identification of all the angles and side lengths in this triangular figure. Note that this identification method will still be effective if used on any other non-right-angle triangular figure, providing that the problem's apex (its included angle G), is positioned above and within the confines of the problem triangle's baseline B, and this baseline occupies a position similar to that shown as line B in the fig. 37 series of triangle sketched examples.

The right hand corner of the roof's included angle has been given the name 'House eaves angle'. It follows that this has been given the initial letter H (for identification purposes).

The horizontal beam shown at the base of (and in one's mind's eye, supposedly supporting the main roof structure) is therefore given the initial letter B for 'beam'.

The included angle shown at the left-hand corner of this roof structure is given the rather unusual (but easily remembered) name, 'Wonky angle', this being due (in this particular case) to its unusually shallow angle. (However, the magnitude of this angle can, in practice, be found to be any angle, up to a possible 89.9degrees). This angle has therefore been given the initial letter W for its identification, (note that this is also the first letter used in the naming of this unequal angle triangle calculation method).

To reiterate, the fig. 37g example shows a miniaturized and sketched view of a simulated gable end of the roof of a terraced house. The view illustrates, in particular, the roof's gable end and shows in the example the three known side lengths being measured in mm, these being 282, 51.1980303, and 327.451187; these side lengths are used in the following example calculations shown in fig. 37, 37a, b, c, d, and e.

The roof's three sides and three angles have been given their separate individual letters in order to identify them positively in the actual positions they occupy in the problem triangle being worked on.

We must first make a large sketch of the problem triangle on a separate piece of paper, and write in all the identifying letters: W, *p*, G, *s*, H, and B, in their correct positions. We must also include in the sketch all of the three known lengths of sides, also in their correct relative positions.

[It may be of interest to know that the individual descriptive lettering used for each side and each angle used in this Combination Probe memory aid, and in the fig. 37 to 37e sketched examples, have been carefully chosen to represent the first letter of descriptive words used in the house building industry.)

Using a careful selection of words has allowed these specific identifying letters to be fully understood also by the speakers of German, because this selection provides the same meaning as the identical initial letters used in the German language.

This useful similarity is shown in the following examples:

The English-named 'Gable angle' can, in German, be translated as 'der Giebel Winkel'; the name 'slate' translates as 'schieferplatte'. The name 'House-eaves angle' translates as 'Haus Dach Traufen Winkel'; the name 'Beam' translates as 'Balkan'. The name pent roof' is translated as 'pult-dach', and lastly, the name 'Wonky angle' is translated as 'Wack-E-Lig Winkel', confirming that the Combination Probe's instructions and its triangular shape, can be easily understood and used by students who are mainly speakers of the German language.]

When the student is making a start on the calculations necessary to obtain the unknown angles in the problem non-right-angle triangle (as shown in the fig.37 to 37e examples), we should refer to side 3 of the Combination Probe for its guidance (these instructions are also shown in fig. 6b and 6c). It is very important that we follow the instructions and carry them out in a correctly strict sequence. (These instructions are also duplicated on the fig. 37f page, which contains the actual cutout calculation aid and is contained within paragraphs' D, E, F, and G.)

To begin making the triangle calculations, it is imperative that we first finalize the actual position and orientation of the 'unequal angled triangle's baseline with the actual triangle being worked on. (The position of this baseline is also shown in fig. 6 b, and 6c, where it is used as the aid's base line B, and in figures 37 to 37e, and in fig. 37f as the baseline used for the calculations. Note that when this calculation system is used for other unequal-angled/non-right-angle triangle calculations, it should be realized that any newly worked-on problem (because of the possibility of variance in their orientation to the problem triangle), will need to be rotated to the standard (correct) attitude, as shown in the fig. 37, to 37e examples, before starting the calculations. To reiterate, it is always the case that line B is used as the established horizontal baseline for this particular calculation, as it is shown horizontal, in the fig.37, to 37e calculation examples.

We can now commence the calculation by following the calculation instructions D, E, F, and G given in fig. 37f, (also shown in fig. 37g). These formulas are now used to calculate the fig. 37, to 37e, examples, (and displayed in figures 6b, and 6c).

The calculation sequence you are about to undertake should be entered into calmly, diligently, in a relaxed frame of mind, and certainly this calculation should not be entered into with any undue haste. It should be understood that this could possibly be the most difficult calculation sequence undertaken so far in this book, and your concentration levels should be kept as high as possible throughout the whole calculation sequence. The intricate moves given to the 'scientific calculator's keys to make the changes between its memory and its visual display can prove to be rather disconcerting to the beginner. It may therefore require several attempts at this whole calculation before the final result will be found to be correct. There can also be a block that can occur on completion of the formula required to obtain the angle G; this is due to a marked tendency for the student to fail to include the necessary and important following sequence INV, COS. This, however, must be used in order to convert the so-called gobbledygook number shown in the display, into the much sought-after angle G, in the preferred decimal degrees required. Similarly, after using the correct formula for finding the included angle of H, the student must not forget to include the important following sequence INV, SIN, to complete the calculation of the required angle H, in decimal degrees. It may therefore be necessary for the student to run through the whole calculation sequence several times before he or she can become fully acquainted with the various accurate calculation sequences required.

STEP ONE.We are instructed first to find angle G (as shown in the example 'step one' in, fig. 37a).

One should note that the figures shown inside brackets in the formulas, must always be calculated first (this is a perfectly normal mathematics convention), and one should always ensure that the result of the calculation made above the division line will later be divided by the result of the calculation made below the division line (this is a very important fact and must be complied with).

We must now make a large sketch of a similarly shaped triangular figure at a similar orientation to the problem triangle being worked on while using its side B as its baseline. It is also an advantage if we copy out the appropriate formula needed to obtain angle G, angle H, and angle W on the same piece of paper, (a simplified version of the angle G formula, when spoken aloud, would be '(s squared minus B squared), plus p squared, divided by the result of two times p, times s). It is most important that we identify each of the problem triangle's sides and angles by using its written identification lettering, G, H, and W for its included angled corners, and its s, B and p for identifying its appropriate sides. We must also enter the dimensional length of each known side in its correct position. The calculations can now proceed by using this previously sketched example and by using fig. 37a as a reference.

We start the calculation by entering into the calculator,the length of the first known side (this will be the first of the bracketed sequences shown in the 'above the division line' formula (s squared – B squared) as follows: we enter the length of side s, 51.1980303: we then multiply this by itself by using the key sequence, '×, =', giving us the figure 2621.238307 (now shown in the display);. We follow this by pressing the XM key (this automatically transfers the displayed figure into the calculator's memory).

We now enter the known length of side B by entering 327.451187. This is multiplied by itself by using the key sequence, '×, =', giving us 107224.2799 (now shown in the display); we now enter the following sequence used to swap the current display with the calculator's memory, to complete the calculation: 'INV, XM, -, RM, =, giving us – 104603.0416.

(Note that the display is now showing a minus quantity); we follow this by pressing the X M key, (thereby transferring the displayed figure into the calculator's memory.)

We now enter the length of side *p*, by entering 282, followed by the sequence 'x, =', giving us 79524 shown in the display. This is followed by the sequence, '+, RM, =', giving us -25079.04156 showing in the display; we follow this by pressing the XM key, (transferring the displayed figure into the memory). (Note that this is also a 'minus' number.)

We have now completed the 'above the division line' part of the calculation, but we should note that we have only so far completed the first part of the calculation sequence required to calculate the magnitude of angle G.

The calculation must now be continued onward by taking the next step, which involves completing the 'below the line' part of the calculation.

We therefore enter the sequence, 2, x, 282, x, 51.1980303, =, giving us 28875.68909. This is followed by entering the sequence 'INV, XM, ÷, RM, =', giving us – 0.868517509. This calculation is now continued by remembering to enter the sequence 'INV COS', to give us 150.2868189, the included angle of G in decimal degrees.

We have now calculated the included angle (in decimal degrees), of angle G (but this is only the first part of the full calculation required to find the unequal-angled triangles' other two angles that will now need to be calculated). The angle G we have just calculated, should now be written into the hand-drawn sketch as 150.286819 degrees as shown in step one fig. 37a. We then press the XM key to retain this figure in the calculator's memory.

STEP TWO.

We now carry on with the next part of the calculation by referring to the fig.37b and complying with the instructions required to find the included angle of H (this is also shown on side 3 of the Combination Probe).

We must now copy out the formula required for obtaining angle H, on the same piece of paper, in order to familiarize ourselves with the calculation sequences required.

We therefore continue the calculation by using the next given instructions to find angle 'H, which are, 'SIN.H = (SIN G) ×, p, ÷ by B,' as follows (note that we always calculate the bracketed part of the formula first). (A simplified version of this formula (when spoken out loud) would be: "Sine H equals Sine G multiplied by p, divided by B'. Having noted that SIN G is bracketed, we therefore use the existing display showing 150.2868189, followed by pressing the SIN, key, giving us (0.495658487 in the display). We now enter the sequence, ×, 282, =, giving us 139.7756935; this is now divided by the length of side B, by entering the sequence, ÷, 327.451187, =, giving us 0.426859633. We must now remember to press the keys INV SIN, to give us 25.26842901 degrees now shown in the display,; this now gives us the included angle of H, in decimal degrees]. We now enter this into the memory by pressing the XM key and add this into the sketch as shown in 'step two' fig. 37b.

We have now reached the stage in the calculations where we now actually know both the included angle of G (150.2868189 degrees), and the included angle of H (25.26842901 degrees). This situation now makes it possible to find angle W by using the following calculation.

Utilizing the retained calculator display of (25.26842901), we enter, + 150.2868189, =, giving us 175.5552479 (degrees), followed by using the XM key,; we now have the combined angles of G and H.

STEP THREE.

[We already know that all triangles must contain exactly 180 degrees (of included angle); we can therefore obtain the included angle of W, by now entering the sequence 180, -, RM, =, giving us 4.44475209 degrees (followed by XM). This gives us the included angle of W, the so-called Wonky angle.

In order to calculate every aspect of this non-right-angle triangle and having found that angle W contains 4.44475209 included degrees, we may now need to find the vertical height of the whole triangle.

[(We should also note that if we are making our calculations on a calculator using a ten-figure display, the last two figures may be found to be at a slight variance to those obtained by obtaining the vertical height if one were using angle H, as opposed to using the angle W in the calculation for checking this vertical height; the difference in this case will be very small, this being approximately 0.000000185 mm.

Because this is such a very small dimension, it can largely be ignored in these calculations. Therefore if after performing a long and involved calculation using angle H and angle W, while using SIN and COS for both calculations, it may be found that the last two figures in the ten-figure display will be at a very small variance between the previously obtained figure, particularly if the calculator being used is unable to produce an absolutely identical result in the last two digits of its ten displayed digits. Of course, in a perfect world, the angles contained within any triangle should always add up to exactly 180 degrees, but in some cases after a long involved calculation, they may add up to, say, 180 degrees plus or minus 0.000000175 mm, for example, and therefore this difference can be ignored for all practical purposes.)

STEP FOUR.

Having previously retained the Wonky angle W (4.444752088 degrees), in the calculator's memory, it is now possible to use this particular corners angle (now located in the position normally occupied by The Angle when one is using the basic Probe and Prompt calculation system), to obtain the vertical height of the unequal angled triangle. To do this, we use the prompt $\angle a$, $b = a \times SIN \angle$. We enter R M, (4.444752088), SIN, (0.077497773) × 282 = giving us 21.85437224 mm; this now gives us the vertical height of the whole triangle.

[For practice purposes, this now makes it possible to use an alternative method to find the vertical height of this non-right-angle triangle, by using in this case the now-known angle H (25.26842901 degrees), and the already known length of side s (51.1980303), and using the basic probe and its prompts, namely $\angle a$, $b = a \times SIN \angle$ for this. We therefore enter the sequence 25.26842901 SIN, (0.426859633) × 51.1980303, =, giving us 21.85437243 mm. We may find that this does marginally differ from the vertical height obtained when using the angle W for this calculation. (Note that if the student has used a different calculator during this particular calculation it may also show that the two end figures may differ slightly. This amount of numerical difference is really insignificant; therefore, for our purposes, this difference can be ignored.)

We can now take the final steps required to complete the full unequal-angle triangle calculation.

We now need to obtain the base lengths of the fig. 37d so-called large triangle and the base length of the 37e so-called small triangle. We do this by converting in our mind's eye, our originally shaped (non-right-angle) triangle figure with its same outline shape now including two right-angle triangles placed back to back; this process will be easier understood if it is first performed mentally, followed by confirming it physically by drawing a vertical line downward from the triangle's apex (the vertex point) of angle G, (this line must be oriented at 90 degrees and square to its' baseline B), as shown in sketches 37d and 37e. (Note that we have now reverted to using the standard lettering and prompts of the original probe's prompts in order to identify these triangles' side lengths and angles for this particular calculation. It will be found easier in this case to rename these two newly introduced right-angle triangles as, for example, the *large triangle* (as shown in fig.37d), and as,for example, the *small triangle* (as shown in fig. 37e).

If we now refer to fig. 37d, (which shows side b of both of the introduced triangles in a vertical position), we can now quote the prompt required to find the base length c of the large triangle (by using the prompt $\angle a$, $c = a \times \text{COS} \angle$, shown on the basic probe at side a of side 1.

We are now in the position where we now know The Angle ($\angle$) of W and the known length of a (282 mm), and wish to find the length of c. We therefore use the given formula; $\angle a$, $c =$, a, $\times$, COS. $\angle$.

We enter The Angle (in decimal degrees) as follows,: 4.44475209, followed by the sequence, COS, (0.996992525) $\times$, 282, =, giving us 281.1518921; this dimension now gives us the base length of the large triangle. We follow this by pressing the XM key (transferring the displayed figure into the calculator's' memory); we also copy in this base dimension at side *c* of our sketched large triangle.

To find the vertical height of the two introduced triangles we use the prompt ∠ *a*, *b* = *a* × SIN ∠.

We enter the sequence, 4.44475209, SIN, (display shows 0.077497773) ×, 282, =, giving us the common basic vertical height of 21.85437224 for all three of the triangles.

We now need to find the base length of the small triangle; we enter the overall base length of the original non-right-angle triangle, by using the sequence 327.451187, - 281.1518921, =, giving us 46.2992949, the base length of *c* of the small triangle.

We have now completed all the calculations required to obtain all that is needed to be known of the fig. 37 problem non-right-angle triangle.

*　　*　　*

If the reader is still unsure of how to calculate the non-right-angle triangle featured in fig. 37g, or finds that his or her calculator cannot perform the rather complicated 'INV, XM, -, RM, =,' sequence that needs to be used in step one of the previous non-right-angle triangle calculations, then this problem can still be solved by using the following rather basic calculation method:-

We first make a large sketched copy of the Wonky Gable House non-right-angle triangle calculation method shown in the fig. 37g drawing; we then use preferably a sheet of A4 paper to provide plenty of space for the calculations. We then identify and insert the letters *p*, *s*, and B in their correct positions, located in the sides and the base of the sketched triangle; this is followed by inserting the letters W, G, and H into their respective positions in the triangle's included angled corners, these positions being W to its left-hand corner, G at the top corner, and H to its right-hand corner.

We now enter the three known side lengths into this non-right-angle triangle; therefore, in this particular calculation's case, side *p* is 282 long, side *s* is 51.1980303 long, and side B is 327.451187 long.

We now write down the formula required to calculate the included angle of G as described in the fig. 37f drawing and also in STEP ONE, of the fig. 37a example, this being:- $(s^2 - \mathbf{B}^{2)} + p^2$, ÷ by the result of the below the division line, $2 \times p \times s$, followed by the sequence INV, COS.

The whole sequence (if spoken under one's breath) will be, (s squared minus B squared) plus p squared, divided by $2 \times p \times s$, followed by the sequence INV. COS.' It should be noted that we always calculate the bracketed part of the formula first (this being normal mathematics convention). It should also be noted that the result of the calculation made above the formula's horizontal division line, is always divided by the result of the calculation made below the calculation's division line to produce a number that must now be followed by the sequence INV, COS, to enable us to obtain the required included angle of **G** in decimal degrees.

By referring now to the formula shown on side 3 of the fig. 37f drawing, we start the calculation by first finding the value of side s squared, we therefore enter into the calculator the sequence 51.1980303 × 51.1980303, =, giving us 2621.238307. We then find the value of side B squared, by entering, 327.451187 × 327.451187 =, giving us 107224.2799, (the resulting numbers found after completing these calculations should now be written down on paper). We then enter 2621.238307 minus 107224.2799, =, giving us -104603.0416.

We then find the value of p squared, by entering 282 × 282 =, giving us 79524.We then enter the previously calculated -104603.0416, + 79524, =, to give us -250790416; this completes the 'above the division line' calculation. We now carry on by calculating the 'below the division line' calculation of $2 \times p \times s$, we enter the sequence 2 × 282 × 51.1980303, =, 28875.68909.

We now enter the previously calculated 'above the division line' number --250790416. (However, to enter this negative number into the calculator, it will in all probability be found necessary to enter this number as a positive number first, followed by pressing the '+/- key', to ensure that the number has now been correctly entered as a negative number. This now-negative number is then ÷ by 28875.68909, =, to give us --0.86851751. This is followed by entering the sequence INV. COS, to give us 150.286819 degrees, the included angle of G.

The formula we use for step two to obtain the included angle of H is (SIN G) × p ÷ by B, as follows: (150.286819 SIN, × p, ÷ B).

We therefore enter 150.286819, SIN, (display shows 0.495658485), ×
282 = 139.7756929; this is followed by ÷ the displayed number by
327.451187, =, giving us 0.426859631. This is followed by the sequence
INV SIN, to give us 25.26842889 degrees, the included angle of H.

In step three, it is now possible to obtain the included angle of W
by the calculation 180 − (G + H).

We therefore use the display-retained number of 25.26842889 +
150.286819 = giving us 175.5552479 degrees, the combined angles of
G and H. We now enter 180 − 175.5552479, =, to give us 4.4447521
degrees, the included angle of W.

In step four, (if required in the final calculation), we can now
obtain the vertical height of the un-equal angle triangle by using the
now-known included angle of W (4.4447521 degrees), and the known
length of its side p (282), in the calculation; this calculation is now
assisted by using two right-angle triangles introduced into the non-
right-angled triangle (in a backto-back or 'side b to side b' situation),
and by now using the original probe's prompts.

We now know The Angle and a; we can therefore use the
prompt $\angle a$, $b = a \times$ SIN $\angle$. We enter the sequence 4.4447521, SIN,
(0.077497774), × 282 = 21.8543723 mm, giving us the vertical height
of all three triangles.

To find the base length c of the *large introduced triangle* (as
featured in fig. 37d), we use the known length of side, p (282), (now
renamed as side a in the newly introduced triangle), and use the
probe's prompt $\angle a$, $c =$, $a \times$ COS $\angle$. We therefore enter 4.4447521,
COS, (996992525) × 282 = 281.1518921 mm, giving us the base
length of side c in the large triangle. We can now obtain the base
length of the small triangle featured in fig. 37e, by entering the overall
base length of the non-right-angle triangle (shown in the fig. 37
example triangles) of, 327.451187, − 281.1518921 = 46.2992949, giving
us the base length c of the small triangle (as shown in fig. 37e).

The above condensed version of the Wonky Gable House method
is used for calculating non-right-angle triangles (in this case, by not
using the memory function of the calculator), and should now meet
the needs of users whose calculators fail to perform the previously
required sequence, INV, XM, − RM, =, in the first original Wonky
Gable House calculation.

Chapter 15

Using the Sine Bar and Its Slip Gauges

The following chapter explains the use of the precision sine bar and its slip gauges to set up the correct angle on the machine's tooling and work-piece as shown in fig. 38.

This combination of precision measuring instruments exists in the tool-making and production engineering industry for use in setting up the required angle on the workpiece relative to its manufacturing tooling. It will be seen in the fig. 38 drawing that it actually aligns the work-piece to an extremely accurate angle relative to the machine's grinding wheel.

The fig. 38 example shows the work-piece inclined to the required 36-degree angle by using the sine bar and its calculated pile of slip gauges to obtain the correct workpiece angle accurately, relative to the machines' work-table and grinding wheel.

This method of setting up the work-piece's angle is fully explained in the following notes that describe in detail how this method is used in the machine shop for the accurate machining of extremely high-precision angled work-pieces.

The use of this sine bar angle-setting instrument, together with its assembled pile of slip gauges will allow even small angular magnitudes to be obtained down to the minimum of 0.001 degrees or up to its maximum of 89.90 degrees.

This sine bar method of setting angles on a work-piece provides a more accurate setting than if one is using the traditional vernier protractor to set the angle.

The example shown in fig. 38 explains the general setup required to surface-grind a work-piece that has been set at the exact angle of 36 degrees with extreme accuracy, by the use of a 5-inch sine bar and a calculated, selected, pile of precision slip gauges.

This method of setting the angle to be machined is often used in both the tool room, and on the research and development machine shop floor, where extremely precise machining of an angle on a component is required.

The sine bar, together with its accompanying selection box of slip gauges, is often used by the inspection department, should a very accurate inspection of an angle be required on a precision-machined component.

In this particular case, the inspection department would be using a calculated pile of slip gauges in conjunction with using a finger-type dial test indicator (affixed to a vernier height gauge), where both are being firmly located on a perfectly clean and flat surface table.

The fig. 38 drawing shows this angle machining operation being performed on a work-piece while a surface-grinding machine is being used to perform the operation.

This machine is equipped with a magnetic chuck, and with the chuck switched on, it will securely hold the cast-iron supporting angle plate, which is in turn supporting the workpiece component by the use of two clamps. The magnetic (or electric) chuck is therefore designed to secure the angle-plate firmly in place for this machining operation. One should note that non-metallic components, such as those made from aluminum, brass, bronze, and possibly some of the stainless steel materials, will also have the need to be held similarly via a cast-iron angle plate to enable them to be securely held and machined. This is due to their lack of containing sufficient magnetic iron in their molecular structure making them virtually non-magnetic and therefore unable to be held directly on a magnetic or electric chuck for this reason.

The sine bar and slips combination can also be used for the accurate setting of angular work-pieces relative to their tooling on other work-shop machines (these being typically, the jig-boring, jig drilling, or milling/drilling machines).

The appropriate precision sine bar and tooling required for this operation includes the previously mentioned five-inch instrument (or its' metric equivalent), an angle plate, (complete with its necessary clamping arrangement), and an accurately madeup pile of precision slip gauges The height of this pile will have been previously carefully calculated and selected from the workshop boxed set of gauges, to provide the exact and precise 'pile height' on which the sine bar's top roller is located.

This chosen pile of slips is then wrung together, to form what is known as a solid support pile.

This is then placed with the pile's top surface face positioned in support of the upper roller of the sine bar. Its lower-end roller is supported on the flat datum face of the magnetic (or electric) chuck's surface.

This precision setup for machining the angle will ensure there will be an accurate grinding of the 36-degree angle on the workpiece (provided of course the correct height of individual slips have been accurately selected to form the exact height of the calculated pile).

The lower datum face of the workpiece is located in close contact with the sine bar's top datum surface; the assembly is then securely clamped to an angle plate, to prevent any movement during the machining operation. The whole assembly is then positively held in a fixed position by the activated (that is, switched on), magnetic or electric chuck, affixed to the reciprocating bed of the surface-grinding machine.

A similar assembly to this would of course be securely bolted down to the machine's' work-table if one were working on a milling machine, (this precaution must be taken because of the increased motive force generated by the milling cutter when it is removing the excess material during this particular operation). It is also very important when surface grinding that the grinding wheel has been dressed to sharpen and 'true up' its cutting surface, leaving it perfectly sharp and concentric, by the use of a diamond dresser, preferably before the setup is started. It is advisable to do this first, as this operation will be found awkward to perform after the set-up has been completed.

The top face of the work-piece is then ready to be precision ground to the required 36-degree angle, by the combined action of the reciprocating machine's table and the revolving grinding wheel that is removing the excess material from the work-piece down to its required finished dimension.

It is the normal practice, in the industry, for a machinist to own his own sine bar (this having been previously checked for accuracy by the work-shop's standards room, who will have checked that it possesses sufficient accuracy to be used in the firm's tool-room). This instrument may have been manufactured to the old imperial standard dimension of 10.000-inch center-to-center (roller) distance (for use on large work-pieces), or as shown in the fig. 38 example, to the more popular 5.000-inch center-to-center distance on the instrument.

It should be noted that the machined accuracy built into the sine bar, namely its center-to center roller diameter's accurately spaced distance, is of course critical, for any discrepancy in the instrument's calibrated 5.000-inch center-to-center roller distance dimension, would seriously affect the accuracy of the calculated angle, and subsequently any work-piece measured or machined by its use. However, if the sine bar when scrupulously checked by the standards room is found to be very slightly less or more than the normal 5.000-inch dimension, then provided the exact dimension between rollers is actually known, made a note of, and used, the instrument can still be used with accuracy, provided this new checked roller-to-roller dimension is used in all the sine bar trigonometry calculations.

The sine bar instrument is therefore universally considered by the workers in the industry to be extremely precise, it being manufactured very accurately, particularly with regard to its center-to-center roller distance and the accuracy of its roller's diameters).

This built-in accuracy will therefore allow it to be relied upon to produce very accurate angular settings to the intended work-piece.

It should be noted that these so-called rollers are permanently fixed to the sine bar's main body. They do not rotate, but effectively allow the bar's bottom or top roller to pivot around to the required angle on the flat datum surface they are placed upon.

If necessary, there may be need to convert this imperially dimensioned 5.000-inch center-to-center (between rollers) dimension, to its metric equivalent, for use with metric-dimensioned slip gauges, and for use during metric-dimensioned calculations.

This conversion is carried out by dividing the existing 5.000-inch dimension by 0.03937, to obtain its' metric equivalent of 127.000254 mm. We can then use this metric dimension in calculations to obtain the overall dimension required for the pile height when using metric-dimensioned slip gauges, to secure a correct angular setting.

However, it may be found that in smaller workshops, the employed production team may only have the availability of imperial-dimensioned slip gauges (this being mainly due to slip gauge sets being very expensive to purchase, (at, say, £500 per set). Occasionally therefore there will be the need for the machine shop personnel to have the ability to use the old imperial gauge sets (these will be marked in imperial dimensions), by converting the imperial dimensions into metric dimensions to enable the calculation of the correct pile height (in metric dimensions), in order to obtain the correct angular setting of (in this case), 36 degrees.

Referring now to fig. 38, it will be seen that in order to calculate and set the required angle of 36 degrees, we must now use the Probe and Prompt $\angle\, a$, $b = a \times$ SIN $\angle$, to obtain the correct pile height.

This prompt is more easily understood (and remembered), if it is committed to memory as, "'I know The Angle (36 degrees), and the length of side a,'" (the center-to-center 5.000-inch dimension between the rollers) of the sine bar.'

Therefore, to obtain the total slip gauge height that will involve the combined individual gauges' thickness that is necessary to produce the 36-degree inclination of the sine bar, we must therefore enter the sequence 36, SIN. (calculator's display now shows 0.587785252) ×, 5, =, giving us the required pile height of 2.938926261 inches (in imperial dimensions),.

It should have become clear to the reader by now why this instrument is called a sine bar, for its use normally requires access to the stored table of sines, in the calculator's' permanent memory that allows this particular calculation to take place. In the past, (that is, before the introduction of electronic calculators), the work-force were forced to use a printed book of tables that contained sine bar constants, to obtain the required slip gauge pile's dimension. The student should note that these particular tables are only published and printed in the form of the required 'degrees, minutes and seconds, and not in the more useful decimal degrees (these being much more compatible, when using scientific calculators on practical trigonometry calculations).

The engraved size shown on each individual slip gauge is of necessity given to four places of decimals.

It is general practice, therefore, when complying with the engraved sizes provided on 'slip gauges' to obtain the sine bar height, (when using imperial slips), to use only the first four digits after the decimal point of the number required.

The reason for this is that the calibrated dimensions engraved on the actual slip gauges are identified to a maximum of four places of decimals that indicate their size.

When one is called upon to make up a correct slip gauge pile height, the student should be made aware of the basic principle needing to be known, to facilitate the correct calculation sequence of the slips' required, before one makes the choice of each individual slip gauge from its box in order to obtain the required height of the pile.

Having decided the correct height required by calculation, it is recommended, (as in general practice), that one selects the lowest denomination of slip gauge from the pile first. We do this is in order to make the initial reduction to the required pile height in small and easy stages. This then allows the student to leave the final slip gauge or gauges required to finish the pile, to be selected in large denominations for convenience during one's last choice.

Referring to drawing fig. 38, we start by making up the calculated pile using either the exact one or the nearest-dimensioned calibrated slip gauge to the one required, available in the box; we do this to remove the smallest 0.1009-inch dimension from the total 2.9389 inch slips required to complete the pile.

We must also bear in mind at this stage in the calculation, that the smallest available slip gauge required in the set, is marked 0.1009 inches. (Note that if this is not available, it is possible to use a combination of 0.1005-inch and 0.1004-inch slips (to give us 0.2009 inches), thereby reducing the next possible slip pile required to 2.7380 inches instead of 2.8380 inches, in order to keep the calculation consistent with the required pile height.

Gauges with a very small, thin dimension, (with the exception of protective slips currently used in corrosive and hard-wearing situations), are usually not included in the set. The reason for this is said to be their inability to remain consistently perfectly flat and accurate, in all working conditions.

To reiterate, this first selection will remove the 0.1009-inch dimension from the total required slip pile's calculated height. (See previous note if this slip gauge is not available.)

Having availed oneself of this particular slip, and having subtracted its engraved figure from the original 2.9389 inches, we are left with 2.8380 inches of pile to be further dealt with.

We can now carry on through the procedure required to select the next suitably sized slip gauge to reduce the pile further.

We therefore select the next smallest available slip gauge (from the box), to remove 0.1380 inches from the worked-on pile.

This is then subtracted from 2.8380, leaving 2.700 inches of pile remaining.

We now select the next available slip (0.700-inches), leaving us with the 2.000-inch slip to be used as the last 'large slip to finalize the completed pile.

Once this selection process is completed, the slips can then be wrung together to form the final solid pile height. (It is a wise precaution at this stage to double-check the completed height of the slip gauge's pile's dimension by using an accurate micrometer, to ensure that the dimension is correct.

[One should note that slip gauges are consistently more accurate than the actual calibrations provided on the checking micrometer's barrel. In fact, slip gauges are often used by inspection departments to check and re-calibrate micrometers and other measuring instruments for possible inaccuracies in their scales readings, because of the possibly excessive wear on their anvils and measuring surfaces.)

However, when working to metric dimensions and metric slip gauges, there will be the need to convert the existing imperial 5.000-inch sine bar (center-to-center roller distance) into its equivalent metric center-to-center dimension in order to make the calculations easier.

We do this by dividing the 5.000-inch center-to-center distance of the existing sine bar by 0.03937. This gives us the figure 127.000254 mm; this dimension will later be called the probe's side a (the longest side of the probe, in the following triangle calculations). We must now find the exact slip gauge height combination (this time by using metric slips) to obtain the correct angular setting for the angle of 36 degrees.

We now need to consult the probe and choose the displayed prompt $\angle$ a, b =, a × SIN $\angle$.

Because we already know The Angle ($\angle$), and side a, and wish to know the length of side b (the required total slip gauge pile height); we now enter into the calculator the sequence, 36, SIN, (display shows 0.587785252) ×, 127.000254, =, giving us 74.64887634 mm. (the required slip gauge pile height in millimeters that provides us with the 36-degree inclination).

This ten-figure calculated dimension will be found (in practice), to be rather too long to be used in total; it will therefore need to be very carefully and meticulously shortened, (in this case rounded up), to a representative figure that contains just three decimal places. This will give us 74.649 mm; (we do this to comply with the limited engraved sizes of available slips in the set).

It is general practice to always select the smallest available slip gauge first, (in this case, it will be 2.009 mm). This number is then subtracted from 74.649 mm. to give us 72.640 mm.

We then subtract the next available slip (2.500 mm,) from the remainder, to give us 70.140 mm. We follow this by subtracting the next selected slip of 2.140 to give us 68.000 mm; we then select and subtract the 25.000-mm slip, to give us 43.000 mm;. We can now select the appropriate slips from the box to make up the 43.000-mm remaining slip height dimension required (in practice this could be 3.000 mm, plus 40.000 mm, (if a slip gauge of 43.000 mm, is not available). The combinations used to accommodate each individual slip gauge size can be varied according to the actual contents of the boxed set.

With the pile now complete, it can be wrung together to form a solid and usable calibrated support pile for use with the sine bar.

Generally, the gauge blocks (or slips) are supplied in boxed sets. They will have been checked very accurately during manufacture, and will have passed a very strict accuracy test. A certificate will have been supplied in the box to this effect, giving their exact sizes to very close tolerances of 0.000001 inches, (a millionth of an inch for imperial sets), or, for metric sets, to a tolerance of 0.0000254 mm. (in microns). Imperial-dimensioned boxed sets contain either 36 or 81 pieces, (depending on their price and the extent of their intended use. Metric sets are normally available in sets of 47 or 88, pieces.

There are two main qualities of slip gauges available; they are supplied in either the top-quality inspection grade (mainly used in the standards room, where they are stored at a constant temperature of 68 degrees Fahrenheit, in order to maintain their engraved calibrated size at this temperature; they are used in this case mainly as reference gauges for checking workshop-grade measuring equipment, (micrometers, for example). They are not usually available for use in the manufacturing workshop, but are retained exclusively for inspection use. Workshop-grade slips are generally used on the production workshop floor. They can also be used for checking the machined width of precision slots etc. often required in components during their manufacture. These workshop-grade slip gauges are used elsewhere in the factory for accurately setting up machines and work-pieces for production purposes. The workshop slip gauge sets are of course also made from good-quality hardened steel, but are manufactured for production purposes and possibly to a slightly lower state of absolute perfection, (hence the lower price). The working surfaces of all qualities of slip gauges have been lapped, honed, and polished to a very fine surface finish in order to obtain an exact calibrated size. The nominal outside (rectangular section) dimensions of these gauges are all basically similar in size at 1/4 by 3/8-of an inch (32 by 9.5 mm.). They will have all been honed and checked to comply with their extremely accurate (engraved) dimensions.

It is of course their flatness and perfect surface finish that contribute to their accuracy in use. This dimensional accuracy allows a selected pile of slip gauges to be wrung together to form the required pile height dimension used to obtain the required angular setting. The overall dimension of this pile of new gauges (when wrung together) should not possess any accumulative errors. It is only well-used slip gauges that are found to have accumulated dimensional errors caused by wear while they are being constantly used in their respective piles.

It should be noted, however, that when the slip pile is wrung together, the individual slips are held together in close contact merely by the exclusion of air from their smooth, lapped contact surfaces, and not (as sometimes thought), by magnetism.

However, after prolonged use, and over the course of time, the slip gauges will inevitably become slightly worn on their precision mating surfaces. This minute loss of accuracy is usually dealt with and taken into account, by a regular (six-monthly), inspection update, of all the 'firm's' slip gauges. This task is usually carried out by the firm's standards room inspection department, who take responsibility for the accuracy of the company's inspection equipment.

This practice of precision checking carried out by the inspectors in the factory's standards room will also use precision measuring instruments called comparators and standards-room-quality slip gauges for their checking procedures.

The inspectors concerned will note down any errors found in each slip gauge and will supply a correction list (to be later supplied with the fully checked gauges in the box). This will include details of all the minute errors found for each particular gauge block.

This correction list or wear factor list must be taken into account when making all sine bar calculations. This is done by adding each of the discovered wear factors found for each gauge used, to the calculated height of the required slip pile.

The slip gauges are individually identified by engraved figures (in decimal notation) on their mating surfaces; (in the case of the large gauge blocks, this is usually on their sides); the extent of these engraved figures is limited to four places of decimals in the case of imperial sets, and to three places of decimals in the case of metric sets. It will now become obvious to the student that care should always be taken to protect the delicate contact faces of the slip gauges from any damage, in order to maintain their accuracy.

Experience has shown that if the gauges are handled by persons who have so-called rusty fingers (this is where the perspiration from the operator's fingers is found to be of a corrosive or acidic nature), then it is advisable to adopt certain protection procedures, for without adequate protection, the corrosion problem can cause quite serious rust damage to form on the precision contact surfaces of the gauges, thereby reducing their accuracy, their useful working life, and their ability to be wrung together to form the necessary solid, accurate pile of gauges.

The extent of this corrosion problem can be minimized by the use of a rust-preventive lanolin lotion (usually supplied in a tube with the box of gauges), this to be applied to the hands prior to using the gauges.

It is now possible to purchase complete sets of slip gauges made from a ceramic material. These do not suffer from a corrosion problem. They are more expensive, however, to purchase, but still possess a similar accuracy to the steel ones. They are also non-magnetic; this is considered an advantage, for with this feature, they will not require de-magnetizing during use, as is often necessary when using steel slips, in order to prevent an accumulation of 'metallic swarf being attracted to their polished steel precision faces, with possibly detrimental dimensional result.

Chapter 16

Explaining how the sine vice work-holding device is used in industry

A continued explanation of the various precision methods used for setting up the angles on machines and their work-pieces, in the engineering workshop.

There follows a description of this alternative work-holding device.

This device is used quite often on the work-tables of precision machinery in the machine shop to obtain extremely accurate machined angles on the production work-piece.

The sine vice is a precision work-holding vice capable of high-precision angular settings of the workpiece relative to the machine's cutting tool.

This extremely accurate work-holding device is used mainly for the precision machining of tool-room-quality components that have the need to possess extremely accurate machined angles on their work surfaces.

Its advantage over using a normal machine vice is that the work-piece can be accurately tilted and locked exactly to the required angle to be machined.

This angular tilt is performed by using the support of a calculated pile of slip gauges. These are introduced between the vice's top roller and its base datum face to ensure the accuracy of its set angle. Its working principle is similar to that used for a conventional sine bar.

The manual setting of its angle of tilt is very similar (in set-up) to that used in the previously described sine bar / magnetic chuck method, (where a selected column of slip gauges is used in a support pile to accurately establish the angle required.

The sine vice is, however, much easier and more convenient to use in practice, because the precision roller mechanism is actually built into the body of the vice.

In use, the correct angular setting will require the slip gauge pile to be calculated, wrung together, and used in a vertical support role to establish the correct angular setting required.

The clamping mechanism, (used to lock the angle to be set), is incorporated into the lower pivot point of the device, which allows easy spanner access for secure clamping.

The design feature built into this accurate angular tilt, is normally based on the imperial five-inch (center-to-center distance) sine bar.

In rare cases, a ten-inch (center-to-center distance) sine bar model is used for setting the required accurate angle on larger components.

Both devices are manufactured with great precision, and they are therefore much more expensive to purchase than the normal, basic, engineering machine vice.

This fact makes this tooling rather special in the workshop, for its precision quality has the tendency to divert its use toward the more important tool-room milling, precision grinding, or for use on inspection work.

Authority to use these particular precision angled vices may require the 'management's' permission before being put to use on normal, run-of-the-mill machine-shop-type work.

*   *   *

fig. 38.
Fig 38.

THE 'SINE BAR' POSITIONED FOR THE ACCURATE ANGULAR SETTING OF 36°;

SHOWING THE SLIP GAUGES USED TO ESTABLISH THIS ANGLE.

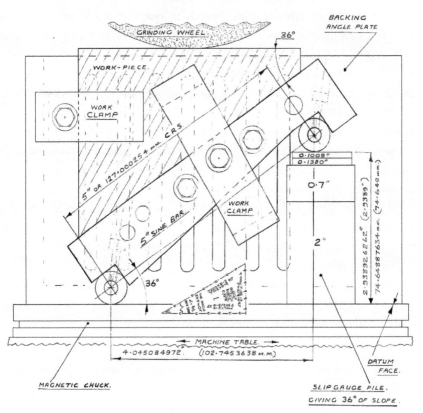

BACKING ANGLE PLATE

GRINDING WHEEL

36°

WORK-PIECE.

WORK CLAMP

5" OR 127·0002·54 m.m. C.R.S.

0·1009"
0·1380"

0·7"

2·93892·6262" (2·9389")

74·648·7634·m.m. (74·649 m.m.)

5" SINE BAR

WORK CLAMP

36°

2"

MACHINE TABLE.

4·045·084972.   (102·745·3638 m.m.)

DATUM FACE.

MAGNETIC CHUCK.

SLIP GAUGE PILE.

GIVING 36° OF SLOPE.

TO FIND HEIGHT OF SLIP PILE REQUIRED
USE 'PROMPT' ∠ a b = a × SIN. ∠.

Fig 39.

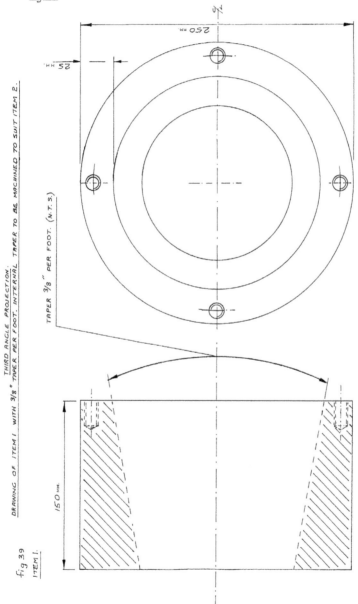

fig 39.

THIRD ANGLE PROJECTION.

DRAWING OF ITEM 1 WITH 3/8" TAPER. PER FOOT. INTERNAL TAPER TO BE MACHINED TO SUIT ITEM 2.

TAPER 3/8" PER FOOT. (N.T.S.)

fig 39
ITEM 1.

250 MM

25 MM

150 MM

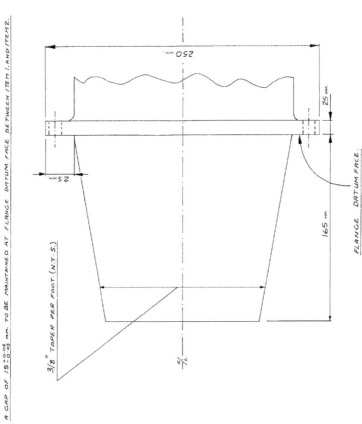

Fig 40.

(header, right) Fig 40.

DRAWING OF SUPPLIED PATTERN (ITEM 2), WITH A 3/8" TAPER PER FOOT (MALE TAPER)

THIS COMPONENT TO BE USED FOR THE MACHINING OF INTERNAL TAPER AND FINAL ASSEMBLY WITH ITEM 1.

A GAP OF 15⁺⁰·⁰⁵₋₀·₀₅ MM. TO BE MAINTAINED AT FLANGE DATUM FACE BETWEEN ITEM 1. AND ITEM 2.

Fig 40.

ITEM 2.

250 MM.

25 MM.

25 MM.

165 MM.

3/8" TAPER PER FOOT. (N.T.S.)

½

FLANGE DATUM FACE.

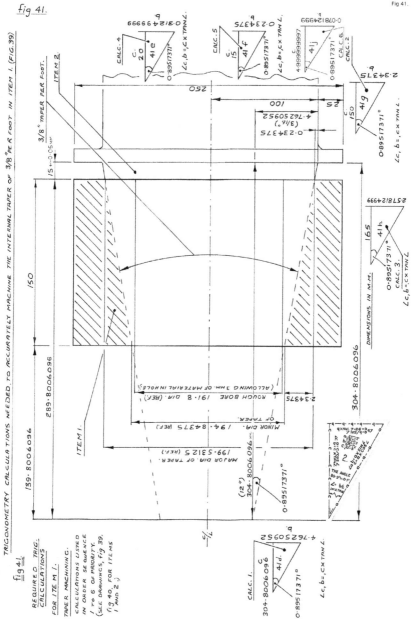

Fig 41.

fig 41.

TRIGONOMETRY CALCULATIONS NEEDED TO ACCURATELY MACHINE THE INTERNAL TAPER OF 3/8" PER FOOT IN ITEM 1. (FIG. 39)

Fig 45.

Fig 45.

THREE METHODS OF TAPER TURNING ON THE CENTER LATHE.

fig 45.

TURNING A TAPER USING THE OFFSET TAILSTOCK.

THE ADJUSTABLE ACTUATION ROD.

THE COMPOUND SLIDE HAND-WHEEL.

THE TAPER TURNING ATTACHMENT.

THE CROSS SLIDE HAND-WHEEL.

TURNING A TAPERED BORE USING THE TAPER TURNING ATTACHMENT.

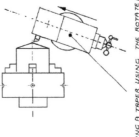

TURNING A TAPER USING THE ROTATED COMPOUND SLIDE.

TITLE
THREE METHODS OF
TAPER TURNING ON THE
CENTER LATHE.

DATE 27-6-14.
DRN. G.N.R.
DRAWING Nº   fig 45.

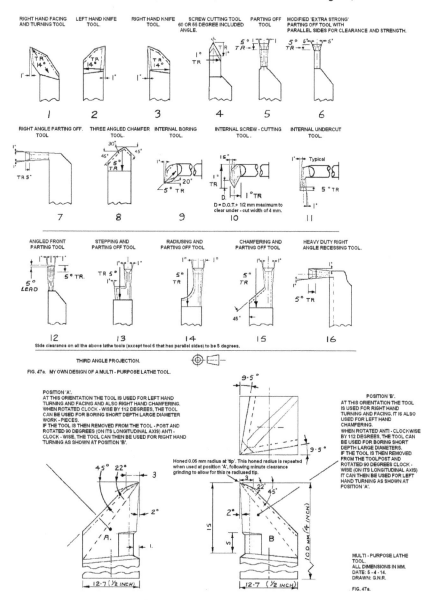

FIG. 47a. MY OWN DESIGN OF A MULTI - PURPOSE LATHE TOOL.

POSITION 'A'.
AT THIS ORIENTATION THE TOOL IS USED FOR LEFT HAND TURNING AND FACING AND ALSO RIGHT HAND CHAMFERING. WHEN ROTATED CLOCK - WISE BY 112 DEGREES, THE TOOL CAN BE USED FOR BORING SHORT DEPTH LARGE DIAMETER WORK - PIECES.
IF THE TOOL IS THEN REMOVED FROM THE TOOL - POST AND ROTATED 90 DEGREES (ON ITS LONGITUDINAL AXIS) ANTI - CLOCK - WISE, THE TOOL CAN THEN BE USED FOR RIGHT HAND TURNING AS SHOWN AT POSITION 'B'.

Honed 0.05 mm radius at 'tip'. This honed radius is repeated when used at position 'A', following minute clearance grinding to allow for this re radiused tip.

POSITION 'B'.
AT THIS ORIENTATION THE TOOL IS USED FOR RIGHT HAND TURNING AND FACING. IT IS ALSO USED FOR LEFT HAND CHAMFERING.
WHEN ROTATED ANTI - CLOCKWISE BY 112 DEGREES, THE TOOL CAN BE USED FOR BORING SHORT DEPTH LARGE DIAMETERS.
IF THE TOOL IS THEN REMOVED FROM THE TOOLPOST AND ROTATED 90 DEGREES CLOCK - WISE (ON ITS LONGITUDINAL AXIS) IT CAN THEN BE USED FOR LEFT HAND TURNING AS SHOWN AT POSITION 'A'.

MULTI - PURPOSE LATHE TOOL.
ALL DIMENSIONS IN MM.
DATE: 5 - 4 - 14.
DRAWN: G.N.R.

FIG. 47a.

MATERIAL - HIGH - SPEED TOOL STEEL.

Chapter 17

Fig. 39 Precision taper turning, on the center lathe

It is generally accepted in the engineering industry, that if an issued drawing shows a fractional dimension on a components drawing, then the dimension requiring to be machined (because it is being given in fractional notation), will not require the absolute precision in its manufacture because of this so-called fractional method of dimensioning.

It is therefore quite the usual practice for a fractionally dimensioned component to be acceptable provided it is within the drawing's stated limits of plus or minus *one sixteenth of an inch (1.5 mm)* of the drawing's given dimensions. A component, manufactured to these particular limits, will, when checked by the inspection department, normally be passed as acceptable.

In the engineering industry, there is always the requirement for the maximum speed of production at the lowest possible cost; it is therefore considered unnecessary for any component to be manufactured to a greater accuracy than the drawing-stated limits provided on the drawing, to save valuable man-hours of production time.

However, there are inevitably some so-called unofficial exceptions to this particular rule, as explained in the following example that actually requires a *very accurately machined component* to be produced, while its taper angle has been given in this case as *'taper inches per foot'* *(this being in fractional dimensions)*. The following situation highlights this anomaly quite clearly. A situation often occurs on the shop floor when no drawings exist to be issued by the drawing office; therefore, in this case the machinist is given just verbal instructions from either the charge hand or the foreman regarding the degree of accuracy the component requires. These verbal instructions are quite often given in a hurried way, and an example of this is in the following instructions given by the foreman to a machinist while he is receiving his next job; these were as follows:

"'This part, indicated by the foreman pointing to the object, (later called item 1 in fig. 39] needs to be bored out to the same taper as that pattern component over there, [said while pointing at the component, now called item 2 in fig. 40], it has a taper of 3/8 of an inch per foot". "You must leave a 15-mm gap of plus or minus 0.05 mm, between the existing flange [of item 1] and the end face [of item 2] when they are finally assembled together". "The timekeeper has given you 2 1/2 hours to finish this job".

I think the reader will agree that these are very minimal instructions with which to undertake this type of precision work, but a skilled toolmaker should be capable of carrying out these instructions without too much difficulty.

A student, however, on being given these same instructions, would certainly feel that he has been thrown in at the deep end and would tend to flounder at the prospect of completing this particular task to the required accuracy in the time available.

For this reason, and to make these machining operations more easily understood by the student reader, the author has prepared three drawings that display the fore-man's verbal instructions. These drawings are numbered fig. 39 item 1, fig. 40 item 2, and fig. 41, which contain the required calculations, the finished assembly drawing, and include the required small triangular calculation sketches itemized as, calc. 1, 41d, calc. 2, 41g, calc. 3, 41h, calc. 4, 41e, calc. 5, 41f, and calc. 6, 41j. The series of machining operations required and the necessary practical trigonometry calculations needed to obtain a precision mating of the two components' internal and external angles are shown in order to arrive at the final accurate spacing of the two components' flange and end face to comply with the foreman's verbal instructions. The required machining operations are described in minute and intricate detail in the following taper-turning operations that are being performed on the center lathe.

The machining operation required is being performed on the internally tapered component featured in the fig. 39 drawing to allow it to fit the taper of the supplied male tapered pattern item shown in the fig. 40 drawing. The fig. 39 item is to be set up and securely held in the four-jaw chuck of the center lathe.

It should also be pointed out that apart from the machinist being told what is required by the foreman, the tool-maker (being a skilled machinist), is rarely told much else, and is expected to do everything required (including the trigonometry taper-turning calculations required) to complete the work-piece without any further instructions.

The author's supplied drawings fig. 39, fig. 40, and fig. 41, include the informative triangle sketches produced in fig. 41 with the object of highlighting the actual accuracy that will be required in this particular machining operation. It should be pointed out at this stage that the fig. 41 drawing explains in practical terms the difference between the open limits spoken of by the foreman of 'three eighths of an inch taper per foot' to the tight limits of plus or minus 0.05 mm, which refers to the precision gap required between the two items. This particular turning operation obviously breaks the general rule (in the use of fractional and so-called less-important dimensions of accuracy), to the present requirement of replacing them with the required precision dimensions that are now actually needed to complete the work-piece to the required accuracy. This new requirement to manufacture the component to provide a precisely dimensioned gap, when the original verbal instructions given by the foreman were for using the wide tolerance of 'three eighths of an inch taper per foot' indicating that the tolerance could be *plus or minus one sixteenth of an inch*, and this should be definitely watched out for when one is working on the shop floor in most of the factories throughout the whole of the U.K's engineering industry.

What the student must realize is that *in this particular case*, the actual requirement will be for a very accurate and precise taper-turning operation to be carried out, in order to make both components fit accurately together, while also providing an *exactly dimensioned* gap to be maintained between one component's flange and the other component's end face. This is a very important fact and should be noted.

The setting and adjustments that need to be made to the center lathe for the manufacture of this particular component will rely on the lathe's taper-turning attachment's angular settings being aligned with *extreme* accuracy); they will therefore need to be performed very accurately, *to a much more precise level* than that envisaged by the initial *fractional taper dimension* given verbally, by the foreman (there being no official drawing issued for this particular operation). This situation is of particular concern regarding the student's interpretation of the accuracy required for the machining of this component. The actual setting up of the taper angle on the center lathe must therefore be undertaken with the utmost care, even though the instructions would have initially indicated that this was to be a fractional, and 'less important' taper angle.

Therefore, when the machinist is in this situation, it is advisable to be very cautious of any verbal instructions received, and to realize that the angular settings made to the lathe must be of an absolutely precise nature, to ensure that the finished product will be machined with sufficient accuracy to pass through the strict inspection process that will inevitably follow by the inspection department.

The advantages gained by using the Probe and Prompt's assistance, for all the trigonometry calculations needed to obtain this correct angular setting, are given in the following meticulous in-depth explanation of the whole calculation and lathe setting-up procedure.

The student should therefore realize that the following explanations and sketches involve a very long (but necessary) calculation and setting-up process.

To assist in solving this particular problem and (in this case) by adopting the process of leaving no stone unturned, we utilize a calculation process known as extrapolation (commonly called 'to extrapolate'), pronounced 'ex-trap-o-late'.

This calculation process is based loosely on the assumption that the trends followed inside an original range of calculations have the distinct ability to be continued outside it to obtain additional required dimensions.

The use of this method allows the student to estimate something unknown from the facts or information already known and readily available.

The fig. 39 (author-produced) drawing shows the part to be taper-turned (which is now called item 1).

The fig. 40 (author-produced) drawing shows the supplied male pattern component detailed later in the fig. 41 calculation drawing. This component contains a male taper with a taper angle of 3/8 of an inch taper per foot. The final machining operation must therefore enable item 1 (fig. 39), to be machined to possess an internal taper that is a perfect fit and sufficiently accurate in angle and dimensions to enable both items to fit perfectly together while accurately maintaining a precision gap of 15 mm plus or minus 0.05 mm (0.0019685 inches) between the flange of fig. 40 (item 2), and the end face of fig. 39 (item1), on their final assembly together. To allow the student to get some idea of the close limit of fit required, the author wishes to point out that the human hair (when measured by micrometer) will normally be 0.05 mm (0.0019685 inches) in diameter; this minute dimension emphasizes the fact that 0.05 mm (when multiplied by two), will provide us with the whole of the 'window of tolerance (of 0.1 mm or 0.003937 inches), that must be maintained in order to produce a correct fit. It should now be realized that this is a very tight-limit tolerance to comply with.

The fig. 41 (author-produced) drawing shows the trigonometry calculations required to obtain the exact angle required (in metric taper dimensions), to enable the accurate machining of the internal '3/8 of an inch taper per foot' bore in item 1(fig. 39), for it to receive the supplied male taper of item 2 (fig. 40). The fig. 41 drawing contains the calculated basic dimensions that were also physically taken from the supplied pattern (item 2 in fig. 40), to enable its existing taper to accurately match the machining of item 1, where (as previously stated) an accurate gap must be established and maintained between the flange of item 2 and end face of item 1; this gap is to be 15.00 mm + or − 0.05 mm, (0.0019685 inches.) when they are finally assembled together. (This particularly close tolerance would be considered by most machinists to be a *very tight limit* and extremely difficult to accurately achieve.).

If we now refer to fig. 41, we use the angular similarity between the six small sketched / inverted right-angled triangles to solve the trigonometry problems found that are beyond the extremities of the original fractional dimensions previously given verbally by the foreman.

The accurate results obtained by using these calculations must be obtained before the machinist can finally set the correct angle on the center lathe's *taper turning attachment*; it will now be realized that the setting of the angular scale on this particular taper-turning attachment *must be of a very precise nature.*

Of course, the general run-of-the-mill short tapers that are normally machined on components can usually be undertaken quite easily on the center-lathe by utilizing the rotated compound slide method (as shown in fig. 45), its tool and tool post in this case being adjusted to half the required included angle stated on the issued drawing, and this is obtained by utilizing the engraved graduations on its angular scale to obtain the correct angular setting.

However, the student should note that in most taper-turning cases, the correct angular setting of the compound slide must be carefully adjusted to read only half of the included angle indicated on the component's drawing.

Should the angled surface to be machined on a work-piece (as in this particular case), be longer than the compound slide's overall length of travel, then to machine the long taper required by our fig. 39 drawing, there will now be the need for the component to be accurately *precision taper-turned* by using the *taper-turning attachment* (affixed to the back of the lathe's longitudinal and cross-slide saddle assembly). This mechanism is initially actuated to perform the taper-turning operation by first removing the cross-slide's locking security bolt, followed by attaching its operating rod securely to the back of the lathe's bed. This then permits the lathe's angle-turning mechanism to be activated; it should be noted that this angled slide mechanism will now be working independently of the normally used longitudinal and cross-slide's movement. Also note that the now-being-used taper-turning attachment's angled slides should be well lubricated before being used.

The angular setting of the taper-turning attachment must be adjusted so that its scale now reads exactly half of the included angle actually shown on the drawing. In this particular example's case, the drawing calls for *3/8 of an inch taper per foot*; therefore, the actual taper angle setting that needs to be set *on the lathe's taper-turning attachment* should now be set to *half* this angle, i.e. *3/16 of an inch taper per foot*).

As previously stated, to fully activate the taper-turning attachment (as shown in fig. 45), the machine's taper-turning *adjustable actuating rod assembly* must be securely clamped to its receiving casting fitted to the rear of the bed of the machine to finally bring the lathe's separate precision'angled slide into use.

Problems can arise when the machinist is seeking to interpret and set the precise angle required to be taper-turned onto the *taper-turning attachment's scale*, due quite often to its scales being calibrated generally to *half-degree* divisions that are inadequately precise for this precision use.

This imprecise graduation of its scales will therefore cause difficulty when attempting to set the exact angle accurately in the required digital degree format on its scale by eye. This will apply particularly when the required angular setting required falls between the existing half-degree graduations on its scale, or, due in our particular case, the required angle has (of necessity) been calculated in *decimal degrees*, where these particular graduated lines are not engraved on the *taper-turning attachment's* angular scale.

It follows that, in use, the correct angled set of this mechanism, (when finally set correctly and clamped), will now actually guide the whole of the angled (and separate cross-slide assembly, including its integral cross-slide, hand-wheel, scale, and cutting tool), to traverse the work-piece at the exact set angle, thereby allowing the boring tool to progress along an angled cutting path that is totally independent of the normally used longitudinal traverse of the machine's existing parallel bed's V ways.

As previously stated, to fully explain how this accurate angular setting on the lathe can be achieved, one should note that the normal (drawing quoted) taper angle of *three eighths of an inch taper per foot* actually refers to the *included* angle of the intended work-piece, and does not refer to the actual angle needing to be machined on or in the work-piece.

The machinist must therefore complete the following rather extensive calculations, by using practical trigonometry, with the help of the probe's prompts, to establish the exact angular setting required on the taper-turning attachment that will allow the work-piece to be accurately taper-turned to the exact requirements of the verbal instructions previously received from the foreman, and to comply with the newly calculated figures contained within the six triangular sketches shown in fig, 41-d, e, f, g, h, and j.

The fig. 39 drawing-stated taper angle of 'three eighths of an inch taper per foot' is therefore halved to provide the required three sixteenths of an inch taper per foot (this being exactly half of the drawing quoted *included* angle); this will then give us the actual taper angle needing to be set on the machine to produce the correct internal taper required in the fig.39, (item 1) workpiece.

This newly quoted 3/16th of an inch taper per foot taper dimension must now be converted into a metric dimension to aid the following calculations.

It should be noted that it is quite the normal practice in the U.K for the engineering draftsman to quote the drawing's taper angles in the pre-1970s imperial dimensions of feet and inches, using the so-called 'taper per foot' method of indicating the taper angle required. (It is understood that this situation still applies to issued American drawings, where imperial (inch) dimensions are still in use.)

If we now refer to fig. 41d calc.1, this calculation begins by halving the 'three eighths of an inch taper per foot' dimension, to the 'three sixteenths of an inch' taper per foot' required for setting the lathe's taper angle. We do this by converting the given fractional dimensions into the required metric decimal dimensions to allow further calculations to be carried out, we initially do this by entering (into the scientific calculator) the following sequence:

Enter, one, divided by sixteen, =, giving us 0.0625 (the decimal equivalent of one sixteenth of an inch); we then multiply this figure by three to give us 0.1875 inches (the decimal equivalent of three sixteenths of an inch). We then convert this imperial dimension into its *metric* equivalent by dividing it by 0.03937, (the author's usual method of converting an imperial dimension into a metric dimension), to give us 4.762509525 mm.

We now need to deal similarly with the 12-inch dimension, by entering into the calculator, 12 divided by 0.03937, thereby converting the 12-inch dimension into its metric equivalent of 304.8006096 mm.

If we now refer to fig. 41, it will be found helpful to the student if he or she studies in turn, the six sketches shown that consist of inverted right-angle triangles, oriented into positions where The Angle end of each sketch (and of the displayed probe) are pointed to the left and toward the problem angle to be machined (these sketched triangles include, calc.1, 41d; calc. 2, 41g; calc. 3, 41h; calc. 4, 41e, calc. 5, 41f, and calc. 6, 41j). (It will also be seen that a photocopy of the Probe and Prompt has shown oriented into its correct attitude for angle and side length calculation guidance purposes).

The actual Probe and Prompt aid being used by the student should now be oriented into a similar upside-down attitude (that is, with its The Angle end (or sharp end) pointing to The Angle end, similar to the sketched triangles.

We now consult the probe and make a selection of, and note that the required prompt to use for finding The Angle (shown at the sharp end of 41d), is $\angle$ =, b/c, INV, TAN (this is the prompt normally used for obtaining The Angle in any right-angle triangle, and in this case, we are using decimal degrees).

We do this by entering the sequence shown in fig. 41d, 4.762509525, ÷ 304.8006096, =, (display shows 0.015625) followed by INV, TAN, giving us the angle of 0.89517371 *degrees* (this is the actual angle required to set the taper--turning attachment on the lathe (it should also be noted that this angle is slightly *less* than *one degree*.)

As previously stated, when one is attempting (in practice) to set this exact angle on the taper-turning attachment's angular scale, it will be realized immediately that this specific angle is not compatible with the graduations engraved on the angle-turning attachment's angular scale.

As a result of observing these rather 'indecisive' graduations on the lathe's scale, we then discover that for this particular job we will be unable to finalize our angular setting in the required *decimal notation* but must first make a preliminary *estimate* of its approximate position on its scale, and lightly clamp this *provisionally set* angle of *just less than one degree.*

For use in calculations, we must now make a physical check of the actual length of the fig. 39 work-piece; we do this by using a vernier caliper, in order to ensure that its 150 mm overall length is exactly correct. This will then make this component's length suitable for use during the following 41g calculation when positioned at its side *c*.

To perform this calculation, and to simplify the setting of the lathe's angle attachment, we now enter into this triangle sketch, the *now-known* angle of 0.89517371 degrees, and designate its length of side *c* to be 150 mm. (this being the actual measured length of the work-piece as shown in fig. 39).

We now need to find the length of the triangle's side *b* (which is now designated to be the machine's so-called offset dimension), by calculation; we do this by selecting the appropriate prompt, $\angle$ *c*, *b* = *c* × TAN $\angle$.

We enter the sequence, 0.89517371, TAN, (display shows 0.015625) × 150, =, giving us 2.34375 mm, (the calculated length of side *b* of the 41g triangle); this now gives us the total amount of machine offset required on the *taper-turning attachment*, while it is being checked by a DTI (or clock) when the tool is traversed over the 150-mm longitudinal length of the work-piece.

By utilizing these two dimensions, 2.34375 mm, and 150 mm, we can now finally check (and correct) the previously *estimated* and provisionally partially set' lathe's setting angle on its taper-turning attachment.

The hand-wheel controlling the saddle's movement, is now rotated to move the saddle (including its tool and its taper-turning attachment assembly) along the bed of the machine to its left in order to check that this setting is correct.

Note that when the *taper-turning attachment* has been accurately set, the tool (in its tool-post) will be seen, not only to move to its left, but also to move gradually transversely across the lathe and away from the operator on the diagonal path of 0.89517371 degrees (slightly less than one degree), while the lathe's tooling assembly will be obtaining a cross-slide movement of 2.34375 mm and the lathe's total longitudinal slide movement will be 150 mm.

It will be found to the student's advantage during this initial setting-up stage of the lathe, if a gap of (say) 5 mm has been left between the left-hand end of the work-piece and the chuck's front datum face; this gap will then allow the internal boring tool's sharp pointed end to be viewed through this gap on it reaching the extreme end of its tapered travel, where (if the lathe is equipped with a digital readout facility), this tool's position can be used (with the help of a handheld lens) to check that the final digital setting-up process is correct.

(The ideal boring tool for use in this operation should be a tool that is very similar to fig. 47, drawing number 9, but will to advantage possess a more robust shank design possessing a diameter of (say) 18 mm, in order to provide sufficient rigidity for the following taper-turning operation; its front cutting edge must also be examined and checked for sharpness, and to ensure that it possesses the vital minimum cutting radius as shown in fig. 47. The tool must of course also be fitted into its tool-post at its exactly correct cutting height.

The specific digital readout measuring equipment being used on the lathe will provide an accurate indication by using its digital scale, and this indication will very accurately check the actual movement of each of the lathe's longitudinal and transverse slides in both its normal longitudinal, and its normal transverse direction of travel. The set calculated readouts of 150 mm and 2.343 mm are then used to confirm that the *final* setting angle of the *taper-turning attachment* is correct. This set angle, when finally checked in its correct position, is then locked by using the taper-turning attachment's locking nut.

A practical method of checking the lathe's set taper turning angle, can be performed (when found necessary), by the accurate and careful positioning of the internal lathe tool's cutting point, firstly at the *exact* beginning of the workpiece (this position being zeroed on the digital readout), and checked by the use of a magnifying glass), followed by the tool being seen (through the gap previously provided at the chuck end of the work-piece) to have reached the *exact* end of the 150-mm-long workpiece (while also being checked by using a magnifying glass) and noting there is a correct reading of 2.343 mm shown on the digital readout.

In engineering terms, the actual surface being machined is called the internally tapered bore'.

While still referring to drawing fig. 41, it will be seen that it shows all the details and the dimensions required to accurately machine the work-piece to complete this taper boring operation.

The initial bore of the work-piece will of course require initial preparatory drilling and boring out to be completed first, (these being its center drilling, undersized drilling, roughing out turning, followed by its preparatory parallel boring stage).

This machining must produce an initial parallel diameter that will be approximately 3mm, smaller in diameter than the required minor diameter of the tapered work-piece (this diameter is calculated and shown detailed on the fig. 41 drawing as the 'rough bore' diameter', of approximately 191.8 mm. The diameter of this 'rough bore must be calculated by utilizing the overall dimensions given in the fig. 39 and fig. 40 drawing. (The so-far unknown rough bore's diameter will need to be calculated as one's initial calculation step.)

To explain how we can obtain this rough bore's diameter and how we perform the necessary calculations to obtain it, we will first need to make a sketch in the form of the triangle shown in calc. 3, 41h, with the length of its triangle's side c being given as 165 mm long (similar to the fig 40 drawing of the supplied male tapered pattern component), with its The Angle shown at the previously calculated angle of 0.89517371 degrees.

Now, by using the prompt $\angle$ c, $b = c \times$ TAN $\angle$, we enter into the calculator the sequence 0.89517371, TAN, (0.015625) ×, 165, =, giving us 2.578124999mm. This is the total amount of offset that *would* have been needed on the taper turning attachment *if it were used* over the full *165-mm* length of the internal tapered part of item 2, as shown in drawing fig. 40, *but* to machine the taper required in the fig. 39 drawing item 1, we will now need to obtain the total offset required for the component in the fig. 39 drawing, which is exactly 150 mm long.

We therefore use fig. 41g, calc. 2, triangle sketch with its side c now being 150 mm long, and use the prompt $\angle$ c, $b =$, $c \times$ TAN $\angle$. We therefore enter the sequence 0.89517371, TAN, (0.015625) × 150

= 2.34375 mm, giving us the *total offset* required for setting the taper-turning attachment for the component we are about to machine.

The 250-mm overall diameter dimension is taken from the supplied pattern workpiece (in figures 40 and 41); this dimension is therefore reduced in diameter by 50 mm (2 flange radial dimensions of 25 mm), to give us the 200-mm calculated diameter of the large end of the supplied taper. We now use this dimension as the starting point for the calculations needed to establish the large end's diameter of the inside female taper in the fig. 39 work-piece being taper bored, which is slightly smaller. It will be seen in the fig. 41 drawing that there is a very important requirement for a 15.00-mm (plus or minus 0.05 mm gap) to be maintained between the two components when finally assembled together on completion.

To obtain the initial rough parallel bore's dimension, we now refer to calc. 3 in fig. 41, in the small triangular sketch 41h, and multiply its calculated side *b* dimension of 2.578124999, by 2, to give us 5.156249998, followed by entering XM (to store this dimension in the calculator's memory). We then use the previously calculated pattern's major taper dimension of 200 mm by entering the sequence 200, -, RM, =, giving us 194.84375; this is now the theoretical final diameter of the small end of the internal taper to be machined in the fig. 41drawing. To obtain our required *roughing diameter*, we now subtract 3 mm from the 194.84375 mm diameter to give us 191.84 mm diameter, allowing us to leave sufficient material retained in the parallel bore to be machined away to produce the final tapered bore. The theoretical diameter of 194.84375 mm should then be achieved after boring the components taper, thereby complying with the fig. 41 drawing.

To machine this *parallel roughing bore*, we are now required to temporarily de-activate the taper-turning attachment's operating rod mechanism by replacing its locking bolt to enable the lathe to be used for just straight turning; we then start the basic drilling out and roughing process of item 1 by initially using a center drill held in the drill chuck, and by using at least 800 rpm spindle speed while also using coolant, followed by using a larger-diameter drill of, say, 1 inch (25.4 mm) diameter (using a spindle speed of 135 rpm) with coolant, to fully drill through the component, this to be followed by completing a normal parallel internal boring operation, until the approximately 191.8-mm rough parallel bore diameter has been reached.

Having completed the 'roughing out' straight boring operation to its calculated 'roughing roughing-out diameter of 191.8 mm, we must now go through the process of re-activating the taper-turning attachment's operating rod mechanism, and removing its 'straight turning' locking bolt to enable the lathe to revert to using its previously set taper angle. We can finally check out the required accuracy of the lathe's set taper-turning angle, by checking the lathe's digital readout scales. These must be zeroed at the exact point where the tool enters into the work-piece's bore. This exact position can be checked for accuracy by using a 10 × magnifying glass. Note that this check must be done at both the entry end and the exit end of the work-piece, until the digital read-out scales indicate that a transverse movement of exactly 2.343 mm has taken place over the exact measured 150 mm longitudinal length of the work-piece, which will now confirm to us that the correct offset dimension is being used.

When this is found to be correct, the final clamping action is then made by tightening the taper-turning attachment's locking nuts.

This correlation between the two dimensions should occur over the exact length of the 150-mm-long work-piece.

If a digital readout facility is not fitted to the lathe, then an accurate angular setting can be achieved by using the following practical method.

We first set the taper-turning attachment's scale by eye to the approximate offset angle required, (in this case, this would be set to an angle that is slightly less than one degree and lightly clamped).

A dial test indicator (known in engineering circles as either a clock or a DTI), is used. This is preferably fitted to a magnetic base, and is secured to the lathe's saddle casting in a suitable position that allows its preloaded probe to be in contact with the tool-post during the hand traverse of the saddle; this must be started with the tool's point at the exact start of the intended taper, (by using a magnifying glass for this check). We must now verify that during the saddle's 150 mm longitudinal movement to its left while using a preset 150 mm telescopic gauge held between the saddle's button and the lathe's longitudinal end stop that this move is coupled with the calculated offset dimension of 2.343 mm being indicated on the dial test indicator's precision dial at the glimpsed end (through the gap) of the tool tip's longitudinal movement of 150 mm along the bore of the work-piece.

Alternatively, establishing this exact 150-mm longitudinal movement of the tool,can be achieved by using a calculated built-up pile of slip gauges (checked by a micrometer to verify that it contains exactly the required 150 mm dimension), this pile is initially held horizontally between the machine's moving saddle button's edge, and its longitudinal 'bed end stop' to ensure that the exact 150-mm longitudinal distance is first set, then removed and traversed by the saddle and tool.

To re-iterate, when the two dimensions of 2.343 mm and 150 mm are seen to co-incide exactly over the 150 mm length of the work-piece (checked by the use of a magnifying glass), the taper-turning attachment can then be finally set and fully clamped to allow the taper-turning operation to take place.

For safety reasons, when using either of these methods to obtain the correct angular setting of the taper-turning attachment on the lathe, it is always important to ensure that the *used* portion of the *tapered* travel along the *angled slide* of the *taper-turning attachment* occurs at a point midway along the total travel of the *attachment* (this is a very important check), and must be done to ensure there is sufficient unimpeded run-in and run-out of the turning tool along its angled slide, both on its approach to, and immediately following the required machined tapered length of the workpiece. This must be done in order to avoid any possibility of the tooling being allowed to reach the end of the taper-turning attachment's total travel before the whole length of the worked-on tapered portion has been completed.

The foregoing has now dealt with the actual setting-up of the taper-turning attachment required for the precision taper-turning operation.

The student should have also previously carried out the preliminary calculation phase in order to obtain the correct roughing-out bore diameter (shown in the fig. 41calculations as 191.8 mm), to enable him or her to start the initial preparations required to internally bore the taper required in the fig. 39 and 41 drawing.

If we now refer to the working dimensions provided in the fig. 41 drawing, it should be noted that the finished width of the precision gap to be maintained between the supplied male tapered flange's datum face (on item 2), and the outer end datum face of the internally tapered work-piece in item 1 in fig. 39, is to be 15.00 mm plus or minus 0.05 mm, therefore the taper-turning dimensions must be very accurately machined in order to achieve this dimension on their final assembly together.

The fig. 41drawing shows all the trigonometry calculations required that will allow the accurate machining of the internal taper of 3/8 of an inch taper per foot, in item 1 (fig. 39). The machinist must also make a check of the lathe tool's setup in its tool post to ensure there exists a 6 mm clearance gap between the gripped end of the boring bar's shank and its holding tool-post, *after* the cutting edge of the tool has been traversed the full length of the internal bore of the work-piece *by hand*, by using the saddle's hand-wheel. The tool's extended shank must then be accurately marked on its top surface to indicate to the machinist the actual full length of the taper being turned, while still allowing the 6 mm required tool clearance relative to its tool-post. It is also advisable to allow the tool, when under power, to have a clear run-up clearance area *before* its cutting edge actually starts cutting the 'taper being machined; this is to allow its taper-turning attachment's 'backlash (and the backlash present in the lathe's drive gears) to be taken up before the actual cut is started. The left-hand end of the work-piece must also be closely monitored to check that the position of the warning mark previously made on the tool's shank is in its correct position to ensure that the tool's cutting tip is never allowed to over-run and reach the face of the lathe's holding chuck.

To start the actual taper-turning operation, the parallel internal bore of the work-piece is sparingly painted with micrometer blue; the point of the tool is then carefully positioned to just touch the entrance to the parallel bore, then with the lathe switched on at a spindle speed of approximately 120 rpm, the tool should then be seen to leave a minute witness mark by removing a minute quantity of blue from this inner surface, and the 'transverse digital readout should now be set to read zero. The cross-slide's scale should also be set to read zero, with its cross-slides' backlash already taken up in an anticlock-wise direction.

(If the material being machined is steel, then coolant is used; if cast-iron material is being machined, then no coolant is required.). The first cut of 0.5 mm is now put on, by turning the cross-slides' hand –wheel anti clock – wise by this exact amount, followed by operating the lift-up *turning* lever to make the first (rather short) tapered cut, of the considerable number of passing cuts required.

Having now performed a whole series of passing cuts along the work-pieces' full length of taper, by having used, say, seven passing cuts of 0.5 mm depth of cut, followed by a penultimate cut of 0.265mm in depth, we will have now reached the stage where we have machined out almost all of the excess material from the *almost finished* internally tapered bore, but are still retaining a minute (theoretical) amount of 0.078 mm, of material still left in the tapered bore. This minimal amount of excess material has been left to allow for the taking of the final finishing cut that should then theoretically obtain the required 15.000-mm plus or minus 0.05 mm gap required between the two components when they are finally assembled together.

At this time however, there is an urgent need for the machinist to check the two components *for a correct and accurate fit.*

With the lathe now switched off, and with the tool and the taper-turning assembly temporarily moved out of the tapered bore and to the right to allow the male pattern component (item 2) to be offered into the now almost finished internally tapered component (item 1), in order to check the present accuracy of its fit, and the dimension of its gap relative to its flange and end face.

Should we now find that the gap now existing between the two components when fitted together, is 20 mm (this being 5 mm wider than required on the finished assembly), we must now use the Probe and Prompt system together with trigonometry calculations, to calculate the exact amount of anti – clock – wise outward movement of the cross – slide and tool (in mm), required to correct this. We do this by using a minute anti-clockwise rotation of the cross-slide's hand wheel (this being indicated on its scale) (or in its digital read out, its number change) to finally allow the male tapered component (item 2), to move 5 mm further into engagement to its left in item 1, and arrive at the required finished 15.00 mm + or – 0.05 mm gap that needs to be maintained between the two components' flange and end face when finally assembled.

To check the exact amount of material that will need to be removed from the tapered bore to allow this to take place, we now make a sketch of an inverted right-angle triangle in a similar orientation to the worked-on problem triangle as shown in calc. 4, 41e, which simulates the existing 20 mm gap length being represented at its side c while also using its calculated The Angle of 0.89517371 degrees, (obtained from the previous calculation at 41d).

By using the chosen prompt $\angle\ c$, $b = c \times$ TAN $\angle$, we now carry out this calculation as follows:

We know The Angle and c and wish to know b. We therefore enter the sequence 0.89517371, TAN, (0.015625) × 20, =, giving us 0.312499999 mm; this figure now gives us the length of side b in this sketched triangle.

We now sketch another similar right-angle triangle, oriented similar to 41e, i.e. pointing to its left and now shown at calc. 5, 41f. This calculation is needed to establish the correct length of side b in this triangle, when its side c is at the *required* length of *15.00 mm.*

Therefore, using the prompt $\angle\ c$, $b = c \times$ TAN $\angle$, we enter into the calculator the sequence 0.89517371, TAN, (0.015625) × 15, =, giving us 0.234375, the length of side b in this triangle.

We now sketch another similarly oriented triangle (calc. 6, 41j), and subtract the length of side b shown in sketch 41f (0.234375), from the length of side b shown in sketch 4,41e (0.312499999), and complete the calculation as follows: We enter the sequence 0.312499999, –, 0.234375, =, giving us 0.078124999 mm giving us the theoretica*l* finishing depth of cut required by the boring tool when using the digital readout or the lathe's cross-slide scale to perform this final cut to establish the exact finished size of the tapered bore while allowing the required gap of 15 mm + or – minus 0.05 mm to be accurately maintained between the two components.

The student at this point should be made aware that the calculations we have made up to this point to find the actual depth of cut required to finish the job, must be considered as being purely theoretical because their accuracy will depend to a large extent on the prime condition of the boring tool's cutting edge and tip radius, and where there is the possibility that it has encountered (so-called) tool push-off during the previous boring cuts, and will now be under much less push-off pressure for this finally calculated 'minimum' depth of cut. The tool could, due to the removal of this 'push off' pressure, now be capable of removing this small amount of material from the tapered bore during this final cut, *without* any further cross-slide movement being given at all. It is therefore considered extremely wise, because the workpiece is now in this almost-finished condition, to make the final decision as to whether or not to allow the tool to take this final taper cut with little, if any, indexing outward of the tool at all. Seriously considering this important precautionary step should allow any existing spring in the tool to be used up on this final cut and achieve the final correct result.

Alternatively the machinist can index the cross-slide anticlockwise (while using its scale or digital readout) outward by the theoretical 0.0781mm. in order to achieve this critical components gap's dimension.

It should also be noted that if the tool being used had been allowed to possess a larger-than calculated cutting radius, then the push-off in this case would have been considerably greater; (thereby proving that a very small cutting radius given to the tool is the much-preferred option).

We should also note that completing this 0.078124999 mm (0.00307578 inch) dimensioned last depth of cut only requires a very small anti-clockwise movement of the cross-slide's hand-wheel to allow the boring tool to reach its finished diameter. Therefore, this final movement given to the cross-slides' hand-wheel must be undertaken very carefully in order to avoid the tool exceeding the close tolerance limits required for an accurate fit, as is specified in the original fig. 41 calculated drawing. It is considered advisable, therefore, to double-check the result of the dimension previously calculated by use of the calc. 6 in triangle 41j. This triangle must have an identical angle of 0.89517371 degrees, and its side b will be calculated to be 0.078124999 mm long; therefore, by using this calculation, we can now obtain an accurately checked length of its side c.

To perform this double check, we use the prompt $\angle$ b, $c = 90 - \angle$ TAN $\times$ b. We do this by entering the sequence, $90 - 0.89517371 = 89.10482629$, TAN, (64.00000002), $\times$, 0.078124999, $=$, 4.999999937 mm, thereby proving that an outward anti-clockwise rotation of the cross-slide's hand wheel and its scale of 0.07812499 mm will allow the tool to remove the exact amount of material (that theoretically remains in the unfinished taper) to permit the male tapered component (fig. 40, item 2,), to enter into the work-piece further by the required 5.00 mm, and, in doing so, obtain the required (and expected) accurate gap of 15.000 mm + or – 0.05 mm, between the assemblies' flange and end face.

The following practical method can be used when the amount to be removed from the taper is so very small that it is likely to be exceeded by any further outward movement of the conventional cutting tool and its cross-slide.

This problem can occur if the amount needing to be removed from a work-piece is extremely small, possibly in the region of (say) 0.0127 mm (0.0005 inch) or less. To do this, the author has successfully used the following method to achieve the required result while using a reasonable amount of engineering skill.

There are times in the engineering work-shop where the amount of material that needs to be removed from the precision bore of a component is so small that the lathe tool (due to the amount of push-off or spring inherent in its boring bar), *cannot be relied upon* to remove exactly the amount being indexed outward on the cross-slide's scale to obtain the exact diameter required in the component. In this case, it is possible (by taking reasonable care) to accurately remove this very small amount of material from the bore of a component, by using a strip of one-inch-wide medium emery cloth securely wrapped around the diameter of a slotted, wooden-handled, and supported hand tool. This tool can be easily made from a piece of (say) 25 mm (1 inch) diameter wood (of a material similar to a wooden broom handle); this is cut to about 350 mm (14 inches) long, and provided with a lengthwise hack-saw cut (positioned on its center-line) at its slotted end, of approximately 25 mm (one inch) deep. The slot should be of sufficient depth to take just the (25 mm) width of the normally issued emery cloth strip commonly used in the workshop. This strip of emery cloth should then be cut to about 5 inches (127 mm) in length and entered firmly into the slot, with its excess material wrapped tightly around the holder's diameter in the same direction as the work-piece is revolving.

With the work-piece now rotating at a normal turning speed, the handle assembly, complete with its tightly wrapped strip of emery-cloth, is entered into the bore of the component and moved longitudinally along, using a reciprocating motion while its handle is being tightly gripped in both hands of the machinist. It is important that one does not allow the tool's emery cloth surface to dwell in any one place or allow its abrasive surface to extend more than half its width at either end of the component's bore. By using this method, followed by making several accurate measurements and checks of the diameter being honed, using (in the case of honing a parallel bore), an internal micrometer, or a precision taper gauge if honing a taper, it will now be found possible to establish a parallel or tapered bore to the exact dimension required.

Chapter 18

Checks and adjustments often required when setting up a center-lathe

This chapter contains many of the preparatory checks and adjustments found necessary when setting up a center lathe for precision turning, screw-cutting, or parting-off operations. These checks and adjustments are designed to ensure that accurate components can be produced on the center lathe.

Section 18/1
An introduction to how the day-to-day engineering problems can be solved when undertaking general machine shop lathe work

This chapter contains many important factors the student should study and consider acting upon that relate to screw-cutting, turning, boring, and parting-off operations on the center lathe; the script also includes some very important and necessary maintenance checks that are often neglected, and it should be pointed out that there will be the need for these checks to be carried out by the machinist prior to using the machine for precision work.

There is a well-known saying, often used in the armed forces and in the engineering industry, which states, 'Forewarned is forearmed'(which indicates there is a distinct advantage in knowing what might happen before it actually does happen).

The following notes and suggestions are based primarily on this particular saying, and these suggestions will include extensive screw-cutting advice, lathe safety precautions, machine accuracy checks, cutting tool procurement, preparation, and sharpening, and helpful tips on lathe tool off-hand grinding and its sharpening techniques. These notes have been prepared to assist the trainee engineer and the student sufficient to allow them to become totally familiar with the wide variety of practical and theoretical steps needing to be taken before it will be possible to become a fully fledged machine shop center-lathe turner or tool-room engineer.

Also included are a host of popular machining techniques practiced by the author and by his experienced colleagues who are currently working, or who have worked in the tool-room and in the research and development departments of the engineering industry, while also working on the production machine shop floor. Helpful notes are also included regarding the most important checks generally found necessary when setting up a center-lathe for a straight turning, parting off, or a screw-cutting operation.

The information gained during these rather lengthy and extremely thorough checking and setting-up procedures, will allow the budding machinist to resolve many of the problems that are likely to occur, before they are actually being experienced.

The author would like to take this opportunity of conveying the following important practical advice to all toolmakers, lathe operators, and machinists.

The lathe you are about to be working on may not belong to you, but it will allow you to earn your living. Treat it as kindly as possible; it is to be cleaned and oiled daily and the machinist should familiarize himself with all of its peculiarities. Its repayment to you will be trouble-free and untiring service, coupled with years of accurate and consistent output. However, if this machine is neglected, then imprecise work and numerous mechanical problems that occur will usually be a result of your neglect.

This chapter describes the experiences and the problems that will inevitably occur in practice, particularly when the machinist is involved in setting up a lathe for a turning or screw-cutting operation. The explanations are given in minute detail and cover a wide variety of tooling problems as and when they occur; the notes then go on to explain how best to deal with and solve these potential problems.

In true life, these particular problems occur quite often, and usually at the most awkward moment in the setting-up procedure. To truly reflect this situation, the in-depth explanations given, will be dealt with immediately, without delay, and before continuing with the original explanation process; they will explain the main screw-cutting problem first, and will return later to the initial subject being explained regarding general center-lathe turning.

These breaks of continuity in the general flow of information on screw-cutting or turning, will allow the newly found problem to be investigated thoroughly and in depth, by using the simple process of leaving no stone unturned, in the final explanation given.

The author has adopted this approach to solving the problems that do occur, in order to make the student fully aware that these problems do always exist, and that they will be encountered on a daily or minute-by-minute time scale. These recurring problems can often cause the machinist to be seriously sidetracked away from the most important task of producing the actual work-piece as quickly and accurately as possible, and will often be undertaken while trying to work out a correct sequence of operations in the stipulated time available, during the normal everyday workshop experience.

This deviation from the immediate task in hand, occurs quite frequently on the work-shop floor, and usually occurs when the machinist is deeply engrossed in setting up a lathe for machining a component that is ultimately found to require several extra calculations to be made, or the use of special tooling to be manufactured or procured, before the final production sequence of operations can be undertaken. The student should therefore be made aware, (rather sooner than later), that a substantial amount of forward thinking and planning must be undertaken by him or her in order to establish the correct sequence of operations that will need to be adopted for the manufacture of a particular component. He (or she) will soon discover that very few turning jobs are sufficiently straightforward to allow the use of just their basic skills, and the basic tool settings necessary, to produce a fully completed work-piece. He or she must therefore run through in one's mind the whole series of turning operations required to finish the job, memorize them, and carry them out in their entirety in the correct sequence, in order to complete the fully machined work-piece. The planning sequences adopted must encompass all the intended operations required, and must also avoid if possible any waste of time in the use of unnecessary extra setups that could possibly be required to complete the work-piece. The student will discover that it is much more convenient to first plan the production of the work-piece by going through all the moves mentally, so that it is possible to get the job completed in just one operation, without upsetting the work-piece's accurate location in the lathe' chucks jaws by turning it around and re-chucking it to complete the component.

If it is found necessary to remove and reverse the work-piece in the chuck to 'finish turn' the previously held end of the component, it will be found to take much longer to finally complete the work-piece. However, this theoretical plan can be seriously thwarted, should the planners of the establishment decide to supply the machine shop with steel billets that have already been cut to an exact length obtained from the drawings' dimensions. This is a case where the planners (*who have only rarely been promoted from skilled turners* and who have a strong wish to save money on the cost of materials), are reluctant to realize that the machinist will often need to secure the component in the chuck by utilizing an extra length of material with which to actually hold the billet, prior to finish turning all its diameters followed by parting off the finished component.

By using one's own serious planning efforts in order to seek and adopt a one-operation sequence of operations (if at all possible), it will then be realized that this will result in all the diameters of the work-piece running true and concentric to one another; this will be due to all the component diameters being machined at one precise setting. (This cannot of course be the case, if the work-piece has been removed from the chuck, turned around, and replaced in the same chuck to allow the final machining operation to be carried out using a second operation to be performed on its previously gripped end without the extra task of using a clock gauge to ensure that it is again running truly in the chuck.

It will also be found advantageous if the student learns the art of using both right and left hand lathe tools to enable him or her to complete all of the operations required to finish the workpiece in just one operation, prior to the work-piece being finally parted off from its holding material.

One of the uncertainties the author encountered while gaining his initial practical experience in the machine shop and tool-room was 'How is it possible to securely hold all the shapes and sizes of work-pieces needing to be turned, while still finding the time between all these different jobs to learn the very important skills of tool and drill sharpening?'

Because of this age-old problem, the author has included ample advice on how to acquire these skills during one's daily work-shop experience.

It is of course a well-known fact that the newer the lathe one is using, the more capable it is of performing very accurate work. The ultimate degree of accuracy produced by a new machine is of course legendary, and is often achieved with little need on the part of the machinist to make any further remedial adjustments to its slides and working parts to obtain the required accuracy.

An older machine can of course be made to perform very accurate work, but to do so, it will be found to require rather more checks and adjustments to be made to it by the machinist in order to arrive at a similar degree of accuracy. It should therefore be realized that a considerable amount of skill and experience is needed on the part of the operator to keep an older lathe in a workable, pristine, and accurate turning condition. The difference in performance accuracy between an old and a new lathe will undoubtedly be caused by the accumulated wear in its machine slides, its spindle support bearings, and by the amount of its cross-slide and compound slide's backlash that will normally be found in the screw threads of its lead-screws. A new lathe's bearings and slides will of course be in perfect condition, because the machine will undoubtedly have passed through a very recent accurate parallelism test before it is allowed to leave the manufacturer's factory.

Many factories, however, are faced with the problem of having to use their older, well-used machines for their research, development, and production purposes,; this is mainly due to the lack of sufficient capital investment being made to replace these machines over the past years. Therefore, at the present time it is rare for a factory or its machine shop to possess any new precision machines.

The student, on arriving at his new place of work for the first time, should be advised that there often exists a certain pecking order among the skilled personnel already working in the machine shop and particularly in the tool-room. This does unfortunately create a situation where the newest and best machines are often reserved for the use of the longer serving fully experienced employees. It is quite the usual practice, for a newcomer to be assigned to a well-used, well-worn machine, which, because of its long-time use on the shop floor, will of course have been operated extensively by everyone else in the wor-shop, who will therefore be fully aware of its merits and its possibly inconsistent characteristics typical of an older machine. These wear characteristics may include a less than-perfect work-piece-holding capability of its chuck (this means that when the work-piece is installed in its three-jaw chuck, it will not have the ability to run perfectly true); this problem can also result in the lathe possessing a poor parting-off performance and an inability to safely perform accurate and consistent parallel turning operations, without certain adjustments being made to its precision slides or its tail-stock location to obtain it. The machine may also possess a tail-stock that, because of wear, is inconsistent in its ability to maintain true work-piece parallelism at all times, particularly after the tailstock has been moved along into a different position on its bed to suit a longer work-piece. It should also be pointed out that these machines' failings are fully expected to be overcome by the newcomer during his first few days of indoctrination into the workshop environment. It will be to the student's advantage therefore if a working knowledge of how to rectify any current lathe problems *without complaint* and without the need to seek fellow machinists'advice on the matter, will also confirm in the minds of your new colleagues that this new employee does actually possess the wherewithal to sort out the machine's problems himself.

Every effort should therefore be made, (and certainly before voicing complaints to your new colleagues, to rectify (as best one can) the problems found on the lathe, to allow it to perform sufficiently accurate work.

In most cases, a combination of checks and minor adjustments, carefully performed by adjusting any possibly loose jib strips or loose compound slides, while also ensuring that all the machine's slides still have full freedom of movement, usually eliminate most of the tool chatter problems that would normally occur. These remedial adjustments will then enable the lathe to perform with greater accuracy, and the lathe will also be found to possess a more secure holding capability for the turning or parting off operation to the required precision standard.

The cutting angles ground onto one's own existing lathe cutting tools may also benefit from some slight modifications. This can be achieved by a process of beefing them up, in order to improve their ability to withstand the additional (and possibly intermittent) loads generated while performing the *critical* parting-off operation. This improvement to the machine's rigidity will then allow the tool to cut more efficiently on an older machine. This improvement will of course depend largely on the amount of wear present in the lathe's working slides, spindle bearings, and working surfaces.

To quote a practical example, when a machinist is using a parting-off tool to part off a component on an older machine, it is often the case that a chattering noise will occur during the operation, because of the accumulated wear in the unadjusted lathe's slides or the accumulated wear in its main bearings. To mitigate this problem to a certain extent, the author has found it an advantage to increase (where possible) the available strength of the parting tool in use, by modifying its overall shape and its clearance angles. We do this by making the extended part of the tool more rigid than its original tool's design. Fig. 47 item 6, shows how this can be achieved by relocating (by using off-hand grinding techniques) the tool's inherent weak point, situated at the tool's butt-welded joint with its shank, held in the tool post. We can modify the strength of the tool by literally moving this weak point forward and nearer to its cutting edge, so that it now occupies a position located much nearer to the tool's front cutting edge and correspondingly further away from its vulnerable and weak welded shank joint near its tool post. This extra rigidity can be obtained by grinding vertically the sides of the tool's extended section, (behind its front cutting edge), particularly in its ground side relief clearance area, transforming it into a vertical parallel form, which produces much stronger straight sided clearance 'cheeks', with its new relief clearance taper now located much nearer to its cutting edge. This modification now allows the extended part of the tool's overhanging section to receive a much stronger support, as it now possesses stronger parallel sides. This parallel section, will now extend from its welded shank junction, up to its new position lying approximately 10 mm (on small parting tools, 5 mm), back from its front cutting edge. This modification to the parting-off tool will be found to increase the rigidity of its unsupported stick-out length from its tool post quite markedly. The old weak point in the tool will now be positioned well in front of its normally taper sided shank's weak junction point. This modification will increase the front cutting edge's rigidity, and therefore prevent possible chatter (or possibly a broken tool) occurring during the parting-off process (which can be very embarrassing among work-shop colleagues).

It is also very important to point out to the student that the graduations supplied on many of the older lathes' cross-slides are calibrated quite differently from those fitted to the modern and newer design of lathe. The older lathe's scale graduations are calibrated to indicate to the operator the exact inward movement of the cross-slide and its tool and, in so doing, fail to take into account the fact that the cutting tool will actually be removing material from *both* sides of the turned workpiece, thereby reducing the turned diameter by *twice* the cross-slide's scale's indicated amount. For this reason, it has become general practice for the operator when using these older types of lathe, to mentally halve the actual depth of cut required by the work-piece, and actually index inward on its scale only half of the calculated dimension required,; this then allows the lathe to turn the work-piece down to its required diameter.

For example, if we have just turned the work-piece's diameter down to 25 mm, and are now required to reduce this diameter smaller to obtain 24 mm diameter, the operator must mentally halve the actual 1 mm required, and index inward by using the cross-slide's scale just half of this amount, (that is, 0.5 mm) to enable the tool to turn the work-piece down to the required 24 mm diameter.

Alternatively, if he is using a *newly* designed lathe's cross-slide's scales, it will be found that they are calibrated to indicate to the operator the total amount of material being removed from the work-piece. Therefore in this case, when working on this new machine, if the machinist needs to reduce a previously turned diameter of 25 mm down to 24 mm diameter, the cross-slide hand wheel and its scale will need to be turned the full indicated amount (i.e. 1 mm) in order to obtain the 24 mm turned diameter required.

To make a reliable check of the actual calibrations that exist on your particular lathe's cross-slide's scale, you can adopt the following procedure:

The cross-slide's hand wheel is first turned clockwise by one complete revolution (to eliminate all the inherent backlash existing in its lead screw); its scale is then set to read zero. A line is then marked (using pencil) on the top of the cross-slide's fixed *and* moving castings to produce a horizontal penciled line. The cross-slide's hand wheel is then turned one complete revolution clock-wise and a similar pencil line is again marked on the two castings. The distance between the two marks is then measured by using an accurate rule (or a vernier caliper). This measurement will then reveal not only the pitch of the cross-slide's lead screw, but also the amount the tool and its tool-post have progressed inward during one complete revolution. If for example, the distance between the two penciled lines now measures 5 mm but the total calibrations shown on the cross-slide's scale indicate 10 mm, then this will indicate to the operator that the lathe is of the modern design, and is calibrated to remove the exact amount from the work-piece that is indicated on its scale from the *diameter* of the workpiece.

If on the other hand the two penciled lines indicate that 5 mm has been traversed by one revolution of its scale, and the graduations on its scale only total 5 mm, then the lathe is of the older design, and the amount of material being removed from the work-piece's diameter by this tool, will be twice the amount indicated on its cross-slide's scale, therefore in this case, the calculated amount of material needing to be removed from the workpiece must now be mentally halved, and only half of the engraved figure shown on its scale should be indexed inward to obtain the desired turned diameter.

The spaced graduations on the lathe's cross-slide's indexing scale, relative to the actual movement of its slide, can also be checked and memorized by using the probe of a magnetically based dial test indicator (the spring loaded probe of the instrument must first be preloaded by one revolution of its needle seen indicated on its scale). It must then be placed in a position touching the tool-post and its scale zeroed, the needle's exact movement is then noted relative to the cross-slide's graduations, when the hand-wheel is being turned in a clockwise direction. This check will confirm the exact amount of movement of the tool-post (and its tool) relative to the actual reading shown on its cross-slide's index scale when compared to the dial test indicator's scale. By using this visual, practical method of checking, it will confirm in the machinist's own mind, the actual value (relative to the diameter of the work-piece being turned) of *each graduated line* and each *space* on the cross-slide's index scale. It is most important (and this cannot be overstressed), that the diametrical value to the operator of each graduation line relative to its next graduated line shown on the cross-slide's index scale must be positively and exactly identified, and a mental note made of the value of each line and each space (in terms of the diameter being turned on the work-piece); this physical check and its confirmation must be performed before the machinist can consider himself capable of using the lathe for precision turning operations.

Carrying out this checking and confirming exercise will avoid any unfortunate and expensive dimensional mistakes being made during any of the following turning operations.

To reiterate, the dimension being traversed relative to the work-piece that exists between the cross-slide scale's graduations, can be accurately checked by using the needle indications of a fixed dial test indicator's probe when it has been preloaded against the tool-post, while the cross-slide's hand wheel is being turned slowly clock-wise. At this time, one can also check the amount of back-lash being present in the cross-slide's lead-screw by noting the amount of lost motion (or slack movement) discovered when the cross-slide's hand-wheel is then turned anti-clockwise, and noting the amount of lost motion on the cross-slide's scale that occurs before the dial test indicator's needle begins to move in its opposite direction. This checking and memorizing of the cross-slide scale's values will be found to be a most worthwhile exercise, particularly before the machinist makes his first attempt at performing very accurate precision internal and external turning operations.

Section 18/2.
Work holding devices explained

Those who intend to become engineering machinists, or toolmakers, will often be using the center-lathe to do most of their work; it is therefore very important that they are made fully aware of the many work-holding devices currently available for their use.

They consist of, primarily, the much-used three-jaw chuck, used for securely holding either round or hexagonal shaped materials, followed (in importance) by the four-jaw chuck, used for holding not only round but irregular, square, and other oddly shaped sections of material. This is followed by the faceplate (used for machining clamped-on irregular-shaped work-pieces or castings that need to be bored, faced, or screw-cut internally or externally), followed in importance by the catch-plate with its accompanying lathe carrier, used to provide the necessary motive drive for turning longer workpieces when held in the situation called 'between centers', while obtaining its motive drive from its lathe carrier.

Section 18/3
Removing and replacing the gap in the machine bed of the center lathe

The existing lathe bed casting on a tool-room lathe is often designed to have a removable section (called a gap),; this is positioned below the chuck end of the main lathe bed structure and supplies the extension of its bed's V ways. This separate precision casting has been machined in situ during the V way and the precision grinding of the main bed during the lathe's manufacture. The two castings are accurately located together in their true positions by the use of dowels and retaining bolts,; these then enable the gaps' easy removal and accurate replacement. The gap is normally retained in situ for most of the machine's daily turning and screw-cutting operations. However, if found necessary, this gap can be removed should the lathe be required for machining large or unwieldy castings that require a larger than normal swing radius at their periphery to enable the work-piece to revolve over the now much larger clearance area. Its removal will provide the additional clearance necessary to allow not only a larger diameter faceplate to be fitted, but will also create sufficient room to accommodate large and bulky clamped-on work-pieces to be fitted to the tooling's front face that require machining.

Section 18/4
The setting up of a workpiece on a lathe's faceplate

When setting up a work-piece on the lathe's faceplate so that it runs concentrically, it is possible to use as a guide (by eye), the numerous concentrically engraved circular grooves that exist on its front contact face in order to establish that the work-piece is being held in its correct position so that it runs truly relative to the lathe's center-line. In the case of a circularly marked-out work-piece, these previously marked-out circular lines, can be used in conjunction with using a sticky pin (explained later) to perform this concentricity operation, or alternatively, by using an existing surface machined diameter or bore of the work-piece to check that it is running truly, relative to the lathe's center-line, by using a DTI dial test indicator) for this task. It is therefore possible to position the work-piece in one of the above ways to obtain the necessary true running concentricity required for the accurate machining of the component.

The marking-out of large work-pieces needing to be bored or turned (if found to have two adjacent machined square faces), can with advantage be located and clamped against a separately held 90-degree angle plate, which in turn is secured by bolts to the face-plate's front face. However, if the clamped-on workpiece possesses a machined diameter, then this surface can be utilized to aid its centralization by using the needle's fluctuations of a fixed dial test indicator (DTI), whose probe is held in contact with its hand-operated revolving diameter. However, if the assembly about to be turned is of irregular shape, it is quite possible that this can cause a weight distribution problem by it being out of balance. In this case, an additional balance weight must be selected (by its weight and size) and securely clamped to the face-plate in an opposing position that is sufficient to balance the work to be used as a counterbalance. This balance weight should be independently assessed as having sufficient weight, and be positioned to provide the necessary balance to the whole assembly. It may also be found necessary to bolt on a work stop, carefully positioned to prevent the work-piece being moved out of position by the tool's influence during heavy cutting operations.

Because of the high centrifugal forces generated during fast rotation, it is advised that only relatively low spindle speeds should be used when using this particular balanced setup, and care should be taken that the cutting tool is not allowed to come into contact with any of the securing clamps or fixation bolts during the work-piece's rotation. The correct balance of an assembly can be checked by firstly selecting a neutral gear, followed by spinning the faceplate and its workpiece assembly, by hand, and noting the lowest point where it actually comes to rest. When correctly balanced, the whole assembly should be seen to come to rest at any random position. Failure to balance the assembly correctly can cause serious vibration problems when switching on and running the machine. In extreme cases, an out-of-balance workpiece can be considered dangerous; therefore, any out of balance must be rectified immediately by rebalancing the whole assembly correctly.

For accuracy reasons, extreme care must be taken when re-placing the previously described lathe bed 'gap' into its original dowelled and seated position. On large heavy lathes, this may be considered to be a two-man job (unless there is a crane available), due to the gap casting's heavy weight. The seating area around both components, (including the area of its locating dowels), must be scrupulously cleaned of all contaminants before an accurate replacement can take place. After replacement, a check must be made by using a small straightedge, positioned over the two newly aligned V surfaces, to ensure they are lined up and seated correctly. The probe of a clock gauge affixed to the tool post can also be traversed over the surfaces of the two castings while using longitudinal movements of the saddle and noting any discrepancy in the dial's needle readings during the traverse between the two components.

After lubricating the lathe's flats and its V ways, the saddle should then be traversed over the original gap to ensure there is free movement to and fro across the gap. A perfectly smooth traverse will indicate that the gap replacement has been carried out satisfactorily.

Section 18/5
Using the collet chuck supplied with the lathe

Tool-room lathes are quite often provided with a full set of interchangeable work-holding collets. These are used as an alternative holding device in place of the normal chuck for securely holding round stock materials, particularly when there is the requirement for the gripped diameter of the material being machined to run perfectly concentrically with the diameter being turned or bored. This will then establish accurate and concentric machining. The interchangeable metric or imperial bore sized collets are held internally in the lathe's spindle's nose, by use of its *collet chuck adapter*. During the preliminary setup for using a collet, the existing chuck is removed, and placed in a safe place where it cannot become contaminated with dust or metal swarf. The lathe spindle's internal taper area must then be scrupulously cleaned out to remove any existing contaminating particles. The adapter and collet are then fitted. The collet is then tightened onto the perfectly round stock material by using its long, hollow, and threaded, *draw sleeve*; this is operated by its hand-wheel, situated at the left 'outer entry of material' end of the lathe's spindle. The operator must however take extreme care in selecting and using only the appropriate sized collet that is found compatible with the diameter of the stock material being used. If a collet of the wrong diameter is used to grip a differing diameter of stock material (whose diameter is incompatible due to its differing diameter), then the collet can be seriously strained during the tightening process,; therefore, this practice must of course be avoided at all costs. Collets are also available for securely holding hexagon or square stock materials, although these particular shaped collets are normally confined for use on production capstan lathes. This design of lathe often have a very long and noisy automatic stock feed capacity of up to 6 meters, the material in this case being automatically fed through a protective tube by use of a push piston being operated under air pressure into the machine's spindle and through its collet. The fed material finally contacts a pre-set length stop after its journey through the collet chuck, (this is a rather noisy operation as the feed of material is normally operated while the lathe is still running and its open chuck is still rotating).

Section 18/6a
How to manufacture a homemade internal collet chuck, on the center lathe

There will be occasions when it will be found necessary to accurately hold a work-piece (or a batch of work-pieces) by using their precision machined bores to enable a second machining operation to be carried out on their (still to be machined) outside diameters. When in this situation, it will be found necessary to make your own internal spigot chuck to hold the component. To do this, it will be necessary to turn a very accurate shouldered spigot diameter onto a spare piece of mild steel material, that contains a larger diameter than the work-piece; this is preferably centralized and held in the four-jaw chuck (for increased rigidity). The spigot's diameter must be turned to an identical outside diameter to fit the work-piece's internal bore. This precision-turned diameter must be extremely accurate and to a critical diameter that is just sufficient to allow the work-piece to *slide* on.

A large-diameter shoulder must also be allowed to exist on the holding stock material to enable the work-piece to be abutted against it in order to locate the work-piece longitudinally with accuracy. The work-piece to be machined can then be accurately located longitudinally against this shoulder while also being located accurately on the precision spigot's diameter to enable it to run perfectly concentrically.

This homemade collet chuck (made from this spare piece of material), will require the following extra work to be carried out on it to enable it to grip the work-piece securely and accurately.

It will be necessary, (depending on the work-piece's bore diameter, and the necessary diameter required for its supporting spigot), to choose a suitably sized countersink tool to countersink the outer end of this proposed collet chuck to a diameter that will suit the stock-sized head diameter of a selected Allen-headed hexagon type countersunk steel screw. A suitable tapping drill will then be required in order to drill the tapping sized hole in the spigot's outer end, followed by this being tapped with the compatible internal thread to enable the chosen countersunk screw to be screwed into its end. (It is important that the chosen screw's head diameter is slightly *smaller* than the work-piece's bore diameter in order to allow sufficient diametrical clearance between the spigot diameter and the component.). The depth of thread in the spigot should be tapped slightly deeper than the length of the selected screw (to allow for the screw's end clearance when it is fully tightened to enable the spigot to enlarge by a sufficient amount to grip the work-piece). The spigot must be modified by using a handheld hack-saw, to cut (while still being held in situation in the chuck), two accurate, but short, longitudinal slots at 90 degrees to one another forming a longitudinal cross, to a depth of approximately a third of the gripping length of the spigot, (this slotted length will of course depend on the actual length or thickness of the work-piece being held). This is followed by de-burring any sharp edges from the slots and the spigot's exposed diameter.

In use the work-piece to be machined is then entered fully onto the previously turned spigot's diameter, with its back face located against the (now existing) machined shoulder, followed by the fitted counter -sunk screw then being tightened. During this tightening process, the screw's tapered countersunk head will cause the spigot's split countersunk end to expand by a sufficient amount to grip the workpiece securely.

The previously awkward task of accurately machining the outside diameters of a batch of work-pieces *to run concentrically with their gripped diameters* can now be achieved with confidence, and accuracy.

Having finished the batch of components and before removing this homemade collet chuck from the lathe's holding chuck, it will be found advisable, (in order to enable it to be re-set accurately again at a later date), to mark the position of the held end of the collet chuck relative to the lathe's holding chuck, by using a carefully positioned and indented center punch mark on both items,; this will then establish their correct re-setting positions. This accurate repositioning of the collet chuck in its holding chuck will then enable its spigot diameter to run truly, while being used on a later batch of components.

Its true running can of course be checked and adjusted if necessary by using a dial test indicator.

Section 18/6
Using the fourjaw chuck fitted to the lathe, and explaining its main use

Experience has proved that the four-jaw chuck is the most useful work-holding device used on the center lathe. Not only is it considered to be the best and safest device for holding the workpiece for parting off a component (because of its longer jaw length and possessing four jaws instead of three), its versatility can also be used for securely holding a wide range of oddly shaped work-pieces (with the exception of course of those of hexagon shape that will require the use of a three-jaw chuck or the use of a suitably shaped hexagonal collet mounted in its appropriate collet chuck). The variety of materials possible to be held securely will include not only all the common round materials, but also those of square, flat rectangular plate-work, and other awkwardly shaped materials found difficult to hold for machining in any other work-holding device.

Its jaws can also be reversed to allow a positive grip to be obtained on large rectangular shaped castings, flat plates that require surface machining, large-diameter flanges, various drilling operations, internal screw-cutting, or the boring of castings needing to be carried out. However, with the chuck's jaws extended far out from the chuck's outside diameter to allow them to grip large irregular-shaped components, it is always a wise practice (before switching on the machine, and after setting up), to check whether the jaw's outer tips or the work-piece's outer corners are liable to contact the bed or the machine's V ways during their rotation, especially when working on large-dimensioned workpieces.

For reasons of safety therefore, having secured an irregular-shaped workpiece into position in the chuck, and before switching on the machine, one should always select a neutral gear, and turn the chuck by hand to check whether sufficient clearance has been provided between the bed of the machine, the work-piece's outer corner surfaces, and the chuck's extended jaw corners.

The four-jaw chuck is considered by the majority of tool makers to be the most versatile and safe of all the work-holding pieces of tooling; this is because of its very safe and accurate gripping potential. It is therefore considered to be the most useful work-holding device used on the center-lathe. The work-piece's gripped datum faces or its varying diameters, can be checked for the necessary true running by using the indications shown on a dial test indicator's needle with its probe in contact with the workpiece's diameter, while the chuck is being turned by hand. The needle's fluctuations will indicate which individual jaw will need to be adjusted to centralize the work-piece. The work-piece can then be 'persuaded' to move across the chuck's front face by the small adjustments made to each independent jaw until a satisfactory centralization of the work-piece has been achieved. However, care should be taken not to allow the work-piece to actually fall out during this process. This can happen if it is momentarily left insecure by being insufficiently gripped by its other jaw during this centralization procedure.

Its four jaws possess (quite surprisingly to the student), a longer grip length than the jaws of its three-jaw counterpart; they therefore provide considerably safer support to the workpiece, particularly when engaged in a parting-off procedure, or when working on large-diameter workpieces that may possibly possess an overlong unsupported 'stick out' (of material) protruding from the chuck.

The four-jaw chuck's disadvantage is that virtually every work-piece will need to be centralized by adjusting its independent jaws before accurate machining can commence. This is done by using either a clock gauge (dial test indicator), to check the concentricity of round components, or a homemade sticky pin, used to centralize a circularly marked-out plate held within its jaws. The four-jaw chuck also possesses the advantage that it can be used to deliberately offset a component's center line in order to produce an off-set turned diameter or bore. An example of this type of operation would be when one is machining an offset diameter on, say, a crankshaft.

Section 18/ 7
Using a homemade sticky pin to centralize the workpiece in the four-jaw chuck

A sticky pin is a very useful (homemade) centralizing device; it is comprised of a common sewing needle (or a similarly shaped and sharpened steel panel pin), held captive within a small quantity of plasticine (or modeling clay) to provide its necessary adhesion to the tool or to an accessible part of the machine.

With the lathe switched off, the 'sticky pin' assembly can be stuck conveniently to any adjacent tooling,; this enables its sharp point to be used to centralize (by eye) a previously circularly scribed line on the workpiece, while the whole setup in the chuck is being rotated slowly by hand. This is of course with the proviso that the work-piece has previously been marked out with an accurate and to advantage large 'scribed diameter' line. An (additional) scribed circle, performed outboard of the required marked-out circle around the hole about to be drilled or bored, will be found advantageous because it is of larger diameter and much easier to pick up, by the sticky pin's point. The individual jaws of the chuck are then carefully adjusted to move the work-piece laterally, by minute amounts within its jaws, to allow full centralization of the work-piece to take place. Extreme care must be taken to keep sufficient tension between two of the chuck's opposing jaws in order to prevent the work-piece actually falling out because of the workpiece being momentarily unsupported),; fallout of the component can cause possible damage to the lathe's bed and to the workpiece. (It is always considered good practice, therefore, to use a flat piece of plywood or similar material on the top surface of the lathe's bed for its protection. Therefore to accurately set up a work-piece to run concentrically by the use of a sticky pin, the work-piece must be carefully persuaded to move across the face of the chuck in minute amounts by using the thrust of one jaw and by relaxing the tension of its opposing jaw sufficient to allow the sticky pin's point to gradually line up with the larger, circular marking-out during the chuck's slow rotation by hand, and also by using one's critical viewing to gradually reposition the workpiece in small stages to allow its circular marking-out to eventually be running concentrically with the 'sticky pin's point. (For extreme accuracy, a magnifying glass can be used to establish absolute true concentricity of the work-piece's scribed line relative to the sticky pin's point, and to the center-line of the lathe's spindle.). With care, this positioning can be surprisingly accurate. This accurate centralization of the work-piece will then allow the component to be accurately turned, drilled, bored, or screw-cut.

This will be due to the fact that it is now known that the work-piece is running concentrically and perfectly in line with the machine's center-line.

When centralizing large, heavy components, it is a very wise precaution to always use the tailstock's sliding barrel's outer end face to support and lightly clamp the end face of the work-piece in order to prevent the possibility of it actually falling out,(this can happen if this precaution is not taken).

The four-jaw chuck's flat front face is also provided with a series of shallow grooved concentric circles,; these can also be used as an aid to setting up the work-piece initially to establish that it is running reasonably true and concentrically by eye.

Section 18/ 8
The following notes describe a safe method of securing irregular-shaped components in the center lathe's four-jaw chuck, also a method used to machine those difficult to-to-turn flame-cut flanges accurately while using the center lathe.

If the work-piece to be machined consists of a perfectly square solid material, this material can be accurately adjusted to run concentrically in the four-jaw chuck by using the assistance of an accurately positioned cutting tool's point, then noting the reading of the cross-slide's index scale when the tool's point is in contact consecutively with all four (of the vertically positioned) faces of the square workpiece. The chuck is rotated 90 degrees by hand (in turn) to allow each of the work-piece's four front face positions to be in a vertical plane when this is being performed.

The external tool's point is positioned (using the cross-slide's hand wheel and its scale's reading), to contact, in turn, each of the four (vertically aligned) faces of the work-piece, while noting the cross-slide's zero scale's reading, and making the required quarter-turns of the holding chuck to arrive at each of the tool's contact points. The cross-slide scale's reading is noted and an estimate is made as to which jaw needs to be tightened (or loosened) to obtain a true secure centralization. These operations can be progressed along by adjusting the appropriate jaws until the cross-slide's scale readings are seen to be virtually identical when contact has finally been made on all four faces of the work-piece. The identical readings obtained will then prove that the square work-piece is now truly centralized.

In practice, the chuck's individual jaws are adjusted to displace the work-piece laterally by a sufficient amount to establish that the cross-slide's index scale does finally indicate there is an identical reading at all four 'tool's point to work-face' contacts. When all tool contact readings are found to be identical, all four jaws are then fully tightened to finally secure the work-piece. Identical scale readings will confirm that the work-piece is now running concentrically and is in line with the center-line of the machine.

Machining Flame-Cut Flanges

If the machinist is machining a batch of (say) large 500mm diameter flanges that need to be turned, and the lathe's capacity is considered sufficient to cope with holding this particularly large diameter, the problem then arises regarding the machining of the material itself. Note that the finished machined flanges will eventually be welded onto the ends of large diameter heavy-duty pipes used for either external heating transportation or oil pipe-lines.

The plate material supplied by the flame cutters to the lathe will probably be initially cut from large steel flat plates by using a flame-cutting torch to produce the required number of steel blanks. These flame-cut blank flanges are cut out from the flat mild steel plate in the areas of their outside diameters and their inner bores. Consequently, the flanges will possess a very irregular flame-hardened crust in all the areas about to be machined. This hardened crust will prove difficult to machine because of its brittle hardness caused by the heat of the flame used to cut them out, coupled with the amount of carbon that is often absorbed by the material during the cutting operation. This results in a very hard and irregular surface being produced that will cause the lathe tool, while under a high cutting load, to suffer initially from a period of intermittent cutting, prior to its cutting edge reaching the crust-free softer core of the work-piece material. This turning job will therefore require the use of plenty of coolant, considerable patience, and the full concentration of the operator, before either its major diameter and its internal bore have been turned to its required diameter by the use of lathe tool number (say) number 1 (in fig. 47), and boring tool (say) number 9 (in fig. 47),; also note that it may require a more robust tools of similar design, to achieve the required dimensions stated on the drawing. The flange's finished internal bore will later be used for resetting the workpiece to run concentrically, by using a dial test indicator, during its *reversed* setup procedure, required to re-centralize the flange in the chuck.

The outside diameter of this rough flange will first need to be held securely by using the four-jaw chuck's *outside* jaws; their gripping surfaces (in this case) are being used to hold just *half of the thickness* of the plate (this must be done in order to allow the flange, after accurately turning its outside diameter over *half its thickness*, and its *bore turned to it's finished full diameter*, for it to be removed, turned around in the chuck, regripped on its *part*-finished outside diameter followed by its bore being re-clocked to be concentric (using a DTI), to enable its previously partmachined outside diameter to be accurately finish turned.

The flange must of course be initially centered relative to the chuck (as best one can), by using a sticky pin, to ensure that sufficient material is still existing in the blank's overall shape, for the tool to finish turn the flange to its full required diameter. The clock gauge's probe is then used to check the flange's possibly existing lateral wobble, by the use of a copper or hide-faced hammer carefully reducing this wobble to an acceptable level by using a few judiciously placed blows on the flange's outer face in the appropriate positions, followed by the chuck's jaws being then fully tightened. (Note that care should be taken to avoid jarring the clock-gauge's internal mechanism (by the action of hammering the work-piece; we do this by carefully easing the probe's point away from the flange's surface while the workpiece is being struck, followed by replacing it carefully in order to check the progress of the possible flange run-out.) The existing flame-cut outside major diameter is then turned by using a high-speed steel right-hand knife tool (number 1 in fig. 47), while using a very low work speed, fine feed, and plenty of coolant. This is a situation where a tool with a ground and honed 1.5-mm radius at its cutting edge tip may be found more beneficial, as this will permit the tool to withstand the extra shock of the intermittent cut that occurs during the roughing-out process, to machine away the hard part of the flange material. We can then revert to using a normal right-hand knife tool with its minimum honed honed radius on its point to finally turn its major diameter (consisting of half the thickness of the flange). Note that the tool when on its final turning phase must finish its cut in very close proximity (but not touching) the hardened chuck's jaws.

Because of the intermittent cut experienced when machining flame-cut, temporarily hardened materials, the choice of using a cemented carbide cutting tool for this particular job may be considered inadvisable because carbide-tipped tools often prove unsuccessful in coping with the intermittent cut without fracturing their glass hard cutting edges.

With the outside diameter of the work-piece now turned down to its finished size over the whole work-piece's thickness, this identical tool can then be used to face off the front face of the flange, (after locking the saddle to prevent any longitudinal movement of the saddle and tool).

To ensure the flange will possess an accurate overall thickness after its machining process, it is considered an advantage for the machinist to use a 90-degree right-angle tool that possesses cutting edges similar to a normal parting tool, as shown in fig. 47 number 16, that is now used as a *back*-facing tool. Its cutting edges are ground very similar to a normal parting tool but its whole cutting area is angled at 90 degrees to its shank; the use of this tool then enables a *back witness face* to be turned on the back face of the flange (in the area situated immediately outward from its now finished bore), before the flange is removed and prior to its turnaround to be reset in the chuck. This inner faced-off area should extend outward from the internal bore of the flange to a diameter of at least 20 mm larger than the bore's diameter. This machined face is then used (after the flange has been turned around) to accurately align the flange (to eliminate any wobble), allowing its face to run true for the second setup, by using a dial test indicator's needle's fluctuations to obtain the required true running surface,; this will then ensure true concentricity of the whole work-piece. Using this method of centering the workpiece avoids the problem of having to set up the flange to run true by the rather basic and less accurate method of using parallel packing strips placed temporarily between the chuck's front face and the back of the flange being machined, followed by using a copper hammer's blows to establish a flat wobble-free surface. (In this case these packing strips must of course be removed before switching on the lathe; (this is of course very important for obvious safety reasons.). The finished bore of the flange can then be picked up by the finger of a finger-type dial test indicator, to obtain a true running concentricity that will include the whole flange. It should also be understood that the finished flange will eventually be welded onto a heavy-duty pipe, with the pipe being fully entered into the flange's bore to ensure that it is concentric and is located accurately before the welding operation takes place.

To promote a satisfactory making of this weld between the flange and the pipe, the issued drawing to the machinist will probably indicate the need for a welding recess to be machined in (what is now), the flange's back face, by using a tool that has been ground specifically to produce a large heel clearance on tool number 16 featured in the drawing (fig. 47);. This tool's clearances must be carefully off-hand ground so that it possesses a vastly increased clearance radius in the area of its outside clearance heel,; this must be of sufficient area to clear the radius of the major diameter of the machined recess). This welding slot is machined into the flange's rear face to a depth of approximately 5 mm and situated at an identical wall thickness away from the bore's diameter that is designed to fit the outside diameter of the pipe. The outside diameter of this slot will therefore be machined so that it occupies an area situated at (two pipes' wall thicknesses) outboard of the outside diameter of the pipe, resulting in leaving a 'proud' circular ring with a recessed cavity that provides the area that will accommodate the weld of the pipe. The machining of this welding slot's recess will then allow the weld to penetrate throughout two identical thicknesses of material (relative to the pipes' wall thickness) so that it can obtain a perfect weld. By machining this slot's recess in the flange, it will be found to improve the quality of the flanges' welded joint.

Section 18/9
The 'proper way of using the three-jaw chuck without damaging its jaws

The three-jaw self-centering chuck is used mainly for securely holding bright round mild steel or hexagon shaped stock materials that are known to possess a reasonably accurate hexagon 'across-the-flat dimension, or an accurate perfectly round diameter. It is considered bad practice therefore to attempt to grip an inaccurate or black extruded 'unreliable diameter' steel materials' in the three-jaw chuck's precision jaws. These inaccurately sized materials possess inherently poor surface finish and their diameters are notoriously oval in cross-section. Should undue force be used during the jaw-tightening process onto these materials, it will cause unequal strain to be imposed upon the chuck's jaws, particularly when they are being forced to grip oval materials. This type of abuse can be sufficient to cause strain or chuck run-out problems later on when the same chuck is being used to secure perfectly round steel materials on components that have the requirement for accurate and concentric machining, or, alternatively, have the need for a very precise and perfectly round workpiece that needs to revolve with perfect concentricity.

Larger diameter components can of course be held securely and concentrically in the chuck by using the extra supplied set of outside jaws. These enable the chuck to grip larger diameters of work-piece by using the jaw's outer gripping periphery. However, when relying on these outside jaws' gripping areas, extreme care should be taken to avoid allowing excessive overhang or stick-out of the workpiece from the jaws. The machinist must ensure (for security reasons), that the workpiece's gripped inner back face is also firmly butted up against the inside front face of the jaws; this must be sufficient to provide a positive backup to ensure the gripped workpiece's security when under load. (A few judicious blows made by using a copper-faced hammer will ensure this important contact is made.). One should also bear in mind that the outside jaws have a relatively short grip length on the workpiece; therefore, one should be extremely careful to ensure that a perfectly safe grip has been achieved before beginning the machining operation.

When working on tubular or plastic materials, the machinist should take extreme care not to over-tighten the chuck's jaws onto these materials as it can cause distortion either to the plastic material or to the tube's structure, resulting in a poor and unsafe grip. A precision-turned steel plug should first be turned up on the lathe to a diameter that exactly fits into the bore of the gripped end of the tube. Taking this precautionary step then ensures that the jaws can be tightened fully onto the workpiece to obtain a much firmer, safer grip. (When machining plastic materials, the use of soft jaws, that have been machined to suit the exact diameter of the material being machined, is to be recommended.). The use of a precautionary plug in the work-piece's bore, will prevent, to a certain extent, the distortion of the tube, and will also prevent the possibility of an accident occurring should the work-piece be forced out of the chuck by the cutting action of the tool. This situation can occur if an over zealous machinist attempts to take too heavy a cut on a poorly supported workpiece. It is also prudent to bear in mind, when in this situation, that the use of a tool with a larger-than-necessary cutting radius should be avoided because its cutting area's larger surface transfers a much greater load onto the work-piece. A large cutting radius ground onto a tool will massively increase the loading suffered by the workpiece, and this becomes particularly important if one is attempting to machine a poorly or marginally supported workpiece that possesses an excessively long stick-out from the chuck without the support of a dead center. This practice should therefore be avoided if possible at all times. The surface area of a cutting tool that possesses a large cutting radius will impart extremely high loads onto the workpiece therefore in this case because of its extremely large area of contact, the work-piece should always be supported by either a fixed steady (as shown in fig. 48) or a dead center held in the tail-stock,; it is also advisable to use a very sharp right-hand knife tool', honed to possess a very small tip radius for this particular job. This minute radius will impart a much lower loading onto the work-piece, and will prevent these very high loads being transferred onto a poorly supported work-piece, this type of work. This problem can be alleviated to a large extent by always using a very sharp right-hand knife tool, the cutting point of which must be very carefully honed to leave a very small radius of 0.1 mm at its tip, coupled with the machinist taking shallow light cuts to complete the machining of the work-piece.

For safety reasons, any serious overhang of the workpiece from the chuck should always be avoided if at all possible, and in all cases, this overhang should be reduced to the absolute minimum in order to maintain the maximum rigidity of the workpiece and the safety of the setup. This cautionary note will apply particularly if a fixed steady or tailstock center support is not, or cannot be used to support the workpiece in this overextended possibly (insecure) workpiece holding situation. This situation is often given the name (in the engineering trade) of being on a wing and a prayer, and this situation should of course be avoided if at all possible.

If the machinist is faced with the problem of workpiece chatter that occurs when machining a particularly large male radius on the outside diameter of a workpiece, and the tool is seen and heard to be leaving visible 'chatter' marks chatter marks on the workpiece's surface, it is possible to improve the existing poor surface finish by using the following technique.

With the lathe now switched off at its isolator switch, and the lathe's spindle placed into neutral gear, the holding chuck is then turned in its normal direction of rotation by using both hands grasping the T-bar of an inserted chuck key. During this process, the radius of the tool must be very carefully indexed inward by using the cross-slide's hand wheel by a sufficient amount to allow the tool to just touch the workpiece's radius, and by just allowing the tool to remove a minimum of material. After several revolutions of the work-piece, and by indexing the tool very carefully inward, it will be found that the tool will gradually remove the chatter marks. This operation is of course very hard work, but it is possible, with patience and perseverance, to remove the chatter marks from the workpiece by using solely this hand-powered method of turning,; it also has the advantage that it will not introduce any new chatter marks onto the work-piece's surface.

Section 18/10
Instructions advising the proper use of a fixed steady, the traveling steady, the pipe center, and the greased dead center

If a long piece of round material is to be machined (whether it is of solid or tubular section), it is recommended that the machinist uses a fixed steady for workpiece support. It is to be located in its most suitable position along the work-piece's extended length (see fig. 48). This steady must be positioned and adjusted so that it is in close contact with the workpiece's diameter while also allowing the work-piece to be held in close proximity of the tool's cutting edge.

If a greased dead center's support is being used at the tail-stock end of a long solid piece of material being machined, then the steady can be positioned centrally along its length to provide additional support in order to prevent the tool being pushed off when approaching this central area, particularly when turning a longer work-piece's outside diameter. Alternatively, a traveling steady affixed to the tool post can be used to support the whole length of the work-piece while it is being turned. The traveling steady can only be used effectively after a suitable length of concentrically turned diameter has been established at the tailstock's end of the work-piece by using the *greased dead center*'s support. This is then followed by adjusting the travelling steady's support rollers so that they accurately pick up (in a supporting role) the short newly turned diameter. The fixed steady can then be removed to allow the work-piece to be precision turned along virtually its whole length and almost up to the chuck.

In the case of using tubular material, the pipe center is used for end support combined with the use of the fixed steady. The pipe center is of much larger diameter than a normal center and normally uses its own internal bearings to support its larger-diameter revolving tapered center. This larger diameter is specifically designed to support the internal diameter of large-diameter pipework. With this tool installed in the tail-stock, it will provide accurate, safe, and concentric, tubular workpiece support. It will usually be found necessary to install a fixed steady temporarily, in order to support the work-piece to run truly to allow the bore of the tube to receive an internal chamfer at its end, sufficient to enable the pipe to run truly on the pipe center by the use of its now-tapered inner end surface. This chamfer is to be machined immediately before the pipe center is finally installed into the bore in order to provide sufficiently accurate end support. To prevent the tendency of long unsupported and slender work-pieces to whip or bend when under pressure from the tool, a traveling steady (as previously stated) should be employed. For example, if one is machining black bar (which is not at all accurate dimensionally) but has been provisionally center drilled and is being held between centers), it will be found best practice to always use coolant and to turn a short length of the bar at the tailstock end to allow the travelling steady and its rollers to be adjusted to suit this newly turned diameter; the rollers will then support the work on this newly machined support area so that the previously turned diameter can be further extended along to almost the full length of the work-piece (after of course removing the fixed steady).

Section 18/10 a
Drilling and reaming a component on the lathe, by using the tail stock's drill chuck or, alternatively, when using the tapered bore of the tailstock's quill to hold taper shank drills and reamers

With the work-piece gripped in the chuck or mounted on the faceplate, it is important that the drill's point is started truly in the hole to enable its following drill's diameter to run concentrically with the center-line of the lathe. For this centralization process to be accurate, we must first use a center drill; this enables the following main drill to pick up its centralized hole. However, to obtain a perfectly true centralization of a drilled hole relative to a precise outside diameter that has already been turned on the work-piece, it is sometimes necessary to enlarge or true up an originally drilled hole by using a single-point boring tool (see fig. 47 number 9, for an illustrated example of this boring tool). This machining method is often used prior to using a reamer to produce a bore of greater accuracy and to establish its true concentricity relative to the work-piece's diameter. It is important to note that the extended center drill's point is very fragile and if dulled by excessive previous use it can be easily broken off if overloaded, or if insufficient work speed or coolant has been used in the process. (The work speed used should be at a minimum of 800 rpm). If insufficient coolant has been used, or the drill has become blocked with swarf, or an insensitive, or too robust infeed has been used by the operator, it will cause the protruding small diameter of the center drill to actually break off in the hole. As a result of this, the component can often be impossible to re-claim, and in fact, the machinist may have to start the whole of the work-piece's manufacture again, from scratch, so students should be warned of this possible calamity occurring.

The reamer drill being used should be of a diameter that will enable its following reamer to remove a minimum amount of material from the hole. This is generally one sixty-fourth of an inch (0.015625) inches, (0.3968 mm), of material, left in the existing bore for the reamer to accurately ream out of the hole. This operation should be performed while using a relatively low spindle speed of, say 120 rpm and by using plenty of lubricant. This instruction will apply particularly when reaming steel components. However, when reaming cast iron, no lubricant is needed or desirable.

Section 18/11
Using a separate set of soft jaws in the three-jaw chuck

If, however, the machinist is working on a large batch of production turned components that all require the same precision grip on their identical diameters, it will be found convenient, (and speedier), if one uses (if available) special set of soft jaws for this task. This *alternative* set of chuck jaws has the ability to be precision bored out by the machinist in situ (that is, while still installed in the chuck of the lathe). This machined diameter in the soft jaws will now provide an identical gripping diameter that suits the whole batch of work-pieces. This boring operation is always carried out while the jaws are firmly *locked* into position in the chuck, with the minimum amount of material left in the jaws to be bored out.

This special set of jaws is manufactured from soft, *unhardened* steel (as opposed to the normally used hardened type); this softness allows the jaws to be bored out to the diameter of the workpiece while using a normal high-speed steel boring tool.

When machined to the required and correct diameter, the jaws will provide a perfectly curved and accurate gripping surface that is used for repeatedly gripping each workpiece's diameter in the batch, permitting it to run perfectly concentrically in the chuck, thereby allowing a large batch of workpieces to be machined to run concentrically for further work to be carried out on them.

These jaws provide a very accurate method of holding a whole batch of production-turned components that need consecutive machining,; their use therefore enables a second operation to be carried out on a batch of workpieces, with extreme accuracy.

By adopting this particular setup, it will be found very useful should the machinist be called upon to machine a large batch of workpieces that require a second operation to be performed on them,; this is particularly useful if there is the added requirement that all the workpieces' diameter's must run perfectly concentrically with their securely held diameters.

To set up the chuck accurately for machining the gripping surfaces of the soft jaws, we must first establish the exact diameter of the workpiece to be gripped. One particular design of soft jaw set incorporates a machine-bored register on the inner back face of each jaw to allow a large flat and machined mild steel washer to be gripped in this bored register in order to lock the jaws securely to allow the boring process to be carried out.

It is essential that the jaw material is not machined away too deeply in its gripping surface region in order to obtain the required diameter. It is most important therefore to adhere to this precautionary advice in order to prolong the usable life of the soft jaw set.

Soft jaws can also be used to hold larger-diameter workpieces accurately and securely by using the alternative set of supplied *outside* soft jaws. These jaws can also be machined to provide the necessary gripping surfaces.

To machine the outside set of jaws, they will need to be bored out to a diameter that exactly fits the workpiece's diameter that is to be gripped. This operation is preceded by securing a piece of spacing material, this being usually a piece of bright round mild steel (the diameter used must have previously been accurately calculated to provide only a small amount of material needing to be machined out of the jaws to obtain the required gripping diameter). This spacer is secured in the chuck's smaller jaw opening and tightened, to lock the jaws effectively. This spacing material's diameter is most important. It must be calculated carefully so that its gripped diameter will permit only the bare minimum of material to be bored out of the gripping jaws' surfaces in order to produce the exact diameter required to grip the work-piece. This spacing material must of course be taken out after machining the gripping bore to its correct diameter. As previously stated, this minimal removal of jaw material must be carried out very carefully in order to prolong the continued useful working life of the soft jaws.

Section 18/12
Familiarizing the student with the task of removing, and replacing, the jaws of the three-jaw chuck

It is very important that the student learns the simple task of removing and replacing the three-jaw chuck's jaws. This operation must be specifically learned to allow the student to make the change of using either the lathes outside or its inside set of jaws. At first sight, this may seem to be a very elementary task, but many students find they have problems when trying to perform this operation successfully, particularly when they have not been given sufficient instruction on the task. They quite often get the jaws completely out of their correct sequence, resulting in the bore of the jaws running completely out of concentricity with the center-line of the chuck and of the lathe.

To perform this task correctly, the lathe must first be isolated electrically by switching off the isolating power switch (usually situated on the machine shop's wall).

Select a neutral gear, and rotate the chuck by hand until number one jaw (or its jaw's slot, if replacing the jaws), is into a position that is uppermost;. (Note that the chuck's jaw's receiving slots, and the chuck's individual jaws, are always numbered in the sequence 1, 2, and 3, to facilitate their correct replacement,; the student should also note that each lathe chuck is unique and is manufactured with its own dedicated set of jaws. Its jaws will have an identical engraved identifying number to those marked on the chuck. This is done in order to prevent them becoming mixed up with the jaws of other chucks, which of course cannot be used as they will be incompatible if used in a different chuck. If wrongly fitted, they will prove to be extremely inaccurate because their gripping surfaces are machined to suit a completely different chuck.).

If the chuck's jaws are already in place, then the chuck key should be entered into a convenient key access socket, and the key rotated anti-clockwise several turns, causing the jaws to move outward fully to their outermost position, whereupon they will finally become disengaged from the chuck's retaining rotating scroll thread, enabling them to be removed. Just before this point is reached, care should be taken to ensure that the chuck's two bottom jaws do not accidentally drop out and cause damage to either the jaws or to the V ways of the bed of the machine. A slat of plywood, placed across the lathe's bed and its V ways will prevent this damage occurring, that is, (if the two lower jaws have not already been caught in the hand).

If we now look deeper into the chuck (from its front face), we will see (just in front of its internal back face), the spiral rotating scroll thread, onto which each of the jaw's matching threads are independently latched when the jaw's gripping action is being carried out using the chuck key.

The chuck key is now turned clock-wise to drive the spiral scroll thread in an anticlockwise direction until the situation is reached where its leading spiral-threaded nose is observed to occupy a position that is just approaching (but-not entering), number one jaw's entry slot. Number one jaw is then offered into this slot, and light downward pressure is applied to the jaw to ensure that its base surface is registering onto the outside face of the spiral curve of the scroll thread. The chuck key is then turned clockwise by a sufficient amount to allow the rotating scroll's point to enter into, and be seen to pass through this number one jaw, (but not to extend very far beyond it).

The key is then removed, and the chuck is rotated by hand in order to locate number two jaw's slot uppermost. The chuck key is entered into its nearest socket key-hole and is again turned clockwise, rotating the scroll's point to approach (but not enter) number two jaw's entry slot. Number two jaw is then offered into this slot, and light downward pressure is maintained on the jaw to ensure that it is held in continued contact with the scroll thread. The key is then turned clock-wise by a sufficient amount to allow the leading point of the scroll thread to enter, and be seen to pass through this number two jaw, (but not to extend very far beyond it). The key is removed and the chuck is turned by hand to position number three jaw's slot into its uppermost position, the chuck key is entered into its nearest available socket key-hole and turned clock-wise until the point of the scroll thread is seen to approach (but not enter) number three slot. Number three jaw is then offered into this slot and light downward pressure is maintained to establish continuous contact with the outside diameter of the scroll thread,; the key is then turned clockwise until the point of the scroll thread is seen to enter and pass through number three jaw.

The chuck key can now be turned clock-wise by a sufficient amount to cause all three jaws to be seen to move inward in unison (to a point where the outermost face of all three of the jaws are seen to match exactly with the chuck's smooth outside diameter). This correct matching of the jaws' outer surfaces to that of the chuck's outside diameter will indicate that they have been entered correctly and will confirm that the chuck is now ready for use.

If, however, the jaws are still found to be incorrectly matched and one or two of the jaws are not actually lining up with the chuck's outside diameter, then the whole process will need to be repeated again from the starting point, until a satisfactory matching has been achieved. A final visual check of the chuck's accuracy, is to grip a piece of bright round mild steel in the chuck's jaws, tighten them, and note whether it is running concentrically when the chuck is being slowly turned by hand.

A very wise procedure to adopt when preparing a three-jaw chuck's jaw opening to accept a piece of material being offered in for turning, is to always start off the procedure with the jaws completely closed. The material should then be offered to the chuck's opening very carefully so that the material is only entered **'as the jaws are being carefully opened to suit the diameter of material being offered'**. *Note that the jaws should only be opened by a* **sufficient amount** to allow the material to slide into the jaw opening before being finally secured by using the chuck key.

This may seem to be a rather simplistic exercise, but there is a good reason for adopting this practice.

The author has been present on many occasions when a student has offered a piece of material into a lathe's three-jaw chuck without giving the matter sufficient thought, and not realizing that the chuck's jaws have been allowed to have too wide an opening, the student has then offered the material into the chuck, tightened up the jaws, and caused the material to be inadvertently gripped (or jammed) off center, in between two of its jaws, resulting in the material being held dangerously and completely out of center relative to the center line of the machine.

Without the student realizing that the material is now being held firmly in an off-center and dangerous position, he has switched on the lathe, and allowed the off-centered work-piece to rotate, strike, and virtually destroy a previously serviceable lathe tool before his presence of mind has allowed him to stop the machine.

By allowing this accident to happen, he will certainly have learned the hard way, and this occurrence will certainly have dented his confidence somewhat in his aim to fully acquire the art of center lathe turning.

Section 18/13
Safety checks that are required to be undertaken on the center lathe and its equipment, by the operator before using it to perform a precision machining operation.

Important safety instructions that warn the machinist of the *dangerous practice* of leaving the chuck key in the lathe's chuck. This occurrence can happen when the machinist's attention is diverted elsewhere at this crucial time.

It is always advisable to make the following check of the lathe and its equipment, to confirm that it possesses no mechanical defects or faults, prior to it being used for precision straight turning, parting off, or screw-cutting operations.

Any faults found, if not remedied, can seriously affect the accuracy of the work produced by the machine.

For safety reasons, one should note that the chuck key (used for tightening and releasing the workpiece from the lathe's chuck) at all times be considered a *potentially dangerous piece of equipment*; it should always be removed from the chuck immediately after it has been used. The chuck key should *on no account* be left in place in the chuck while leaving the chuck unattended for any reason whatsoever, (this safety advice is often forgotten or ignored by the student when the chuck key is being used to adjust the stick-out dimension of the material that needs to protrude from the chuck during the lathe's setting-up phase). All personnel in the machine shop should be encouraged to follow this advice, this being the best and safest practice to adopt. Ignoring this rule can be the cause of a serious accident or serious injury to occur in the work-shop, particularly with regard to the machinist, or to a passer-by who may be unlucky enough to be struck by an ejected chuck key immediately following the machine being switched on. Considerable damage can also be caused to the actual key itself by its shank being actually snapped off at its weakest point, usually at its junction with the squared-off end of its round shank, while suffering the impact of striking the machine's bed during its initial rotation. Damage is also likely to be caused to the precision bed and V ways of the lathe should they be struck by an ejected chuck key. This accident is quite likely to happen if the lathe has been inadvertently switched on *with the key still in the chuck*; therefore, this safety rule must be strictly adhered to for the protection and safety of all those working in or passing through the machine shop.

There follows the author's recommended setting-up procedures. These must be carried out before using the center lathe for any precision turning, boring, screw-cutting, or parting-off operations.

These setting-up procedures will also be found necessary prior to screw-cutting the fig. 42, illustrated example of a male, (external), 60-degree metric coarse screw-cut thread, and these procedures will also be applicable when the machinist is engaged in just a straight turning operation.

A basically similar setting-up procedure can be used when setting up a screw-cutting tool on the center lathe, for either a 60-degree included angle external American thread, or for a 60-degree included angle external metric thread, except that one screw-cut thread's design will be using imperial dimensions of pitch and diameter, while the other will be using metric dimensions of pitch and diameter.

The text also mentions briefly, how one can (in times of emergency, and particularly if a 60-degree included angle screw-cutting tool is not available), to put to use a 55-degree included angle (Whitworth) external screw-cutting tool (ground specifically for cutting English imperial threads), to cut either the previously mentioned metric 60-degree external thread, or an American 60-degree external thread.

In this case, it is most important that the machinist uses the angle approach method of screw-cutting (in place of using the normal 'straight plunge' method of screw-cutting). The difference in the tools, cutting angles will (in this case) be taken up by the deliberate angling (by the machinist), of the lathe's compound slide to the required 60 degrees. This slide's angular setting then allows the difference between the 55-degree cutting tool's angle and the 60-degree cutting tool's angle to be adjusted using a screw-cutting gauge to align *the left-hand side* of the tool's 55-degree cutting edge so that it now aligns with the left-hand side of the 60-degree setting gauge's cutout V, and ignoring the minimal angular gap that will now exist between the cutting tool's right-hand side cutting edge and the right-hand side of the V cutout in the gauge. The tool will now cut a 60-degree included angle thread.

Section 18/14
Useful tips on center lathe turning, and the repair work to the center lathe that is often undertaken by the machinist, including the important safety precautions that must always be complied with.

The following, preparatory work is often undertaken by the individual machinist, before the lathe is considered to be in a fit condition to cut a precision thread on or in the work-piece. The following notes apply particularly when one is intending to screw-cut the fig. 42 example drawing of a male (external), right-hand, 24 mm by 3 mm pitch, metric coarse screw-cut thread.

The reader is also advised to comply with all the following advice regarding the safety precautions the machinist should take before using the center lathe.

Firstly, one should always wear safety glasses for eye protection against grinding dust, ejected liquids, and metal particles. Secondly, he should refrain from wearing a necktie unless it can be tucked away safely into the shirt in order to prevent it becoming entangled with the machine or the workpiece. If he or she has long hair, it should be tied back for safety reasons, or use a hair net. If the machinist has decided to use the favored 1-inch wide emery cloth strip to clean up a revolving workpiece, then for safety reasons, it should first be cut into very short (say) 4 inch (100 mm or less) lengths that are just sufficient to prevent the fingers from becoming entangled between the strip and the work-piece by them being drawn into the revolving component.

It is not generally realized by students, that when metal is being turned on the center lathe, the metal chips (or swarf) produced by the action of the tool cutting the metal work-piece will often change its color because of its rapid change in temperature. This is caused by the heat generated during the turning process. Should the color of the metal chips be seen to turn to a bright blue, (when cutting steel), this will usually be caused by a lack of coolant being applied to the tool and to the work-piece, or to too high a spindle speed being used. The swarf (or metal chips) produced in this case will have reached a temperature of at least 300 degrees centigrade, making them capable of causing serious burns to the operator should they be ejected and allowed to contact the skin. This 'very hot swarf' problem occurs more frequently when the machinist is using cemented carbide-tipped tools, as these are often used dry and at a considerably higher spindle and work speed than would normally be used if one were turning steel using high-speed steel tools that of course require lower work speed and a plentiful and continuous supply of coolant.

High-speed steel lathe tools should always be used at a much lower work speed than cemented carbide-tipped tools, in order to prolong their useful working life. High-speed steel lathe tools do have the disadvantage that if used dry and at too high a work-speed, they are prone to burn out through either the lack of coolant being used, or too high a work speed or feed being used. This rough usage of the tool will cause more frequent re-sharpening being necessary, and can, in exceptional cases, lead to a complete burnout and the ultimate scrapping of the tool.

Fortunately, the high-speed steel tools do have the advantage that they can be re-ground to the required sharpness by using the operators' off-hand grinding techniques by the use of a pedestal-type tool grinder,; this machine is capable of maintaining their sharpness and overall shape by the use of its conventional grinding wheels. The high-speed steel tool's cutting edges are also less prone to chipping (or contact damage caused through accidental contact with hardened steel, (a problem often found prevalent when using the very brittle cemented carbide-tipped tools).

When carbide-tipped tools are being used dry at high spindle and work speeds, they generate extremely high working temperatures in the work-piece, and produce potentially dangerous curled lengths of very hot swarf; this situation must be avoided by the operator for safety reasons. It is also a known fact that due to the excessive heat being generated in the work-piece while using tipped tools for the machining operation, it can cause expansion damage to the accuracy of the three or four-jaw chuck's parallel gripping surfaces. In the worst case, this can result in the chuck's jaws becoming permanently over-strained (or in engineering speak, bell mouthed), making them incapable of safely or accurately retaining future work-pieces over their *whole* gripped length. This generation of very high working temperatures in the work-piece during the machining process, should always be avoided if at all possible, in order to prevent the possibility of this type of damage being caused to the precision gripping action of the chuck's jaws.

It is also considered most desirable for the machinist to go through the following lathe checking procedures, particularly if the machinist is not fully acquainted with the machine in question.

This should be done in order to ascertain, firstly, that the machine you are about use is in good condition, and most importantly, that it has been left, (or handed over to you by the previous user) in a safe and perfectly usable condition.

In practice, it is not unknown for important items such as retaining bolts, chuck retaining camlock socket screws, and other important lathe equipment to be found insufficiently fastened, and protective guards have been moved out of their correct positions. One should be very careful therefore before taking over an unfamiliar lathe from a previous machinist, to make sure that everything is satisfactory, by going through a thorough check of any obvious items that could require certain adjustments being made, before the lathe can be considered safe for use by you.

Section 18/15
Checking the accuracy of the lathe's parallelism relative to its tail-stock, also checking the cleanliness of the chuck and its vital 'spindle nose' precision mating surfaces, should the chuck be found to be not running true

After giving this particular lathe a full inspection, it may be discovered that its previously installed three-jaw chuck is found to be running in anout-of-true manner, and that the lathe's tailstock is also found to be set over into an off-central position relative to the center -line of the machine. If the lathe is used in this condition for turning a long, tailstock-supported work-piece without further remedial adjustments being made to its tailstock or to its chuck's fixation, then the lathe can only produce tapered, and inaccurate out-of-balance work.

This 'running out of true' situation must be rectified immediately, by using the following checking and remedial procedures.

When the three-jaw chuck has already been installed on the lathe, there is no way of knowing how meticulously clean its mating joint faces were when it was originally installed by the previous user. Any small particles of dirt or swarf retained between the joint faces of its locked spindle's nose or its chuck's mating datum faces, (be it of the camlock fixation) (or the 'flange and thread' fixation), will cause the whole chuck assembly and its mounted work-piece to run in an out-of-true and inaccurate manner.

The amount of the chuck's concentric inaccuracy can be verified by checking the visual run-out of its outside diameter (which should be running true if everything is correct), and by checking the concentricity of its jaws while they are gripping a perfectly round piece of test rod material.

For these tests, a (clock gauge / dial test indicator), is used, held by either a magnetic base for convenience of positioning, or with its supporting linkage secured in the tool post. The lathe is placed into neutral gear, and the chuck rotated slowly by hand while the probe of its clock is maneuvered so that it contacts the outside diameter of the chuck, (by choosing a circular area of its outside surface that is free from holes or slots), for the first test of checking the chuck's concentricity while noting the amount of fluctuation indicated by the gauge's needle. (This diameter of the chuck should be found to be running true if it has been fitted correctly.) The clock is then placed with its probe touching the round diameter of the clamped test piece's diameter. We now note the amount of fluctuation of the 'clock's needle' when its probe is touching this diameter while the chuck is being rotated by hand. If the dial's needle is seen to fluctuate wildly to and fro, this will indicate that the test piece is running out of true.

However, if the dial's needle was seen to fluctuate when previously contacting the rotating chuck's diameter, this indicates that further investigation is required. We should always at this stage, place a flat piece of wood underneath the chuck to protect the lathe's bed from abrasion damage. We then remove the chuck.

If the chuck's fixation is of the camlock design, then with the chuck removed, it should be scrupulously cleaned at its fixation joint mating faces, and any 'bruising' found in its seating and mating areas carefully eased away by use of a smooth sharpening oil-stone; we do this to remove any burrs caused by the squeezing effect of the crushing contact of any swarf particles. The chuck should then be replaced and re-checked, by using a similar test procedure.

If, however, the chuck's fixation is of the 'flange and thread' type', (this chuck's design being of the screw-on type), then its flange joint faces should be scrupulously cleaned of any contamination, any obvious burrs removed, and its securing threads thoroughly cleaned out. This internal thread cleaning operation can be accomplished quite effectively by using an old toothbrush. The chuck should then be replaced and the test piece re-installed, where both the chuck and the test piece should then be checked for concentricity by using the dial test indicator method.

If dirt contamination or bruising of the joint faces was the original cause of the chuck's 'out of balance' or 'running out' problem, then completing this exercise with care should cure the problem.

If both the chuck and its work-piece's diameter are now found to be running true after carrying out this check, then the chuck should now be capable of 'turning an accurate and parallel work-piece, providing of course the work-piece is being held solely in the chuck and is not using its tail-stock for support.

Unfortunately, this particular test will only prove that the chuck is capable of producing accurate work if the material being turned is held solely in the chuck, and is not using the tail-stock's support.

The previous confirmed check of the lathe's parallel turning capability will not apply if the work-piece's outer end is now being supported by the greased tail-stock center if it has been positioned inaccurately on its supporting V ways, and is not truly in line with the center-line of the lathe.

The tail-stock's center-line must therefore be repositioned transversely to be exactly in line with the lathe's center-line before true parallelism of a work-piece (when using the tail-stock's support) can be achieved.

To reiterate, when the tail-stock's center is being used to support the outer end of a work-piece, the accurate alignment of the tail-stock relative to the machine's center-line is critical, and must always be checked to ensure that it is perfectly in line with the chuck's center-line before accurate parallel turning can be achieved.

If we now return to the previous parallelism test given to the basic chuck, and find that after cleaning out any contamination, the replaced chuck is now found to be running true, but the test piece it is holding is still found to be running out of true, then this can indicate that the actual chuck's jaws may have become either damaged, worn, or overstrained during its previous use, (this misalignment could have been caused by excessive over-tightening of a work-piece that was previously located at the extreme front of the chuck's jaws (a particularly bad practice, as this can seriously strain the jaws). If this has occurred, the jaws will be less capable of positively gripping the work-piece over their entire gripped length (this being a crucial factor, and it is therefore very important that this potential problem is checked out and corrected by using the instructions contained within the following paragraphs).

This or any other suspect treatment previously inflicted on the chuck's jaws can introduce a condition known as 'jaw bell mouthing' (i.e. making the jaws' gripping surfaces adopt a basically similar but exaggerated shape to the inside bore of a bell). To rectify this problem in a professional way, the jaw's internal gripping surfaces will need to be internally ground (in situation on the lathe),; this operation then enables their internal gripping surfaces to be returned to their normal accurate parallel and concentric gripping condition. This operation is normally performed by using a small internal precision grinding machine mounted in the tool-post, and by using a reciprocating motion utilizing the saddle's reciprocating longitudinal movement, or alternatively by using the parallel movement of the lathe's compound slide to internally grind out the existing bell mouthing of the jaws. This remedial work is usually carried out on the lathe's chuck jaw assembly by the works maintenance department.

However, if this suspected bell mouthing of the chuck's jaws is considered only minimal (i.e. less severe), then a rectification job can be carried out on the lathe by the individual machinist, by using a precision lapping technique. This operation can only be carried out if the lathe's tail-stock has previously been set to be perfectly parallel to the center-line of the machine.

Section 18/16
Rectification, of the condition known as 'chuck jaw bell mouthing' relating to, (in this case) to the rectification of the three-jaw chuck

This rectification job is performed by using a process of lapping in the internal gripping surfaces of the jaws of the chuck with a fine abrasive compound, combined with using a 'lap that is held 'in situation' (while the jaws are still being retained in the chuck). We do this by using a parallel and perfectly round piece of turned brass or similar material, (alternatively, a precision round steel shaft can be used),; this should be about half an inch (12.7 mm) in diameter and approximately 6 inches (150 mm) long. One of its ends is held firmly in the *now perfectly aligned* tailstock's drill chuck. The exposed surfaces (at the lapping end) of the lap are coated with fine valve grinding paste (as used when grinding in car engine valves). This coated end is then entered into and through the lightly nipped jaws of the chuck; the lap must be seen to extend to at least 10 mm deeper than the jaw's full depth.

With the lathe now switched on at a relatively slow, say, 135 rpm spindle speed, and assisted by using an in/out reciprocating motion of this nipped lap by using the tailstock's hand-wheel's rotation in a to-and-fro motion, performed while the lap is in rubbing contact with the inside surfaces of the chuck's jaws. The lap will be seen to hone away the deeper regions of the lightly nipped jaws gradually. By using this action, the lap will be seen to be lapping away the excess material from the innermost surfaces and particularly from the deepest regions of the chuck's jaws. The jaws will of course need to be judiciously re-nipped occasionally (by use of the chuck key when stationary), to ensure they are making continuous contact with the lap. We do this by first stopping the lathe and carefully tightening the lathe's chuck using its chuck key to enable the jaws to just lightly nip the lap. Occasionally, (and with the machine again switched off), several very small light tightening adjustments to the jaws will be found necessary during the lapping process, and the grinding paste will also need to be re-distributed and replenished as necessary. (Note that it is essential that the tailstock has been previously centralized on the lathe's center-line prior to this operation being carried out), ensuring that the lap is actually lining up perfectly with the center line of the machine.) This remedial lapping operation being carried out on the chuck's jaws will eventually produce true jaw internal parallelism. Note that this lapping operation can only be considered complete when the inner gripping surfaces of all three of the chuck's jaws are seen to display a matt-gray surface finish throughout their entire grip length,; this accuracy should be checked later by using a perfectly round and parallel test rod securely held in the tailstock's chuck, lightly smeared with micrometer blue, entered into the full length of the jaws and noting that a full area of blue is being removed from the rod when the lathe's chuck is rotated by hand. When the honing operation has been completed successfully, the hand rotations of the lathe's chuck should show a consistent removal of blue from the test rod's diameter along the full length of the jaws contact area.

All trace of grinding paste must of course be removed from the chuck's jaws or any part of the lathe after the lapping process has been completed. Note that this abrasive material is very detrimental to the wearing surfaces of the lathe should it be allowed to remain.

The majority of machine shop lathes will possess (stored in their tool cupboard), a very useful piece of tooling called a running center. This is used as an alternative method of supporting the tail-stock end of the workpiece for machining. This 60-degree included angle support center, actually revolves with the workpiece and relies totally for its lubrication on its own sealed bearings. This item of tooling is used as a tailstock work support. Its use is similar in operation to the normally used greased dead center and is similarly fitted into the tailstock of the lathe while in use. However, this particular supporting device is at times found to be rather unpopular among some machinists when it is being used to support small-diameter workpieces, because its rather large and bulky bearing housing, tends to get in the way of the lathe's cutting tooling, particularly when it is desirable to have the tool working in very close proximity to the tailstock end of the work-piece, where only a minimum of tool clearance is available. As a result of this problem, this particular item of tooling is often replaced (fairly quickly in some cases by machinists) in favor of using a normal 'greased dead center for work-piece support, which possesses a much smaller diameter.

Section 18/17
Turning very small diameters on the center lathe, which involves them having either supported or unsupported workpiece ends

A particular cutting tool clearance problem can occur when the work-piece being machined is of very small diameter at its tail-stock supported end; this situation can occur when even the normal greased dead center's 60-degree tapered support is found to be much too large and bulky for the cutting tool's clearance requirements.

In this case, half center can be used in its place. This is actually a modified normal center that has been carefully modified by the use of either precision or off-hand grinding so that it possesses a relieved clearance area (on the center's operator's side) of the center's 60-degree included angle tapered support. This area is ground away to provide tool clearance to within a nominal 2 mm of the center's center-line, at the nearside area of its supporting point. This clearance modification then allows the lathe tool's cutting point and its leading cutting edge, to reach, turn, and face off very small diameters of material, without the tool's cutting edge or point impinging upon the dead hard support center, causing possible damage to the tool's cutting edges. This relieved half center then provides the vital end of work-piece support for machining small diameters.

While describing the problem of machining small diameters on the lathe, the machinist, during his normal daily work, may occasionally be supplied with a difficult-to-machine piece of material,; this problem arises largely because insufficient thought is given (on the part of the machine-shop planners), who have supplied very small diameter round stock materials with which the machinist is expected to reduce down to even smaller diameters, without in some cases the component being allowed to possess a drilled center support hole, that normally makes it compatible with a supporting greased dead center. In essence, this means that the required small diameter to be machined on the work-piece will have to be turned, of necessity, *without* the use of a supporting center.

In the following case, the author quotes a difficult turning example where one is asked to use pieces of supplied 6-mm diameter mild steel rod, with a requirement to produce a 3-mm spigot diameter that also needs to be turned down to possess a 5-mm diameter shank, with the material (in this case) being held in a three-jaw chuck, followed by each item being parted off (*these operations to be performed without a tailstock center's support*).

Precision work produced while holding this very small diameter of material will normally be found extremely difficult to achieve because the weak material is pushed off by the tool (where the unsupported piece of material is being pushed away from the cutting tool, because of insufficient end support being given). In this case, the diameter of material supplied to be machined will be found to have insufficient diametrical strength to prevent this push-off problem occurring. It will be discovered that the machining of very small-diameter materials, coupled with one's inability to use adequate end support, will prove to be extremely difficult to achieve because the material itself is incapable of providing sufficient support without using the assistance of what is now the *disallowed* center's support.

The author has found the only way to overcome this small diameter turning difficulty is to replace the supplied small diameter material with much larger diameter of exactly similar grade material of say, 10 mm (or even12 mm diameter),; this larger diameter, while being held in the chuck, will have the ability to maintain sufficient rigidity in the set-up to allow the turning of the required small diameters while still using the holding rigidity of the three jaw chuck.

We must in this case use a high spindle speed of at least 800 rpm. This turning operation must also take place very close to the jaws of the chuck, and must employ the use of a very sharp right-hand knife tool, (see fig. 47 tool (1) or fig. 47a tool B, with its cutting point ground and honed to the very small cutting radius of 0.05 mm at its cutting tip. The machinist will also find it is advisable in this case, to take the full depth of cut required to produce the spigot's smaller diameter of 3 mm in just one cut which actually removes the full and required 3.5 mm, by the process of using coolant, and a very fine feed in order to produce the 3 mm unsupported spigot required. This is followed by taking the next full depth of cut of 2.5 mm (to obtain the 5 mm diameter required) while still using coolant and a very fine feed. This setup then allows the machinist to produce the 5mm shank length required, and provides sufficient residual strength in the setup to allow the component to be parted off successfully.

By using this method of precision turning, it will be realized that the turning load (in this case) is being transferred mainly to the large diameter of the gripped material, thereby avoiding the weaker spigot diameter being pushed off by the cutting tool.

This high loading on the material would of course normally be dissipated toward the very weak 3 mm diameter spigot of the 6 mm diameter (weaker) supplied material.

The use of a very narrow parting-off tool will of course be required. This must be very carefully off-hand ground down to a width of possibly 1.2 mm (or even 1 mm, if further loading problems occur); for this, we must use a high spindle speed of 800 rpm and use a very delicate and progressive in feed speed, while using plenty of coolant.

Using this method will allow the machinist to maintain good diametrical accuracy of the work-piece. Its main disadvantage will be in the increased cost of using the larger-diameter holding material in order to get the task completed accurately, without its use of end support.

Of course, if a 6 mm collet-holding device had been available on the lathe being used, then the operation could have been performed by using the 6 mm issued material, with modified tooling. This operation would then have relied on the superior support of the collet for securely holding the original supplied material.

If one analyzes this situation, it will be realized that the actual turning operation, while using the collet's superior support, will also have been performed much closer to the lathe's main support bearings (this being due to the absence of the three-jaw chuck's longer overall length of 'stick-out from *its* support bearings). It should also be realized that if one had used the original three-jaw chuck with its jaws gripping the original 6 mm diameter work-piece, then the turning operation would have been taking place much further away from the machine's main support bearings because of the chuck's considerably longer stick-out length if compared to the collet chuck. Therefore in this particular case, the extra overhang of the whole turning operation, because of using the longer stick-out length of the normal three-jaw chuck from its main bearing support, would have (in this case) provided a much less rigid support if one were using the lathe's precision tooling to machine the work-piece during this operation.

The importance of maintaining a high standard of cleanliness when working on the lathe.

For reasons of cleanliness and accuracy, one should never insert a male Morse-tapered support center (or a Jacobs drill chuck's male taper), into the receiving internal Morse taper socket of the tail-stock's barrel, without first wiping out the barrel's tapered bore and the external taper of the drill or drill chuck, to ensure that it is perfectly clean before locating this tooling finally into its working position.

Section 18/18
Further important tailstock alignment checks that allow accurate taper-turned work to be produced

A thorough check of the lathe's tail-stock's alignment is a most important necessity.

If the center-line of the tailstock has not been set exactly in line with the *lathe*'s center line (particularly when machining long, tail-stock-supported workpieces), then only tapered work will be produced.

Of course, if a tapered work-piece is required, then the exact angle to be machined on the workpiece's surface must first be established before setting over the tailstock by the required amount, (see fig. 45, 'Turning' a taper using the offset tailstock technique').

It should be noted that only comparatively shallow (and arguably less precise) angles can be produced by using this particular method of taper turning,; one should also realize that the support centers in this case are under further considerable stress because of their 60-degree support angles being compromised and unable to provide a full surface area of 60-degree contact in their support of the workpiece.

In the case of turning an angled workpiece when using the three-jaw chuck's jaws to grip the driven end, and an offset tail-stock's center to support its outer end, the machinist is faced with the particular dilemma of how one safely supports the workpiece without putting additional strain on the chuck's jaws by using this angled setting.

When in this situation, there is the temptation to use a very short grip length of the chuck's jaws in order to prevent any possible jaw straining or bell mouthing being introduced into the set-up by reason of the offset center's angled support.

If on the other hand, the machinist adopts the practice of using the full length of the chuck's jaw's to secure the work-piece's diameter, while he is still using the offset tailstock technique, then he will be deliberately allowing the offset tailstock to introduce sideways strain onto the chuck's jaws, causing not only a bending moment to be introduced into the work-piece (an undesirable feature), but also transferring an additional sideways load onto its supporting center. This extra sideways load, while being transferred to the chuck's precision jaws, will be placing them under considerable strain, extra loading, and will, during prolonged use in this situation, lead eventually to bell-mouthing the chuck's jaws. This practice should therefore be avoided if at all possible to safeguard the precision accuracy and secure holding capacity of the chuck's jaws. (When in this particular situation, it is considered preferable to use the between-centers set-up to machine this type of tapered work-piece, as shown in fig. 45, without using the three-jaw chuck at all.)

The working accuracy of holding the work-piece to run concentrically in the chuck's jaws must always be considered very important. This built-in jaw accuracy must be safeguarded at all costs in order to prolong the continued secure gripping and turning of accurate work-pieces.

A dedicated precision center-lathe machinist will always agree there is nothing worse, (in machine-shop work-holding methods), than a three-jaw chuck that fails to hold a work-piece to run concentrically, for when in this situation it will undoubtedly cause many unnecessary problems to be suffered by the machinist, particularly when he is desperately attempting to machine a precision diameter on the work-piece that must of course be running perfectly and concentrically with its gripped end's diameter.

Every effort should therefore be made on the part of the machinist to protect the original built-in precision accuracy of the lathe's chuck jaws in order to prolong their future accurate use.

In practice it will be found that even if one is using a chuck that is known to run and hold the work-piece to run concentrically, if the workpiece is removed for any reason from the chuck part-way through a machining operation, it may not be found to run perfectly truly or concentrically on being returned into the chuck's jaws.

It is often found in practice that, if the workpiece has been disturbed (that is, by it being removed from the chuck), there is a strong possibility that it will not now run perfectly concentrically when it is replaced back into a *new* position in the chuck.

Should the machinist be faced with this particular lack of a work-piece's concentricity, it is considered a worthwhile tip to always use the same key socket in the chuck (for the chuck key), when removing and reinstalling a replaced workpiece. Therefore, to ensure a positive removal and accurate replacement of the workpiece back into the chuck, it is advisable that the chuck's front face, and the workpiece's adjacent surface should be clearly marked with (say) a scribed mark or a center punch indent, with which to later identify the exact original position the workpiece previously occupied when it *was* running concentrically, *before* its removal. By following this precautionary advice, it should now allow the workpiece to be replaced accurately back into the chuck by using its original position marks.

Most chucks will have the manufacturer's name inscribed on their front face. It is often found advantageous if one actually chooses the nearest chuck key socket to this inscription, in order to positively identify the correct socket to use. It has been found when adopting this practice, that the replaced workpiece will definitely stand a much better chance of running concentrically on its replacement.

Section 18/19
Notes regarding the between-centers method of taper turning (as shown in fig. 45)

There is of course a limit to the angle of taper possible to be turned when the lathe is set up for turning a work-piece between centers.

This particular taper-turning operation is facilitated by deliberately using the offset tailstock technique to turn the desired taper on the work-piece. The body of the tailstock in this case must be set over by an accurately calculated amount to obtain the required taper. This 'set over' angular dimension will of course depend on the calculated length of the work-piece and the amount of 'taper per foot' (in angle) required to be turned. In practice, the tailstock's center-line is set over to half of the total amount of taper required (this being over the full length of the work-piece), and most importantly, this calculation must be over the whole length of the area that needs to be taper turned.

For a more reliable, accurate, and stress-free method of precision turning of angled workpieces, the reader is advised to refer to fig. 39, **'Precision taper turning on the center lathe'**. See also the fig. 45 drawing of a taper-turning attachment, shown prepared for use.

Turning a work-piece held between centers and using a test piece for checking the lathe's tail-stock for parallelism

The name 'center lathe', given to this particular lathe's design was undoubtedly chosen by its manufacturers because it is capable of machining a work-piece held in a machining position called 'between centers'.

To set up a lathe for this method of turning (and also for screw-cutting angled pipe threads), the chuck is first removed,; a catch-plate is substituted, together with a special reducing sleeve assembly complete with its headstock center. This is fitted internally into the spindle nose cavity of the lathe. Care should be taken to clean out the spindle nose cavity thoroughly and the headstock center's surfaces before the assembly is finally entered into the spindle's nose, in order to ensure its concentricity in use. A lathe carrier (designed to transmit the spindle's drive to the work-piece), is fitted and tightened onto the work-piece in a convenient position. The work-piece may need to be machined down to a reduced diameter over a short length to provide sufficient space for the lathe carrier to be fitted (in some cases therefore an extra length may need to be added to the workpiece's overall length to allow for this). This reduced-diameter section should be of sufficient length to allow the lathe carrier to be fitted, and should provide sufficient length and diameter for the proposed taper turned workpiece to be fully machined to its required size. The drive assembly is of course placed at the headstock end of the driven workpiece.

The workpiece will have previously been center drilled in both its ends, to allow it to be located into its correct position of 'between centers'. The outer end of the workpiece will use the tailstock center's support,; its driven end will use the headstock 'center's support. The lathe carrier (or drive dog), will receive its transmission of power from its catch-plate's drive pin. The lathe carrier is with advantage secured to its drive pin by the use of a piece of cord (or similar material) to prevent noisy clatter occurring during use, and this will prevent the drive pin overrunning the lathe carrier (this safeguard is particularly important during screw-cutting operations). Note that if, for work-piece design reasons, the lathe carrier cannot be fitted within the required workpiece's overall length, the machinist will need to add an extra length (of material) added to the work-piece's overall length, to allow sufficient space for it to be fitted,; the extra length required should therefore be decided *before cutting the work-piece material to length, by using a power saw.* This extra material will of course need to be removed on completion of the workpiece by a process of facing off the excess material in order to leave its end completely clear, to comply with the drawing's overall length specification. This extra work will therefore require a second operation to be carried out that requires the workpiece to be reset in the lathe's three-jaw chuck while using a fixed steady's support to allow the facing-off operation to be carried out safely.

*　　*　　*

Accurately turning a between-centers held workpiece.

To turn a work-piece perfectly parallel when the material is being held between centers will require a check of the tail-stock's alignment. This must be done in order to test the lathe's ability to turn a long, accurate, and perfectly parallel workpiece while using the support of the chuck at its driven end, and a greased dead center at its tailstock end. The tailstock assembly can be accurately checked for its correct alignment to the lathe's center line by making very small lateral adjustments to the tailstock's main casting (relative to its base casting), in order to obtain its correct alignment with the center-line of the lathe.

This alignment test can be performed by first selecting a spare (or stock) piece, of bright, round, mild steel, of, say, 20 mm diameter, approximately 12 inches, (300 mm) long, for use as a relatively soft machinable test piece. (Note that the availability of this chosen piece of material when received will probably be in an as-issued state and supplied direct from the metal stores,; it will not normally have been previously machined;.We must therefore make a cursory check of the condition of this piece of material in order to ensure that it is reasonably round, straight, and parallel, by rolling it over a perfectly clean surface plate and noting if there exists any wobble). If it is considered by this test to be straight, the material is then secured in the three-jaw chuck with an initial nominal stick-out (from the chuck) of about 25mm. This exposed end is then faced-off smooth, using preferably a right-hand knife tool (as described in drawing fig. 47 tool number 1),; the resulting sharp corner of the 'test piece should also be chamfered to produce a 1 mm × 45 degree chamfer by using the chamfer tool (shown in fig. 47 tool number 8). It's end should then be center drilled, by using (say) a number 3 center drill, held in the lathe's Jacob's drill chuck, which is in turn secured in the lathe's tailstock. The center drilling operation should always be performed at a high work speed of, 800 rpm and should be supplied with plenty of coolant. This spindle speed should always be used unless of course the material is very tough, where the use of a lower work speed with coolant is advisable. Of course if the material being used is of cast-iron material, then no coolant will be required. This center drilling operation will produce in the workpiece a suitable 60 degree internally tapered (female) work support cone that will be compatible with the male 60 degree tapered and pointed end of the lathe's greased dead center support, held in the tail-stock's sliding barrel (called its quill).

The chuck's jaws are then loosened,; the test piece is then moved outward to leave a nominal 15 to 20 mm retained in the chuck for a suitable grip. The chuck is then carefully secured by being tightened onto the workpiece while the test piece's outer end is being supported in the greased tailstock's supporting dead center (provided of course the tailstock itself has been suitably repositioned and locked into its correct longitudinal position to receive it). Its adjusting handwheel is then turned very carefully clockwise by just a sufficient amount to nip and secure the test piece onto the newly greased dead center, (undue longitudinal pressure should not be used at this time in order to avoid overloading the center), the tailstock's sliding barrel (or quill) locking device, is then secured to prevent further longitudinal movement of the test piece.

It should be realized that a test piece or work-piece can become very hot during heavy machining operations, (this overheating is generally caused by either the lack of the necessary coolant being applied to both the tool and the workpiece, or to the total absence of lubricant being applied to the center's support). It should also be realized that a hot test piece or workpiece will expand marginally in length because of this increase in temperature); therefore, a careful watch should be made of the support center's loading and its lubrication. It may be found necessary to re-adjust the tailstock's center's support from time to time during a machining operation, to relieve any increase of longitudinal pressure caused by this expansion, and to prevent the possibility of the overloaded center becoming burnt out through either excessive pressure or lack of having sufficient lubrication (this is quite a common fault that tends to occur quite frequently among trainees).

Please note that continued information in the use of this test piece check is carried out in Section 18/22.

Section 18/20
Refer to fig. 47.
The authors selection of useful lathe tools.

The author's personal selection of the most useful and suitable lathe tools for use during the various turning operations required.

It is customary for experienced machinists to possess an assortment of sharp off-hand ground' lathe tools, stored in their own tool-boxes; these are primarily used for external turning, internal boring, parting off, chamfering, and external/internal screw-cutting. These retained tools enable the experienced machinist to get started on the job very quickly after receiving the appropriate drawing from the workshop's supervision. Their speed of response to this job change is liable to cause a certain amount of dismay among new students, who have yet to learn how to get up to speed in this quick job-change situation.

The student, while under supervision, may also feel that he is placed at a disadvantage by being allowed to receive only one specific lathe tool from the tool stores at any one time.

Lathe tools are usually supplied on request from the tool stores in a so-called 'new as-issued state'; it will therefore be found that on receipt (especially by the student), they do not possess all of the required precision ground cutting angles, clearance angles, or honed tip radii, that will be found necessary to accurately machine the job in hand, without extra grinding or honing work needing to be carried out on them prior to their use.

It will therefore be found necessary for the student, over time, to cultivate the ability to perform these precision off-hand grinding and honing improvements to the issued tooling by using his or her accumulated experience of off-hand grinding techniques. This includes the sharpening and honing skills required to transform the original tool's as-issued cutting angles, clearance angles, and point radii into more suitable ones for use on the particular job in hand.

It will be found helpful therefore if the student is allowed to accumulate and eventually possess (over a period of time), the basic minimum number of at least six, sharp, hand-ground, honed, and ready-to-use lathe tools in his toolbox, for his immediate use. This will then enable him or her to start the machining of the workpiece without spending too much valuable preparation time on tool grinding and honing before actually starting work on the job in hand. Any spare time a student can accrue in the meantime, will be considered well spent if it is used to practice grind the necessary cutting and clearance angles needed on his accumulating stock of lathe tools, with the aim of eventually improving not only his off-hand grinding techniques, but also the cutting ability and durability of his own set of lathe tools.

The most economical, useful, and durable turning tools are usually produced by using off-hand ground butt-welded high-speed steel stock items (these being considerably cheaper to buy than the solid high-speed steel tool bits. The alternative to this is the 'inserted tip' carbide tools, which are much more expensive to purchase.

If the possessed lathe tools are placed in a relative order of importance and usability in the general machine shop or tool-room environment, they will be firstly, the right-hand knife tool, (used for most straight turning and facing-off operations).

Second is the parting tool, used for parting off the finished workpiece and for machining various external slots, together with an internal undercut tool used for internally undercutting the workpiece.

Third is the chamfer tool, ground with suitably modified multiple cutting angles that are designed not only to remove the sharp corners on the workpiece at an angle of 45 degrees, but also to machine chamfers on the left and right-hand sides and ends of the workpiece. In engineering circles, it is usually accepted that if a sharp corner is intended to be produced on a component, it will be specified on the drawing,; otherwise, all sharp corners whether external or internal, should be generally machined by chamfering at 45 degrees by 0.5 mm in order to avoid one's hands being cut (for safety reasons), and to improve the general appearance of the finished component.

Fourth is the external right hand screw-cutting tool, ground to either an included angle of 60 degrees (for use on metric and American threads), or 55 degrees included angle for use on English imperial threads.

Fifth is the boring tool, used for precision internal boring, or alternatively for boring out existing holes into larger and more precise diameters, on holes that have been previously rough drilled by use of smaller-diameter drills to drill the workpiece initially.

Sixth is the internal (right hand) screw-cutting tool, used for screw-cutting internal threads in the work-piece; this to be ground to either an included angle of 60 degrees (for metric and American threads), or to an included angle of 55 degrees for English imperial threads. Note that internal thread-cutting tools should always be provided with much greater clearance angles than external screw-cutting tools, particularly in the area below their leading cutting edge, this being due to their use in very confined spaces within the internally screw-cut workpiece. These particular clearance angles must be ground away by a sufficient extent to enable its clearance angle to exceed (by at least 5 degrees) the helix angle of the internal thread being cut. (This clearance grinding applies particularly to the area of the tool's underside and to its heel.). We must always ensure that sufficient clearance exists in order to allow the front cutting edge of the tool to be fully capable of clearing the *helix angle* of the work-piece thread during *all* screw-cutting operations.

When internally screw-cutting small-diameter threads, it is extremely important that sufficient clearance has also been provided (by using off-hand grinding) to the *back of and under the tool's extended overhanging shank*, particularly at the tool's leading cutting edge; this clearance must be sufficient to allow the tool to fully clear the helix angle of the thread being cut, plus at least an extra five degrees, (an important consideration). It will therefore be found very important that the tool's cutting edges (including its extended shank) are always ground away with adequate clearance sufficient to prohibit its cutting edge from actually rubbing (or failing to cut),; it is also important that the tool's supporting shank is continually monitored by the operator to ensure that it is always fully clearing the thread being cut during its deepest passing cuts.

Section 18/21
The inspection, sharpening, and honing of lathe tools, and how one obtains a correctly honed radius on the tool, for use on an externally used screw-cutting tool

To perform a basically straight external turning operation, a right-hand knife tool (similar to tool number 1 in fig. 47 and the authors multipurpose lathe tool shown in fig. 47a), is the preferred choice. In practice this tool is normally selected from the machinist's own tool-box, carefully scrutinized for sharpness, (resharpened at its point radius and re-honed if necessary), followed by being carefully fitted into the tool post. A magnifying glass can be used (with advantage) to examine the condition of its cutting edge and it's minutely hand-honed cutting radius. The tool's point height must then be set to exactly the centerline height of the work-piece (and of the lathe), by using the lathe's dedicated tool height-setting gauge (if this tool is available for the lathe you are working on).

The following text describes the importance of using an accurately hand-honed radius at the external screw-cutting tool's cutting point

One should bear in mind that the hand-honed radius carefully produced on the screw-cutting tool's cutting point/radius, not only forms the shape of the *core diameter* of the external thread in the work-piece, but also possesses the exactly honed form that is required at the cutting radius of the screw-cutting tool. It therefore follows that this particular radius should be very carefully hand honed, and scrupulously measured to ensure that it has been honed correctly. It will be found (in practice), that if the exact and correct radius has been honed onto the tool's tip, it will be discovered, during the screw-cutting process that on reaching the end of a screw-cutting pass, and on the finishing cut, that it will leave an exposed and minute *witness mark* on the undercut's diameter at the immediate end of the screw-cut thread. This *witness mark* should occur at this time if the tool has been honed with an absolutely correct radius, (that is of course providing the undercut's diameter has been previously turned to its correct screw-cutting clearance diameter).

It must be stated that during the screw-cutting procedure, the appearance of this witness mark is considered by the machinist to be a welcome sight, as it indicates that the full (calculated) depth of thread has been reached, and the workpiece may now only require possibly one minimum depth passing cut to be made to establish the full depth of thread has now been reached. This situation is followed by using the *thread ring gauge* to check the thread's dimensional fit, which should not reveal any end float or shake being present in the screw-cut thread. A good fit being obtained will signify that the workpiece will now pass through the inspection test (coming later), that the screw-cut component is now considered fit for purpose.

To reiterate, there now follows an important piece of information that highlights the critical importance of strictly maintaining the dimensional accuracy of the cutting radius honed onto the screw-cutting tool's tip.

It should be noted that if the tool has been honed with a *too-large* tip radius, the machinist will discover during the final screw-cutting passes, that on the tool's tip radius reaching the lathe's *scale-indicated* full depth of thread, the thread ring gauge will be found to screw onto the work-piece extremely loosely, indicating that the thread has been cut too deeply. The reason for this, is that the tool, when honed with this too-large radius, and on reaching the thread's full theoretical core diameter, will give the machinist the false impression that the screw-cutting tool has completed its work correctly, but in actual fact, the larger width of the tool's side cutting edges will have (by now) removed far too much material from the side cheeks of the newly screw-cut thread, resulting in the thread being cut *too loosely* (now evidenced by its abnormal clearances).

Therefore, during the last of the screw-cutting passes, when taken by using a too-large radius honed on the tool, the *side cheeks* of the thread will now have been cut slightly wider than required to obtain a precision fit in the thread ring gauge.

[If the machinist is unfortunate enough to encounter this situation, it should now clearly indicate to him that the screw-cutting tool *must* be honed so that it possesses a smaller and *correctly dimensioned* radius at its tip].

It is unfortunate that when the trainee machinist arrives at this situation, he will often carry on taking *several extra* screw-cutting passes of 0.05 mm deep, (reaching far beyond the scale's indicated and calculated depth of thread), to supposedly *increase the tools depth to get the ring gauge to fit*, only to discover (but now too late), that the side cheeks of the screw-cut thread have now been cut much too deeply' (i.e. too wide), making it very unlikely that the workpiece thread will be recoverable because it now possesses a very much oversize and loose fit, thereby making the work-piece probably now to be considered as being scrap.

If on the other hand, the machinist had honed the tool's tip radius very slightly *smaller than its calculated size*, then, on the tool's tip finally reaching the full calculated depth of thread, the tool's side cheeks would not now be removing any excess material from the side cheeks of the thread being cut, and consequently the thread ring gauge would now not screw onto to the thread, because it is still undersize to the drawing's requirements. Therefore in this case, the machinist will need to take several extra passing cuts until the gauge actually fits the thread, and does not exhibit any shake or end float being present. The moral of this story is that it is much better (and safer), to hone the screw-cutting tool with a slightly *smaller-than-optimum radius* which then allows the machinist to take an extra cut or cuts to complete the work, rather than honing the tool with a *too-large* radius that will probably *scrap* the component.

It is also considered a wise practice, during the screw-cutting operation, to make a continuous visual check of the gradual narrowing of the thread's exposed external flat on its diameter that will be visible (in our example's case, at its 23.9 mm diameter); which *should* be 0.50 mm wide when the screwcutting operation has been finally and correctly completed). In this case, if the existing flat is found to be narrower than the required 0.50 mm, and gives the appearance of being considerably *sharper* than the 0.50 mm dimension required, then this will indicate to the machinist that the thread has been cut too deeply and may in consequence be *too loose* for a correct fit. (See the fig. 43 drawing of the thread to see the correct flat dimension of 0.5mm shown on the 24mm diameter metric coarse thread depicted.)

It is also considered best practice, therefore, (for both for the machinist and for the final fit of the screw-cut thread, to always ensure that the exact calculated tip radius has been honed onto the screw-cutting tool's point; as shown in the lower drawing of fig. 43, this tool's tip radius should be 0.1443 × the pitch of the thread being cut, (this tool is also shown in fig, 47 tool no. 4). This radius must be carefully checked in order to prevent a possible scrap workpiece being produced by the use of a *too-large* radius being honed on the tool.

It will be found in practice that it is much more sensible (in the long run), for the student to err towards using a very slightly smaller radius on the tool than was originally calculated, rather than *falling into the trap* of using a *too-large* honed radius for this particular screw-cutting job. The author therefore recommends that the student should adopt this checking procedure during his training in order to prevent the aforementioned problem occurring.

The example 24 mm by 3 mm pitch thread as shown in the fig. 42 drawing will require a radius to be honed onto the screw-cutting tool's tip of 0.4329 mm. The calculation formula used for obtaining this radius (and used for calculating the male radius on all screw-cutting tooling used for cutting metric coarse threads), is 0.1443 × the thread's pitch. Therefore for calculating our example of a 24 mm metric coarse thread, we enter into the calculator the sequence 0.1443 × 3 =, giving us 0.4329 mm. We can check whether the correct radius has been honed onto the tool visually, by using a magnifying glass and comparing the existing tool's radius with the flattened end of a drill (or drill blank) of 0.8658 mm diameter, while it is being used as a comparison gauge.

The tool's cutting points height must also be adjusted to match the exact center-line height of the work-piece. (Note that the dead center's tapered point held in the tail-stock can also be used to verify that the tool's cutting height is correct as this possesses an identical height.

A quick alternative and accurate method of checking the general height of an external lathe tool, particularly when using the right-hand knife or parting tool, is to use them (experimentally), in a facing-off operation on a test piece consisting of a piece of round scrap steel material securely held in the chuck, and using an adequate supply of coolant during the process.

If, during this facing-off/parting-off procedure, either of the tools are seen to be leaving a small pip (or protrusion) of excess material in the central area on the end of the faced-off test-piece (or work-piece), this will indicate that the tool height is set too low. Alternatively, if the tool tends to require extra effort to be applied to the cross-slide's hand wheel when nearing the center of the test piece and also possesses a tendency to be reluctant to cut on nearing the material's center line, then this indicates that the tool is set too high, and must be lowered sufficiently to allow the tool to cut cleanly up to the center-line of the material while leaving no residual tell-tale pip (i.e. protrusion) visible on the material's center-line. This facing-off test can also be used on other tools to establish whether they have been set to their correct cutting height,; therefore, this practical method of testing the hight of the tool can be used on all the used external lathe tools.

For the internal tools, they can be checked for correct cutting height by using the far side of a round test piece, and by interposing (in this example) a 150-mm-long rule between the tool's tip and the back of the round work-piece, followed by checking that the rule remains vertical, thereby confirming that the tool height is correct. If the rule leans toward the operator, then the tool is set too high,; alternatively, if the tool is seen to lean away from the operator, then the tool is set too low. The use of this method will confirm that the internal tools are set at their correct cutting height.

It should also be noted that it is bad practice to allow a cutting tool's point to progress *beyond* the center line of the material when facing or parting off. The student should also realize that the rotating material beyond the workpiece's center line will be travelling in an upward direction on the work-piece's end surface. Therefore if the tool is allowed to go beyond the center-line of the end of a rotating test piece's material, then the tool's cutting edge will not in this case be cutting, but will merely be rubbing the surface, and causing its cutting edge to be dulled, thereby seriously reducing its future cutting ability and its working life between regrinds. The practice of allowing the tool's cutting tip to over-run the centerline of the workpiece should therefore be avoided in order to prolong the working life (between the resharpening grinds) of the tool.

Section 18/22
Continued further use of a test piece (initially partially detailed in section 18/19), to obtain true parallelism of the tailstock relative to the center-line of the lathe

Following the previous guidance notes on the use of the lathe's tooling, the author now returns to provide yet more advice on the recommended method used to obtain an accurate parallelism of the lathe's tailstock relative to the center-line of the machine, in this case by using the *turned test piece* method. We must first center drill the test piece to suit the fit of the tapered supporting greased dead center, and set up the test piece using the dead center's support in the tailstock.

A basic check of the accuracy of the lathe's parallelism can be achieved (when using a parallel and perfectly round test piece,) by positioning the right-hand knife tool's point, (by using hand rotations of the saddle and cross-slide's hand wheels), sufficient to allow the tool's cutting point to contact the test piece's diameter in an area close to the lathe's chuck, followed by adjusting the cross-slide's reference scale to read zero.

The tool's point is then re-positioned (using rotations of both the saddle and cross-slide's handwheels), to contact the diameter of the test piece in an area close to the right-hand end of the test piece, (i.e. adjacent to its supporting dead center).

If on this second contact of the tool with this supposedly parallel test piece's diameter, the cross-slide's scale is again seen to read zero, then this will indicate that the lathe's tail-stock is positioned reasonably accurately.

The statement 'reasonably' accurately,' is used in this case, because the accuracy of the diameter of the chosen test piece's material is so far unchecked and it may be in an as-issued state,; therefore, it should be realized that this stock material may not be perfectly accurate in diameter or straightness at this stage. If a significant difference is found in the readings of the cross-slide's scale between the two tool contact points, then a more accurate method of checking the lathe's parallelism must be undertaken.

To do this, we must now physically turn two spaced diameters on the test piece. One diameter is to be turned near its chuck end and a second diameter is to be turned at its tailstock end while using an identical cross-slide scale reading in both cases.

After turning the two diameters, they are then measured for diametrical similarity by using a micrometer,; the ultimate aim will therefore be (between any intervening tailstock adjustments), to obtain true parallelism of the lathe by a test piece that will contain two identical diameters. It should be noted that it is always more convenient to turn two spaced diameters on the test piece rather than turning one long diameter along the whole length of the test piece, in order to save valuable time.

To carry out this parallelism check fully, the tool's point must first be positioned so that it just touches the diameter of the test piece at the tail-stock's end, and its scale set to read zero, the tool's point is then positioned just to the right of the test piece.

With the cross-slide and tool still set at its zero scale's indication, the tool is now indexed inward clock-wise (by using the cross-slide's hand wheel while noting its scale), by a nominal 0.3 mm, (followed by re-zeroing its index scale). We then switch on the lathe and commence to turn the diameter by using a slow feed to progress the tool to its left, thereby turning the right-hand end's diameter to this new setting. This new diameter is carried along for approximately 10 to 12 mm of the test piece. We then rotate the cross-slide's hand-wheel anticlockwise (i.e. outward) by one complete revolution to clear the test piece,; we follow this by moving the tool, (using the saddle's hand-wheel) to approach within 25 mm of the chuck at the gripped end of the test piece, to allow a second diameter to be turned, (note that when preparing to turn this second diameter, we must always allow the tool to enter into the test-piece gradually while the workpiece is rotating; we do this by indexing the cross-slide inward clockwise to an identical depth setting, (indicated on its cross-slide's scale, to that used for turning its first diameter, (namely zero). This cut is started by turning the cross-slide's hand wheel (almost) one complete revolution clockwise, then slowly indexing inward the last 0.3mm down to its zero point while the tool is moving to its left from its new starting position of approximately 25 mm from the chuck followed by completing the turning of this new diameter to within approximately 10 mm of the chuck. The tool is then withdrawn and the lathe switched off. The two turned diameters are then measured by using a micrometer to compare both diameters for size.

If during this checking procedure, the two diameters are found to measure identical diameters, (which is unlikely at one's first attempt), then the lathe should now be capable of turning a parallel work-piece while the work-piece is being supported by the tailstock's greased dead center.

If, however, the two diameter,s are still found to be dissimilar, then the difference between the two diameters will confirm that the lathe is still not cutting parallel, and that the tailstock is still out of line with the center line of the lathe.

To rectify this tailstock's misalignment fully, the main tailstock casting will need to be moved over so that it will now occupy a correctly aligned position. We do this by physically moving the tailstock's top casting transversely across its lower base casting by a sufficient amount to remedy the misalignment. This relatively small transverse movement of the main tailstock casting relative to its lower base casting, will need to be adjusted either toward, or away from the operator, depending on which of the turned diameters of the test piece measures the larger diameter.

If the tailstock end's diameter is larger, then the main tail-stock's casting will need to be moved (minimally) toward the operator. This lateral movement of the tail-stock will then allow the tool, on its next experimental passing cut, to reduce the tailstock end's diameter of the test piece. The chuck-end diameter of the test piece must then again be turned using the identical cross-slide scale's setting.

Note that the tool must be seen to be removing material from both of the diameters of the test piece to ensure that this new parallelism test is valid. A micrometer check of the two new diameters is then carried out. (The total lateral movement of the tail-stock's casting will be approximately half the difference found between the originally turned diameters).

To reiterate, the ultimate aim during this checking procedure, is to obtain a true parallelism of the test piece or work-piece's turned diameters, by using a sequence of turning operations that will involve the measuring of both of the newly turned diameters, ultimately arriving at both the diameters' having identical cross-slide scale's readings. This is accomplished by making the required sideways adjustments to the tail-stock's upper casting, sufficient to allow the tool to eventually turn two *identical* diameters that will now indicate that the lathe's tail-stock's true parallelism has been achieved. .

The main and base tailstock castings are engraved with minute reference scales at their outermost (right-hand) ends. These engraved scale lines indicate the exact alignment of the machine's center-line relative to both of the individual tailstock castings. These castings were accurately aligned and engraved during the lathe's manufacture, and therefore can be relied upon to provide a reasonably accurate guide for use when initially setting up the tailstock's alignment relative to the lathe's center-line. However, for absolute precision alignment, it will be found that the eventual turning of two identical diameters on a test piece held between centers and using the same cross-slide setting during this process is the most accurate option.

To initiate the correct tailstock's center-line alignment, we will need to complete a series of fine adjustments to the tailstock's lateral adjusting screws, this to be combined with turning the two aforesaid spaced diameters on the test piece. The two tailstock's lateral adjusting screws are situated low down on its casting, one on each side of the main tailstock casting. These screws operate on the principle that while one is used as a jacking screw, (to move the tailstock laterally),; its opposing screw is used as a locking screw to finalize the tail-stock's new position. (This function is of course reversed if one needs to adjust the tailstock in its opposite direction.) The locking security bolt, situated at the right-hand end on the lower casting's underside, is first loosened by a sufficient amount to allow this lateral movement to take place. Very fine adjustments are then made by using very small, (say, approx. 1/8 of a turn), rotations of the adjusting screws, by using a suitably sized hexagon socket-type key. These adjustments must be coordinated with turning the two new diameters on the test piece, followed by measuring their newly turned diameters.

By carefully completing this rather time-consuming process, it is possible (after several adjustments have been made to the adjusting screws, and several exploratory turning cuts have been made on the test piece), to eliminate the 'out of parallelism'of the tail-stock's top main casting, relative to its lower base casting, thereby returning the tail-stock to its truly aligned parallel position on the lathe's center line.

Note, it may be necessary to re-turn and re-check both of the turned diameters several times, before a satisfactory alignment has been achieved.

When the final adjustments have been made and both of the turned diameters are finally identical, the locking bolt under the tailstock's lower casting is then secured.

With the lathe's parallelism check now satisfactorily completed, a tail-stock-supported parallel work-piece (or a parallel screw-cutting operation), can now be accurately undertaken with confidence.

It is often possible (and expedient) to use the actual work-piece to establish the lathe's parallelism, provided of course there is sufficient excess material on its outside diameter to permit the test turnings to be carried out prior to machining the required major diameter of the actual work-piece. The use of this method will avoid the problem that can occur if one uses a test piece to check the lathe's parallelism, and then needs to move the tailstock longitudinally along the bed of the lathe to accommodate a differing length of work-piece. Note that if one is using a worn machine, then any further longitudinal movement given to the tailstock after using the test-piece method to obtain parallelism, can reduce the accuracy of the whole setup if the tailstock's base-plate's V guides are badly worn, as this can cause inaccurate longitudinal or transverse repositioning of its casting relative to the parallel bed of the lathe.

Section 18/23
Selecting an appropriate lathe tool for the requirements of the work-piece

Selecting the lathe's gearbox lever settings required for screw-cutting a metric 3 mm pitch thread on the center lathe, and a general description of how one screw-cuts a component on the lathe

Having completed our preliminary lathe checks and adjustments, we can now progress on to consider the problem of selecting the correct appropriate screw-cutting tool for the job in hand, and to decide on the setting requirements of the lathe that are necessary prior to screw-cutting an accurate external metric thread.

The lathe's screw-cutting 'pitch lever' positions should now be manually selected to comply with the lathe's affixed screw-cutting pitch information plate's instructions, these exact position will then enable the lathe to produce the required 3 mm pitched thread. We must do this in the first instance in order to check and establish that the lathe will actually produce, (for this particular thread) the required pitch of 3 mm between each of its screw-cut threads.

This check is most important at this early stage in the setting up procedure.

However, it is most important that the trainee machinist realizes that the majority of center lathes are equipped with two separate power-feed drive systems (these comprise the normal power feed, (used for powering the automatic turning and facing operations), and the *separately* geared lead-screw drive feed, used for powering the screw-cutting and gear train mechanism). These separate power feed drives should at all times be kept completely separate and isolated from one another, to ensure they are never allowed to become engaged into gear together, for, should this accidentally occur, it could certainly cause serious and expensive damage to the drive gears in the gearbox.

As a safety precaution, and to eliminate the possibility of these two feed drives being engaged into gear together, the gearboxes of most center lathes are designed to possess various safety features to prevent this accident happening. The lathe's gearbox drive systems are usually provided with either a 'neutral gear' selector, (which isolates the 'power feed from the lead-screw feed), or on some lathe designs, a 'slipping cone clutch' mechanism is incorporated into its driveshaft, that is designed to slip should an immovable object be met, or the lead-screw inadvertently engaged while the power feed drive is in use. Alternatively, a sliding hand operated (in or out of mesh), 'dog-clutch arrangement is used, designed to physically isolate the two systems from one another, or lastly, by the introduction of a shear pin, fitted into the lathe's driveshaft mechanism (usually made from soft aluminum), and designed to shear in the event of serious overloading or accidental misuse.

As a warning to all machinists, the author now quotes an episode that occurred in the machine shop where an operator, who, having repeatedly encountered shear pin breakage on his lathe, (because he was seriously overloading the lathe's drive system by taking extremely heavy power feed cuts during a turning operation in order to 'get the job completed quickly), effected his own 'cure' to this troublesome problem of shear pin breakage, by introducing into the power feed drive-shaft a much stronger shear pin, made from tough *silver steel*. This temporarily solved his problem and allowed him to carry on overloading the lathe's drive system to get the work-piece completed in the allotted time. However, having forgotten he had substituted the standard-issue soft aluminum shear pin in the drive system with an illegal (and much tougher) shear pin, he was abruptly made aware while preparing the lathe for a screw-cutting operation, that he had inadvertently engaged both drive systems into gear together, this being announced by the lathe now producing a loud grinding noise emanating from its gearbox, subsequently causing severe damage to the gearbox drive gears.

This severe damage to the lathe's gearbox was caused by his use of a totally illegal tougher drive-pin in the drive system, which of course did not shear when being seriously overloaded during the clash of the two drive systems. His stronger modification to the drive system had solved his initial problem, but had negated the built-in safety factor of using the soft drive shear pin initially installed by the manufacturer of the lathe, designed to shear in the event of any overloading or misuse taking place. The moral of this story is that the machinist should never tamper with, or change, the standard-issue material used in the manufacture of the lathe's drive shear pin. The drive pin in question was designed by the manufacturer to perform the very important task of protecting the gears in the gearbox from any accidental damage.

Screw-cutting on the center lathe, a general description

The author now returns to the continuing task of the setting up and troubleshooting the lathe to allow it to perform accurate screw-cutting operations while using either the straight plunge or the angled approach method of screw-cutting.

If a screw-cutting operation has not been undertaken on this particular lathe for some considerable time, then its vulnerable and exposed lead-screw's drive threads, in all probability, will have been subjected to contaminants produced by the previous host of straight-turning and boring operations. The presence, on the lead-screw of these undisturbed contaminants can cause serious problems to the drive mechanism by preventing the positive engagement of the screw-cutting 'split drive nut's locking mechanism, which can prevent it from securely latching onto its lead-screw. This failure to engage correctly can be caused by the split nut's inner threads becoming clogged with swarf (scrap metal turnings). This area of the lathe should therefore be totally cleared of any accumulated dirt or metal particles, particularly in the area of the lead-screw's threads and its split-nut mechanism, by the use of a stiff brush to clean out its deeply contaminated threads. It is therefore preferable at this stage to have the lathe switched on and the lead-screw in rotating mode while this cleaning operation is being performed with care by the use of a stiff brush, followed by switching the lathe off.

The lead-screw's threads and its right-hand end support bearing should then be lightly lubricated with machine oil.

As a preliminary check and to confirm that the correct gear lever settings for the required thread's pitch have been selected (by using the guidance of the lathe's fitted instruction plate), we follow this by sliding the straight-turning power feed sliding dog clutch, or its 'in/out push/pull lever (or in some cases, using a similar isolating device, depending on the design of the lathe), into its *neutral* position.

We then place the lathe's spindle shaft and the chuck's drive mechanism into gear with its lead screw by engaging the appropriate *lift-up thread cutting lever* (in effect this closes the split nut onto the lead-screw to allow it to take up the drive). One should observe, on starting up the machine (by using the *lift-up starting lever*, and importantly having previously selected a low spindle speed), that the saddle assembly, (including its tool and tool post assembly), should be now moving to its left by the exact and required 3-mm pitch dimension for each single revolution of the machine's spindle and chuck.

The spindle and chuck speed change levers are normally situated on the front or top of the spindle's work-head and are usually distanced from the traverse speed or feed change levers located lower down adjacent to the gearbox levers and its drive mechanism. It is very important that the machinist becomes fully acquainted with the uses of all the individual levers and switches on the lathe, in order to avoid any possible confusion between the spindle's speed change levers, the thread cutting gear change levers, and the tool and saddle traverse speed change levers.

Note that if a low spindle speed has not already been selected, then the lathe should be switched off and the slowest speed available shown on its speed chart should be selected. Note that the lathe must always be stopped before attempting to change its speed change gear, and one should always listen attentively to the noise made by the gears immediately after switching off and as they slow down, until all noise has ceased before attempting to move the change levers,; this must be done in order to avoid damaging the gears by their teeth clashing together. The selected low spindle speed will usually be approximately 60 revs per minute (a low speed is of course a necessary requirement when screw-cutting steel components, particularly when using high-speed steel screw-cutting tooling). The correct choice of the work-speed rotation will ultimately depend on the diameter of the workpiece about to be screw-cut, and the toughness of the actual material being cut (large-diameter work-pieces will require slow work-speeds not only to obtain a good surface finish of the screw-cut thread, but also to prolong the useful working life of the screw-cutting tool. Smaller diameter work-pieces can of course be allowed to have a slightly higher work-speed, because of their lower surface speed relative to the material actually passing the tool's cutting edges. For the screw-cutting of threads, however, the lower range of spindle speeds is to be the preferred option (as will be explained later).

The machining of Cast iron, for example, should be performed dry to prevent moist sludge entering the lathe's precision slides, but when screw-cutting steel, one should always use a plentiful supply of coolant on both the tool and work-piece in order to prolong the between-sharpening life of the tool, and to improve the surface finish of a screw-cut thread.

An additional reason for choosing a particularly low work speed is that later on, when engaged in a screw-cutting pass, the machinist must develop the ability to extract the tool *very quickly* from the workpiece at an exact moment in time and place, while the workpiece is still in rotating mode. Higher work-speeds will make this operation much more difficult to achieve, because the higher work-speed does not allow this operation to be completed safely without the tool damaging the work-piece by overrunning the undercut provided for the thread's end clearance. One way of ensuring that the tool is withdrawn at exactly the correct time and place, is to affix a warning marker (in the form of a strip of drafting tape or similar visible adhesive material), to the outside surface of the holding chuck. This marker when correctly placed will give an early warning to the machinist of the exact moment where the tool *must* be withdrawn as it is seen to pass through his peripheral vision. The method used to accurately fix this marker into its correct place on the chuck's outside diameter, is to perform a dummy screw-cutting passing cut, immediately prior to making one's first serious screw-cutting passing cut. We do this by having the tool positioned close to, but not touching the work-piece during this dummy run. On the tool's point reaching the end of the intended screw-cut thread, and where it has just entered the safety of the undercut area, the lathe is instantly stopped, and the warning marker is stuck in place on the chuck's outer diameter at this exact point. This warning signal, can now give the exact extraction point of the tool from the work-piece, and will be found to assist the machinist greatly in maintaining an accurate tool withdrawal action at the end of each passing cut.

*   *   *

The following notes refer to the problem of removing or eliminating the existing backlash present in the cross-slide and the compound slide's lead-screws before the lathe is used for either straight turning or screw-cutting.

Backlash is the name given to the lost motion (or looseness) found between an ill-fitting or worn nut when screwed onto the male thread of a machine screw. All lathes will be found to have a certain amount of back-lash existing in their cross-slide and compound slide lead-screws; the amount of this will of course vary with the age of the lathe and the wear sustained in its lead-screw's threads, it will be found that even a new lathe will still possess a certain amount of back-lash in its cross-slide's lead screw. This backlash, end float, or lost motion, found in the lathe slide's operation, must be taken up, i.e. eliminated, *in its correct working direction of rotation*, before the lathe and its cutting tool should be used for either external/internal turning or external/internal screw-cutting.

When used for external machining, this is done by turning the cross-slide's hand wheel and its compound slide's hand-wheel carefully in a *clockwise direction* by a sufficient amount to remove any existing lost motion found between its nut and its screw.

However, in the case of performing an internal boring or the internal screw-cutting of a thread, the cross-slide's hand wheel in this case must be turned in an *anticlockwise* direction to take up the back-lash or its lost motion; this procedure must be completed before actually attempting to 'turn the workpiece. In this internal machining case, the rotation of the hand -wheel must be performed in an *anticlockwise* direction by a sufficient amount to remove any existing lost motion found between its nut and its screw.

If we need to accurately check the exact amount of backlash that is present in the cross-slide's lead-screw, we can use a dial test indicator (DTI, or clock), fitted to a suitable magnetic base (enabling the assembly to be magnetically attached to the bed of the machine). The clock should be positioned so that its movable probe is just touching the far side face of the tool-post's surface; its indicating needle should then be preloaded by one revolution followed by its scale being set to read zero. This is followed by turning the cross-slide's hand wheel clockwise by a quarter of a turn,; the cross-slide's scale and the dial test indicator needle's scale should then be re-set to read zero. The cross-slide's hand-wheel is then turned anticlockwise very slowly until the clock's needle is seen to move. The total amount the cross-slide's hand wheel has needed to be turned in this anti-clockwise direction, in order to produce a slight movement of the dial's needle will then be indicated by noting the cross-slide's scale as a reference,; this should now indicate the amount of lost motion, end float, or back-lash that exists between the cross-slide's lead-screw and its nut by the machinist noting the exact movement now shown on the cross-slide's graduated scale.

To further test the amount of lost motion that exists in the cross-slide's lead screw in a practical way, the tool-post can now be grasped in both hands and pulled toward the operator,; the amount of backlash now present in the cross-slide's lead screw will be instantly identified by noting the amount of movement, found by the indications shown on the DTI's scale. The extent of this movement will now emphasize to the machinist the need to back up the lathe's cutting tool always by turning the cross-slide's hand-wheel in its correct clockwise direction for external machining before using the tool to turn the external surface of a component.

To reiterate, when externally turning a component, the backlash that always exists in the cross-slide's lead screw should always be taken up by turning the cross-slide hand wheel in a *clockwise* direction and its scale then set to read zero before taking the first cut. When internally boring a component, the backlash that exists in the cross-slide's lead screw should always be taken up in an *anticlockwise* direction and its scale set to read zero prior to machining the internal surface of the component.

Section 18/24
General notes to assist the machinist, when forced to replace a tool during a screw-cutting operation.

Once the machinist has actually experienced the problem of using a too-high work speed for screw-cutting, he will by now have become fully aware by realizing that a too-high work speed can seriously affect the accurate judgment of the tool's crucial extraction point at the end of each screw-cutting passing cut. Therefore, having reached this stage in a screw-cutting procedure, he will now understand that excessive work speed can be the cause of serious damage occurring to the work-piece if the tool has been allowed to overrun the end of its passing cut by allowing it to pass into and beyond the machined relief undercut provided at this point for the tools thread clearance.

This accidental over-run of the tool can result in extensive damage being caused to the work-piece's approaching shoulder, its tool's vulnerable cutting edges and its cutting tip. This damage to the tool is often caused by the abrupt overloading and shock sustained at the tool's cutting edge as a result of the enforced emergency stop, where its cutting edges can become embedded in the work-piece. This accident will therefore result in the tool becoming deeply embedded in the body of the work-piece.

Therefore, if on completing a screw-cutting passing cut, the tool has been allowed to become embedded into the work-piece because of the machinist's failure to use the 'quick tool withdrawal' method at the correct time when reaching the under-cut area, or his failure to operate the foot-or hand-operated emergency stop switch lever at the correct moment in time, the resulting damage caused by the tool's overrun will (with hind-sight), remind the machinist that the dead stop emergency safety switch, should have been used correctly and without delay, to prevent the accident or damage occurring.

To safely resolve the embedded-tool situation, the lathe must first be switched off,; the machinist should now be made aware that if attempts are made to extricate the jammed tool from the now-stationary workpiece, then the embedded tool's cutting edges are quite likely to become chipped, or the tool's vulnerable cutting point broken off, if extreme care is not taken during its removal.

The machinist should also note that if one has an embedded tool, the snatched method of tool withdrawal should *not* be attempted *after* the workpiece has *stopped rotating*, as there is the definite risk of a broken or chipped cutting tool resulting. It should also be noted that if a cemented carbide tipped screw-cutting tool is in use at this time, then this advice will be doubly important, because of the extreme brittleness of the carbide tool's tip material, it being glass hard, making this particular tool much more prone to chipping during any careless removal than its equivalent high-speed steel tool's less brittle material.

The machinist should also remember at this point, that it is important to note down the exact depth the damaged tool had reached in the work-piece *before* being removed. This should be done promptly by noting down the readings from both the cross-slide and the compound slide's scales before the tool is removed for its close inspection. This noting down of the depth it had reached will be very useful when resetting up the new tool into its partial depth of thread in the work-piece, during the new tools setting-up procedure.

To prevent serious damage being caused to the cutting edge of a jammed screw-cutting tool by careless removal, its securing tool-post clamp screws should first be carefully loosened to relieve all pressure on the tool, its tip, and its vulnerable cutting edges, followed by the tool being very carefully removed for a closer examination.

Should the tool be found to be damaged, it should be very carefully replaced by an identically contoured, honed and sharp screw-cutting tool, and oriented into its correct position in the tool post by the use of a square-type thread setting gauge.

Section 18/24 a
Replacing a screw-cutting tool midway through a straight plunge screw-cutting operation while using the center lathe with a permanently engaged lead-screw, and where (in this particular case), the machinist has made a note of the depth of thread so far reached by the damaged screw-cutting tool before it was removed

If, however, on removal, the tool is found to be undamaged and still in a serviceable condition, it will need to be replaced in the tool-post and reset at its correct orientation, by using a square-type screw-cutting gauge held with its top face edge in close contact with the diameter of the parallel turned work-piece; its cutting edges are then carefully aligned with the angled sides of the gauge's 60-degree precision cutout to maintain the tool's previously set accurate orientation in the tool post. We must also check, and, if necessary, adjust, the cutting height of the replaced tool by using the lathe's height setting gauge. (It should be noted that both of the slide's indexing scales need to be re-set after the tool has been replaced.) The tool's cutting edge's will also need to be repositioned longitudinally relative to the partially cut thread's existing contours by very carefully rotating both the cross-slide and the compound slide's hand-wheels (while the lathe is engaged on an *experimental passing cut*. This tool's 60-degree *angular alignment* in the partially cut thread must be completed before resuming the cutting of the thread. It is also important at this stage to ensure that there is no overhang of the compound slide relative to its support casting, and that the screw-cutting tool itself possesses the minimum of overhang from its tool-post support in order to ensure there is full rigidity in the set-up.

Having fitted the new screw-cutting tool and correctly oriented its 60degree cutting edges relative to the work-piece by the use of a square-type screw-cutting gauge, the compound slide's hand-wheel is now rotated clockwise by a quarter of a turn to eliminate its inherent back-lash and its scale set to read zero, this is followed by now rotating the cross-slide's hand-wheel in a clockwise direction, sufficient to allow the tool's point to actually touch the 23.9 mm outside diameter of the material being-cut; its scale is then reset to read zero. The cross-slide's hand wheel is then turned one revolution *anticlockwise* to allow the tool to fully clear the work-piece, and the saddle's hand wheel is then turned clock-wise to maneuver the tool-s point to its right so that it occupies a position approximately 5 to 6 mm to the right of the work-piece's end. This is followed by turning the cross-slide's hand wheel clock-wise by one complete revolution down to zero on its scale in preparation for starting a new exploratory passing cut.

The partially cut thread is then sparingly painted with micrometer blue, using a brush to avoid any excess,; the lathe is then switched on in its normal direction of rotation (at approximately 60 revs per minute) and a very close study is then made of the front apron's rotating screw-cutting dial. When the machinist's previously selected number (or mark) is seen to line up with the apron's fixed datum line, the *lift-up screw-cutting lever* is operated to start this new experimental screw-cutting passing cut.

While on this pass, the machinist must now carefully maneuver the replaced tool's cutting edges (by using both hand wheels) into the V contour of the thread to allow the moving cutting edges to actually-*pick up* the contours of the previously part finished thread. This maneuvering and centralizing process given to the tool must be carried out and *completed* during this or any following experimental passing cuts. The tool's progress and true centralization in the thread can be checked by scrutinizing that an equal amount of blue is seen to be being removed by the tool from *both sides* of the screw-cut thread and its vital tip radius (when straight plunge screw-cutting), during these experimental passing cuts. Note also that it may take more than one passing cut to establish that the tool is perfectly centralized in the partially screw-cut thread.

[However, in angle approach screw-cutting, it is the tool's left-hand side and its tip that should be seen to be removing blue from the thread because in this case the compound slide has been rotated through 60-degrees for cutting metric threads.]

This picking up of the thread must be carried out very carefully, while the lead screw is engaged, the tool is in motion, and the work-piece is rotating on a screw-cutting passing cut. The necessary longitudinal and transverse adjustments that need to be given to the tool during this pass must be achieved by using *very small* movements of the cross slide and compound slide's hand wheels in order to re-position the tool's angled cutting faces so that they line up accurately within the unfinished thread's contours *during this or any following experimental passing cuts.*

It should also be noted that it is essential that both of the slide's hand wheels are being turned in a *clockwise* direction, *at the point where centralization of the tool in the thread has been achieved.* Both index scales should then be very carefully reset to read zero.

The fact that the machinist has now observed that the cutting edges and the point of the tool are seen to be removing minute quantities of blue and minute slivers of metal from *both sides* and the existing *core diameter* of the unfinished thread during this experimental passing cut will prove that true centralization of the tool in the unfinished thread has been achieved, and that this has occurred while the lathe was under power and on a screw-cutting cut. [On the tool reaching the undercut area of the thread, the machinist should instantly dead stop the lathe's motor by using the emergency stop lever (or foot pedal).

With the point of the tool now stationary in the work-piece and now positioned in the *undercut area*, the tool's full required depth of thread can now be accurately ascertained by turning the cross-slide's hand wheel very carefully and minutely clockwise until the cutting tool's tip is seen (and felt) to be just touching the under-cut's diameter (this under-cut's diameter will have previously been machined down to 20.3 mm diameter, the actual core diameter of the thread being cut). The cross-slide's scale is then marked at this exact point by using pencil, (this pencil mark now indicates the full depth of cut required on its finishing cut to obtain its full depth of thread). The cross-slide's hand-wheel is then turned anti-clockwise by one revolution to allow the tool to clear the work-piece. The lathe's motor is then switched back on in *reverse direction of rotation*, to take the tool back to its start position, where the motor is then stopped by using the hand stop lever. The cross-slide's hand-wheel is then turned clockwise by *almost* one revolution down to a point of approximately half a millimeter *before* the previously set zero mark on its scale. The lathe is then switched back on in its normal direction of rotation to commence the next cut, (the coolant pump is also switched on at this time).

During this passing cut, the tool is indexed inward carefully clockwise (using the cross-slide's hand-wheel) until the tool has reached the new penciled mark on its scale and is again seen to be removing a minute sliver of metal from both sides of the thread,. On the tool reaching the under-cut area, and by noting with a brief glimpse of the 'warning marker affixed to the chuck's periphery, the *quick method of tool withdrawal* is used of one revolution anti-clockwise of the cross-slide's hand wheel, the motor is then instantly reversed to take the tool back to its start point of 5 to 6 mm to the right of the work-piece, followed by the motor being instantly stopped. The cross-slide's hand wheel is then turned clock-wise by one revolution to prepare the lathe for the next cut. The motor is switched back on in its normal direction of rotation but with the addition of 0.05 mm (the depth of the next cut) being indexed inward clock-wise on the cross-slide's hand wheel's scale and marked with pencil. On reaching the end of this cut, the *snatched method of tool withdrawal* is again used as before.

Successive 'cuts of 0.05 mm in tool depth are then used for each passing cut until the '*full depth of thread' pencil mark* (previously made on its scale) is reached, indicating that the tool has now reached its full depth of thread, (this being 1.85 mm).

A thread ring gauge is then used to check that the screw-cut thread now possesses a correct fit without any end float or wobble being present.

Section 18/24 b
Replacing a damaged screw-cutting tool midway through a straight plunge screw-cutting operation, when using the disengaged lead screw technique

If the machinist, is using the straight plunge method of screw-cutting coupled with using the disengaged lead-screw technique and discovers that the screw-cutting tool is damaged or requires replacing or sharpening, it must now be replaced with a new or re-sharpened tool. This must be set at its correct cutting height with the minimum stick-out from its tool -post and oriented correctly by using a square-type screw-cutting setting gauge. To do this, the top datum edge of this gauge must be oriented accurately against the parallel turned 23.9 mm diameter, of the workpiece. The tool's cutting edges are then entered into the 60-degree cutout in the gauge; when found to be correct, the tool-post's securing screws are fully tightened onto the tool's shank. The cross-slide and compound slide hand wheels are then turned clock-wise by a quarter of a turn to eliminate any existing backlash in their lead screws and their scales are set to read zero,; a cursory check should then be made to ensure that the compound slide is not overhanging its support casting. The new tool's cutting point should also be checked to ensure that it is set at its correct cutting height by using the lathe's tool height setting gauge. This is followed by positioning the tool's point, (using the cross-slide's hand wheel) to just touch the previously turned (23.9 mm) of the thread's major diameter. It is most important at this time that the cross-slide and compound slide's hand-wheels are both being turned in a *clockwise* direction (to eliminate their existing backlash prior to the tool making contact with the diameter of the material). When contact has been made, both indexing scales should then be reset to read *zero*. The cross-slide's handwheel is then turned anticlockwise by one revolution to clear the workpiece. The tool assembly is then repositioned back to its starting position of 5 to 6 mm to the right of the work-piece by using the saddle's hand wheel. The cross-slide's hand wheel is then turned clock-wise by one revolution and down to its zero setting. The partially finished thread is then sparingly brushed with micrometer blue, while avoiding leaving any excess in the thread.

The lathe is then switched on in its normal direction of rotation, and the front apron's rotating screw-cutting dial is carefully studied until the machinist's previously chosen line or number is seen to come into alignment with the apron's fixed datum line, (the exact position of this chosen number or mark should also be memorized for future reference). The *screw-cutting lift-up lever* is then operated to engage the lead-screw fully to start a new screw-cutting passing cut. During this exploratory pass, both of the slide's hand-wheels are minutely turned independently, using small amounts, (also ensuring that they are both still being turned in a clock-wise direction to eliminate any backlash being present in their lead screws). These minute tool movements must be of sufficient accuracy to establish that the tool's point and its angled cutting edges are finally centrally disposed in the rotating thread, and are seen to be removing a thin sliver of blue and a thin sliver of metal from the previously (partly finished) thread during this passing cut. (Note that several passing cuts may be needed before true centralization of the tool can be finally achieved.). The lathe is then stopped midway along this pass, and its cross-slide scale is carefully reset to read *zero*. The lathe is then switched back on to finish the cut, and the lathe is then instantly stopped at the point where the tool is exactly positioned to be just in the undercut area,; the cross-slide's hand-wheel is now turned very carefully clockwise by a sufficient amount to allow the tip of the tool to just touch the under-cut's diameter. A pencil mark is then made on its scale to show that this will be the final full depth of thread, (the undercut's diameter was previously machined to this exact full depth of thread diameter, making this diameter identical to the final depth of thread required). It is important that the machinist also now monitors the position of the warning marker (previously affixed to the chuck's outside diameter), *which may now require some adjustment.*

As previously stated, the exact tool's extraction point must now be anticipated and the machinist must also be aware of the possibly new position of the rotating adhesive warning marker, which would normally enable him to judge the exact moment when the tool *must be withdrawn instantly* (by one revolution anti-clockwise). With the lathe stopped at this point, the warning marker can be adjusted into its correct place on the chuck's periphery consistent with the position of the tools accurate withdrawal. The tool is now withdrawn from the work-piece by using one revolution anti-clockwise of the cross-slide's hand wheel, the *screw-cutting knockdown lever* is also instantly operated and the assembly is returned back to its start position of 5 to 6 mm to the right of the work-piece by using the saddle's hand-wheel. The cross-slide's hand-wheel is then turned clockwise down to *almost* the *zero* mark indicated on its scale. The lathe is then switched on in its normal direction of rotation, its coolant is switched on, and the rotating screw-cutting dial on the lathe's front apron is again studied until the chosen mark or number is once again seen to line up with the apron's fixed datum line, whereupon the *screw-cutting lift-up lever* is operated to commence a new screw-cutting passing cut. During this passing cut, the cross-slide's hand-wheel is turned clockwise very carefully down to its zero mark and beyond until the tool is again seen to be removing a very fine sliver of blue and metal from both of the tools cutting edges and its tip. On the tool reaching the undercut area, and also by the machinist observing the new position of the chuck's warning marker as it is seen to pass through his peripheral vision, the *snatched method of tool withdrawal* (of one revolution anti-clockwise), coupled with instantly using the *screw-cutting knockdown lever*, followed by returning the tool assembly back to its starting position by hand using the saddle's hand-wheel.

With the coolant supply now switched on, it will now be possible to complete the required number of passing cuts to finish the job, by using the new partial depth of thread zero setting as a new continuation point, and by taking the necessary 0.05 mm depth of cuts for each of the following passing cuts (these being indexed inward in a clock-wise direction using the cross-slide's scale), until the pencil mark is reached which indicates that the full depth of thread (calculated to be 1.84 mm,) has been reached. A thread ring gauge is then used to check the screw-cut thread for an accurate fit.

Section 18/24 c
Replacing a damaged screw-cutting tool midway through straight plunge screw-cutting operation when using a lathe with a permanently engaged lead-screw, and where the machinist has failed to note down the depth of cut so far reached by the damaged tool before its removal.

If the crossslide's scale's 'depth of thread' indication was *not* noted down when the tool was removed for inspection or possible replacement, a new replacement 60-degree included angle screw-cutting tool should now be fitted into the tool-post, while ensuring that only the minimum of stick-out from its tool post has been allowed. It must be oriented correctly by using a square-type screw-cutting gauge (shown in fig. 49). This must be held with its front datum edge in contact with the turned screw-cutting diameter of the workpiece while the angled cutting edges of the tool are accurately entered into the gauge's 60-degree precision cutout V. The tool's final position when correct is then secured by using the tool-post's clamp screws. The new tool's cutting height should also be checked for height by using the lathe's tool height setting gauge. It is most important during the tool's resetting procedure, that both of the slide's hand wheels are being turned in a *clockwise* direction in order to eliminate the backlash that exists in their lead-screws. Both scales should now be set to read zero, and the machinist should also ensure there is no overhang of the compound slide beyond its support casting, to ensure complete rigidity of the setup. The cross-slide's hand-wheel is now turned clock-wise to allow the cutting tool's tip to just touch the 23.9 mm diameter of the screw-cut thread; its scale is then reset to read zero. The tool is then moved to its starting position of 5 to 6 mm to the right of the work-piece. The thread is now painted sparingly with micrometer blue by using a brush to avoid excess.

The lathe is now switched on at a low spindle speed of 60 rpm, in its normal direction of rotation, the lift-up screw-cutting lever is then engaged exactly where the machinist's chosen line or number indicated on the lathe's front apron's rotating screw-cutting dial is seen to align with the apron's fixed datum line (note that this used position should be memorized while the lathe is on this screw-cutting pass to enable this identical position to be used for subsequent screw-cutting passes).

During this exploratory passing cut, both of the slide's hand-wheels are each turned clockwise independently, by minute amounts, to centralize the tool's point and its cutting edges into the V-grooved *partially cut thread* during the pass', (we must ensure that both hand-wheels are being turned in a clockwise direction at the point where centralization of the tool in the partially cut thread is finally achieved). To confirm the tool is in its correct position, it should be seen to be removing a very thin sliver of blue and a very thin sliver of metal from both of its cutting edges and its point during this or any following checking passes. The lathe should now be stopped halfway along this pass, and its compound slide's scale should be re-zeroed (taking care not to move the position of its hand-wheel).' a pencil mark should then be made on the cross-slide's indexing scale to indicate the depth of thread the tool has so far reached. The lathe is now restarted, and the tool, on reaching the under-cut area of the thread, is stopped instantly by using the emergency stop pedal or lever. The adhesive warning marker previously affixed to the chuck's periphery may now need to be adjusted to a new more accurate position to indicate the exact position where the tool must be instantly withdrawn at the end of a pass.

With the lathe now stopped and with its cutting tool now having just entered into the under-cut clearance area, the cross-slide's hand wheel is now turned very carefully and (minutely) clockwise until the tool's point is seen (and felt) to be just touching the minor diameter of the *undercut* (this diameter being identical to the core diameter of the finished depth of thread). The cross-slide's scale is then marked with pencil to indicate the exact full depth of thread the tool will ultimately reach on its finishing cut. The cross-slide's hand wheel is then turned anti-clockwise by one revolution to allow the tool to clear the work-piece. The lathe is then switched back on in reverse direction (while still leaving the lead-screw engaged), to return the tool assembly back to its start position of 5 to 6 mm to the right of the workpiece, where it is again stopped by using the hand stop lever (or foot pedal).

The cross-slide's hand wheel is then turned in a clock-wise direction almost one revolution down to the first *partial depth of thread* pencil mark, followed by it being indexed inward by a further 0.05 mm to put on the next cut, where its scale is again marked with pencil. The lathe is then re-started in its normal direction of rotation. The tool is allowed to complete this pass, and the lathe is then instantly stopped (using the guidance of the repositioned warning marker) on the tool reaching the under-cut area. The cross-slide hand-wheel is then turned anti-clockwise by one revolution to allow the tool to clear the work-piece and the lathe is re started in its reverse direction of rotation until the tool has returned to its start point of 5 to 6 mm to the right of the work-piece, where the lathe is then stopped. The cross-slide's hand-wheel is then turned clockwise by one revolution down to the new penciled mark that indicates its *new partial depth of thread*, followed by indexing the tool minutely inward *clock-wise* by a further 0.05 mm, to prepare the lathe for the next cut to be taken. The lathe is then switched back on in its normal direction of rotation, the coolant is also switched on to supply both the tool and work-piece, and the screw-cutting passing sequence is then repeated.

These screw-cutting sequences are continued onward by using a series of 0.05 mm depth of cut for each pass until the cross-slide's scale is seen to reach the *full depth of thread' penciled mark* on its scale (the depth of thread reached should now be the required 1.84 mm depth of thread), obtained previously from the tool's contact with the diameter of the accurately machined undercut),; this will now indicate that the full calculated depth of thread has been reached. A thread ring gauge should now be used to check the thread for its accuracy of fit.

Section 18/25
Replacing a damaged screw-cutting tool, midway through an angle approach screw-cutting operation, when using a permanently engaged lead-screw, should the machinist have failed

to note down the depth of cut so far reached by the tool before its removal.

If the machinist is using the *angle approach* method of screw-cutting (that is with the lathe's compound slide already angled and locked at 60 degrees), as shown in fig. 42, with its main lead-screw permanently engaged, and the tool discovered to be damaged, it should be removed and replaced with an identical and correctly re-ground tool, the orientation of its cutting edges should be correctly set up by using the square thread setting gauge (shown in fig. 49) to obtain its correct re-alignment with the work-piece. The machinist must ensure that the replaced tool and its tool-post are being fully supported by the compound slide's casting, and that the tool is only allowed to possess a minimum of overhang from its tool-post; its cutting point should also be set at its correct cutting height by using the lathe's *tool height setting gauge*. It is *essential* that the *compound* slide is not allowed to overhang its support casting; this must be done to ensure that the maximum rigidity of the setup is maintained. It is also important that the compound and cross-slide's hand wheels have been turned clock-wise by at least a quarter of a turn to eliminate any backlash existing in their lead-screws; this is followed by resetting their scales to read zero. If a penciled note was not made of the depth of thread so far reached by the damaged tool on the *compound and cross-slide*'s scales before the tool was removed for inspection, it will be found best practice to adopt the following procedure in order to set up the new tool into its required accurate position, to allow it's cutting edges to pick up the existing partially cut thread while on a new passing cut.

With the lathe now stationary, the *cross-slide* hand wheel is now turned clock-wise until the tool's tip is seen and felt to be just touching the major screw-cutting diameter of 23.9 mm; its scale is then reset to read zero. This is followed by its hand-wheel being turned anti-clockwise by one revolution to allow the tool to clear the work-piece. The tool is then moved to its start position of 5 to 6 mm to the right of the work-piece by using the saddle's hand wheel, in preparation for making a provisional exploratory passing cut. The *cross-slide*'s hand wheel is then turned clock-wise by one complete revolution down to the set zero mark on its scale.

The partially cut thread is then lightly painted with micrometer blue, by using a brush to avoid any excess. The lathe is then switched on in its normal direction of rotation at a spindle speed of approximately 60 rpm, and a close scrutiny is then made of the *rotating screw-cutting dial* situated on the lathe's front apron, until the chosen line (or number) is seen to line up with the apron's fixed datum line. The *lift-up screw-cutting lever* is then operated to engage the lead-screw to start an experimental cut. (A mental note should now be made of the chosen line or number on the dial for future use).

During this passing cut, both hand wheels will need to be turned very carefully and independently in a clock-wise direction to maintain their backlash in its correct direction, while the tool's cutting edges are maneuvered into a centralized position in the revolving and partially cut thread during this passing cut. The accurate centralization of the tools angled cutting edges into the thread may take several passes to complete. The tool can only be considered centralized when it is seen to be removing a thin sliver of blue and a thin sliver of metal from mainly its *left-hand side cutting edge and its point* on this passing cut (this is because we are now using the *angled approach* method of screw cutting, where the removal of metal will be predominantly from its *left-hand side cutting edge and its point*).

On the tool reaching the *under-cut* area, the lathe is instantly stopped by using either the hand lever or the emergency foot stop control. The rotating chuck's warning marker, (previously fixed to its outside diameter), may now need to be repositioned slightly to suit the new cutting tool's position relative to the under-cut's position. With the tool now in this situation, the cross-slide and the compound slide's scales are now set to read zero. The undercut's clearance diameter should now allow the compound slide's hand-wheel to be turned very carefully clock-wise until the tool's tip is seen (and felt) to be just touching the previously turned under-cut's diameter of 20.3 mm, (this being identical to the diameter required by the tool's cutting tip on reaching its full depth of thread). The *compound slide*'s scale is then marked with pencil to indicate what is actually the full depth of thread. The *cross-slide*'s hand wheel is now turned anti-clockwise by one revolution to clear the work-piece, and the tool assembly is returned to its starting point by switching the lathe back on in its reverse direction, until the tool has reached a point 5 to 6 mm to the right of the workpiece, where the motor is instantly stopped by using either the hand lever or the foot stop pedal. The lathe's coolant is now switched on over the area of the work-piece about to be screw-cut.

The cross-slide's hand wheel is now turned clockwise down to its scale's zero mark. The *compound* slide is now turned anti clock-wise by half a revolutionfollowed by being turned clock-wise down to the previously penciled line, and indexed inward minutely further clockwise by 0.05 mm in order to put on the next cut. This is followed by switching the lathe back on in its normal direction of rotation to allow the next cut to be taken.

[Note that because we are now using the *angle approach* method of screw-cutting, the cross-slide hand wheel must in this case always be returned exactly to its zero position before taking a cut. (It should also be noted that the full depth of thread required when using the *angle approach* screw-cutting method will be reached when the compound slide's scale has actually traversed 2.125 mm from its originally set zero position,as shown in fig.42a), (whereas the full depth of thread required when using the straight plunge method of screw-cutting will be indicated when its *cross-slide*'s scale has reached 1.84 mm, also shown in fig. 42a].

A series of passing cuts are now carried out on the work-piece by returning the cross-slide to read zero on its scale prior to each cut being taken, and by indexing inward the *compound* slide and its scale by increases of 0.05 mm, for each passing cut until the full depth of thread (previously marked in pencil on its scale), indicates that the full depth of thread has been reached. A thread ring gauge should then be used to check the thread for correct fit.

Section 18/25 a
Replacing a damaged external screw-cutting tool partway through an angled approach screw-cutting procedure, while using a *disengaged* lead-screw.

If the machinist needs to replace a damaged screw-cutting tool midway through an angled approach screw-cutting procedure while using the disengaged lead-screw technique, it is advisable to use the following procedure to establish that the tool has been accurately replaced and is set perfectly in line with the unfinished screw-cut thread. It is important that the following sequences of operations are used in order to obtain complete accuracy following the replacement of the new tool.

With the lathe now stationary and with the replacement tool accurately replaced at its correct cutting height and at its correctly oriented cutting angle in the tool-post (by using a square-type screw-cutting gauge with its top datum edge touching the 23.9 mm diameter to obtain the tool's cutting edges correctly orientated relative to the work-piece's thread), the tool is now moved away from the work-piece by turning the cross-slide's hand wheel in an anti-clockwise direction to position the tool well away from the work-piece, to avoid accidental contact occurring between the work-piece and other parts of the lathe. The stability of the compound slide should be checked to ensure that its slide is not overhanging its support casting and that the replaced tool is only protruding from its tool-post by the minimum required amount to maintain full rigidity of the setup. (The compound slide's hand-wheel may need be turned anti-clockwise until full support has been achieved, followed by turning it clockwise by (say) a quarter of a turn, and its scale reset to read zero, this will then establish that the compound slide's lead-screw backlash has been taken up in its correct clock-wise direction.) The cross-slide's hand wheel is then turned clockwise and the saddle assembly is maneuvered to allow the tip of the screw-cutting tool to be seen (and felt) to be just touching the 23.9 mm diameter of the thread being cut; its scale is then reset to read zero. The cross-slide's hand wheel is then turned anticlockwise by one revolution to enable the tool to clear the work-piece. The screw-cutting tool is then positioned back at its start position of 5 to 6 mm to the right of the work-piece by using the saddle's hand-wheel. The cross-slide's hand wheel is then turned clockwise by one complete revolution, down to its set zero scale's position.

The unfinished thread is then sparingly painted with micrometer blue, using a brush to leave no excess of blue on the thread. The lathe is then switched on at a spindle speed of approximately 60 rpm in its normal direction of rotation. A close scrutiny is then made of the rotating screw-cutting dial situated on the lathe's apron. The *lift-up screw-cutting engaging lever* is then operated at the exact moment when the machinist's chosen mark or number is seen to co-incide with the apron's stationary datum line. (The exact position used should be memorized and noted down for use later, on the next screw-cutting passing cut). During this exploratory screw-cutting cut, the cross-slide and the compound slide's hand-wheels are now turned very carefully, (keeping them turning in a clockwise direction), to position the tool's predominantly left hand cutting edge and its point to accurately pick up the contours of the revolving thread during the pass.

Because we are using the angle approach method of screw-cutting, it is important that the tool's left-hand side cutting edge and its cutting point are being accurately positioned so that they pick up the partially finished thread during this pass. It is also most important that the machinist is actually turning both of the hand-wheels carefully in a *clockwise* direction at the moment that centralization of the tool in the thread has been achieved; this must be done in order to eliminate the risk of any back-lash being present in either of the slide's lead-screws.

During this passing cut, the tool must be seen to be removing a fine sliver of micrometer blue and a fine sliver of metal from the thread by using (mainly) its left-hand side cutting edge and its point. Its right-hand side cutting edge should be *ideally* merely brushing the right-hand side of the thread without actually removing metal. When full centralization of the tool in the thread has been achieved, the lathe should now be stopped along its present cut, and its *compound slide's scale* marked with *pencil* to indicate the depth the partially cut thread has so far been reached by the tool.

The cross-slide's scale is now set to read zero, and the lathe is switched back on to complete this passing cut, allowing the tool to reach the beginning of the under-cut area (this position is of course aided by the machinist catching a fleeting glimpse of the (now repositioned rotating chuck's tool extraction warning marker); the lathe is then instantly stopped by using either the hand stop lever or the foot stop pedal control. With the tool now in this position, the compound slide's hand-wheel is now turned *very carefully* clockwise by a sufficient amount to allow the tip of the tool to actually touch the undercut's 20.3 mm diameter. The *compound slide's scale* is now marked with *pencil* to indicate exactly where the tool's full depth of thread actually occurs when engaged on its finishing cut; this is followed by turning its hand-wheel anticlockwise by half a revolution, followed by turning it back clock-wise down to its previously set *partial depth of thread pencil mark*. The cross-slide's hand wheel is then turned anti-clockwise by one revolution to remove the tool from the work-piece, and the saddle's hand-wheel is then turned clockwise to return the tool to its start position of 5 to 6 mm to the right of the work-piece. The cross-slide's hand wheel is then turned clock-wise by one revolution down to its set zero mark followed by the compound slide's hand-wheel being indexed minutely inward clockwise by 0.05 mm to put on the depth required for the next cut. The lathe is then switched back on, and a close scrutiny is now made of the rotating screw-cutting dial, situated on the front of the lathe's apron, the *screw-cutting lift-up lever* is then operated when it co-incides with the machinist's exact chosen line or number as it is seen to line up with the apron's fixed datum line. The lathe now takes the next passing cut. When the tool has reached the end of this passing cut, having just entered the under-cut area, the snatched method of tool withdrawal, of one revolution anti-clockwise is made by the cross-slide's hand wheel, coupled with *instantly* operating the *screw-cutting knockdown lever*, to isolate the tool from its lead-screw. The tool is now moved back to its starting position of 5 to 6 mm to the right of the work-piece. A series of passing cuts are now carried out by turning the cross- slide's hand wheel one revolution clock-wise down to the zero mark on its scale, and using the compound slide to index inward a minimal in-feed of 0.05 mm depth of cut for each of the following passing cuts, until the compound slide's scale reaches its full depth of thread indicated by its *full depth penciled line*. The lathe is now switched off and the tooling is moved away from the work-piece, followed by a *thread ring gauge* being screwed onto the thread to check its fit.

If the thread has been cut correctly, the gauge should be able to be screwed on without exhibiting any slackness or end float being present. If it is found to be too tight, then a further screw-cutting pass (or passes) will need to be taken. When a correct fit has finally been obtained on the thread we can then use the screw-cutting tool to independently machine a 30 degree chamfer on left-hand end of the screw-cut thread, by using the screw-cutting tool's right-hand cutting edge, as explained in the following procedure:.The lathe is now switched on and the cutting point of the tool is maneuvered over the under-cut area to enable it to be entered very carefully into this area by using the cross-slide's hand-wheel being turned in a clock-wise direction and down to its scale reading of zero (indicating that the tool is now at its full depth of thread of 1.84 mm), the *saddle's hand wheel* is then turned very carefully clockwise until the end of the thread is seen to be just starting to be machined; this is followed by very carefully moving the tool (using the saddle's hand-wheel) a further 1.5 mm to its right to complete a 30 degree chamfer on the left-hand end of the thread. The cross-slide's hand-wheel is then rotated anticlockwise by one revolution to clear the work-piece. A thread ring gauge can then be used to check the final fit of the screw-cut thread.

Section 18/25 b
Replacing a damaged screw-cutting tool partway through an angled approach screw-cutting procedure, when the lathe is used with a permanently engaged lead-screw, and where the machinist has failed to note down the depth of thread so far reached by the tool, before its removal

If the machinist has failed to note down the compound slide scale's depth of thread reading prior to the tool's removal for inspection or replacement, then the tool's depth of thread settings and the correct position of the warning marker on the chuck's outer periphery will also have been lost by the tool's removal, therefore the tool's full depth and partial depth of thread settings, and its accurate position relative to the thread being cut must now be re-established before one can continue the screw-cutting operation.

In this case, after the new screw-cutting tool has been accurately oriented in the tool-post by using the square-type thread setting gauge's top datum face being held against the work–piece's screw-cutting diameter, the tool's cutting edges must then be aligned into its

60 degree angled cut-out, and the new tool's cutting point must also be set at its correct cutting height. We must also ensure that the tool has only the minimum of stick-out from its supporting tool-post, and that the compound slide is not overhanging its support casting in order to maintain rigidity of the setup.

Having now checked all these points visually, we must now discover the new tool's intended final depth of thread setting, by first turning both the cross-slide and compound slide's hand-wheels in a clockwise direction by a sufficient amount, (usually half a turn), to remove all the backlash existing in their lead-screws; this is followed by re-setting both of their indexing scales to read zero. The cross-slide's hand wheel is then turned clock-wise to position the point of the tool to just touch the 23. 9 mm screw-cutting diameter previously turned on the work-piece; its scale is then reset to read zero. The cross-slide's hand-wheel is then turned anticlockwise by one revolution to allow the tool to clear the work-piece.

Micrometer blue is then applied *sparingly* to the partially cut thread using a brush. The tool is then repositioned at its starting position of 5 to 6 mm to the right of the work-piece by using the saddle's hand wheel, followed by turning the *cross-slide*'s hand-wheel one revolution clock-wise to reach the set zero mark on its scale. The lathe is then switched on in its normal direction of rotation at a low work speed of approximately 60 rpm. An experimental passing cut is then commenced by operating the *lift-up screw-cutting lever* at the exact point where the machinist's chosen line or number (previously selected) on the apron's rotating screw-cutting dial, is seen to line up with its fixed datum line. (A mental note should now be made of this chosen line or number for future use).

During this experimental passing cut, the *cross-slide* and *compound slide*'s hand wheels are turned very carefully (keeping them *always* turning in a clockwise direction in order to continue eliminating their existing backlash), until the cutting edges of the tool are seen to be accurately picking up the contours of the existing partially cut thread.

Because we are now using the angle approach method of screw-cutting, the tool's left-hand side cutting edge and its cutting point must now be seen to be removing a fine sliver of micrometer blue and a fine sliver of metal (while using predominantly the tool's left-hand side cutting edge and its point during this pass). The tool's right-hand side cutting edge should (ideally) be merely brushing the thread's surface and not be removing any significant metal from the thread. (Several passes may be needed to obtain true tool centralization in the partially cut thread.). The tool should be carefully monitored during this passing cut, and on reaching the beginning of the undercut area it must be instantly stopped by using the foot stop pedal or the hand-operated stop lever. The *cross-slide's scale* is then set to read zero, and the *compound slide's scale* is then marked with a pencil line showing the partial depth of thread the tool has now reached.

At this point, with the lathe stopped, the *compound slide's* hand wheel is very carefully turned clock-wise by minute amounts until the tip of the tool is seen (and felt) to be just touching the under-cut's diameter, (a magnifying glass can be used to check this), its scale is then marked with *the second pencil line* to indicate the tool's *full* depth of thread, (to be later reached by the tool on its final passing cut (note that the under-cut's diameter will have been previously turned to 20.3 mm diameter, this being identical to the core diameter of the thread being cut).

The *compound slide's* hand wheel is then turned anticlockwise by half a revolution, and the *cross-slide's* hand wheel is turned anticlockwise by one revolution to allow the tool to clear the workpiece.

The lathe is then switched back on, in *reverse* direction of rotation, to return the tool to its start position of 5 to 6 mm to the right of the work-piece, where the lathe is then stopped by using the hand stop lever or the foot stop pedal.

We have now established the new settings (by the use of two penciled lines on the compound slide's scale) that will now indicate both the *full depth of thread* and the *partial depth of thread*. We must now adopt the following procedure.

The *cross-slide's* hand wheel is then turned clockwise by one revolution down to its previously set *zero* mark; the *compound slide's* hand wheel is then turned half a revolution clock-wise until its scale is again seen to read the previously marked *partial depth of thread pencil mark*. We then slightly increase the tool's depth by indexing inward clockwise a movement of 0.05 mm in order to prepare the tool for its next cut. The lathe is then switched back on in its normal direction of rotation to allow the lathe to take its next passing cut.

On the tool reaching the under-cut area, the lathe is instantly stopped by using the hand stop lever or the foot stop control pedal. [Note that at this point the warning marker on the chuck's periphery may now need to be adjusted into its correct position to give warning to the machinist of the precise moment the *snatched method of tool withdrawal* is to be used. This tool withdrawal is now made, using one quick anti-clockwise rotation of the *cross-slide*'s hand wheel, followed by the lathe's motor being switched off,followed by being switched back on in *reverse direction* to take the tool back to its start position of 5 to 6 mm to the right of the workpiece, where the motor is then again stopped.

The compound slide is now indexed in clock-wise by 0.05 mm. The lathe is now switched back on in its normal direction of rotation, the coolant is also switched on to supply both the tool and the workpiece and the next cut is taken.

This new cut is then followed by using the *snatched method of tool withdrawal* on reaching the undercut area, and instantly stopping the motor, followed by *reversing* the lathe's motor to return the tool to its start position of 5 to 6 mm to the right of the workpiece. The *cross-slide*'s hand-wheel is then turned one revolution clockwise down to the *zero* reading on its scale. This is followed by taking successive cuts of 0.05 mm in depth of cut by using the *compound slide scale*'s graduations, and is continued until this scale is seen to have reached the full depth pencil mark (previously made on its scale; this penciled mark now gives us the finished angled approach depth of thread required of 2.12 mm.

The thread ring gauge is then used to check the thread for correct fit.

If the gauge still proves to be rather a tight fit on the thread, then extra cuts of 0.05 mm depth for each cut will need to be taken, while still using the *compound slides scale*'s graduations as a guide, until a correct fit has been established.

Section 18/26
Choosing the correct work speed for screw-cutting steel components, while also confirming the need to make a trial cut

on the test material to establish whether the correct pitch has actually been selected for screw-cutting the 3 mm pitch of the thread required

Having fully assimilated the foregoing guidance notes regarding tool replacement and any re-sharpening or replacement found to be necessary during the screw-cutting operation, we must now return to a more in-depth explanation of the correct way to select a work speed for screw-cutting, and to finally establish the correct gear lever positions for screw-cutting our fig. 43 example of an external 3 mm coarse pitch metric thread being cut on steel material.

The machinist is strongly advised to refrain from attempting to change from a higher work speed to the recommended 60 revs per minute lower spindle and work speed, *during* the screw-cutting operation. The reason for this is that while one is going through the process of changing the gear lever's position, (the gear teeth in the gearbox will of necessity have to pass through a period of being in a neutral gear); therefore, the previously set mesh of the gears will momentarily become out of step on their re-engagement with the next selected spindle speed, and this can cause the tool's original relationship with its lead-screw to be temporarily lost. This situation can result in a slight misalignment of the cutting tool relative to its previously cut thread when re-engaged.

We must now consider the problem of checking the accuracy of the pitch of the thread to be cut on the lathe. Having selected the appropriate gear lever positions to comply with the engraved instructions printed on the lathe's fixed instruction plate to obtain the required pitch of 3 mm, we can easily check this pitch practically, by the simple process of clamping a piece of test material, usually a piece of round scrap bright mild steel material (its actual diameter must be at least half an inch (12.7 mm) diameter to ensure rigidity of the set-up), into the lathe's chuck. We now select a sharp (and preferably pointed) lathe tool, (it need not necessarily be a screw-cutting tool), and locate this securely at its correct cutting height in the tool post. Now, using the cross slide's hand-wheel being turned in a clock-wise direction, we allow the point of the tool to just touch the diameter of the test piece, we then zero its scale, followed by the tool being moved to the right to clear the test piece. We then turn the hand-wheel clockwise by an additional 0.1mm (indicated on its scale), apply coolant, switch on the lathe, and operate the *lift-up screw-cutting lever*, allowing the tool to perform a mock screw-cutting operation on the material. The lathe tool should now produce a shallow spiral groove along the material's surface; this is allowed to progress along over a nominal distance of (say) 12-15 mm, where the lathe is then stopped.

Having now progressed the tool along (under its own power) for several revolutions of the test piece, the resulting center to center distance between each screw-cut groove can now be carefully checked and measured to establish the required pitch accuracy.

This check can be made by using either a vernier caliper gauge (with its gap set to 3.0 mm), or a thread pitch gauge (the preferred option) (shown in fig. 49), using its 3 mm pitch measuring blade for comparison with the pitch just obtained, or by comparing the center distances achieved with the graduations of a precision rule; alternatively, the pitch can be checked by placing a known thread of the required pitch, in close proximity to the work-piece and comparing the two thread's pitches by using close scrutiny by eye.

Having now confirmed that the pitch of the thread the lathe is about to cut is correct, the test piece can now be removed and replaced with the piece of stock material selected for use as the screw-cut workpiece. Note that we should provide sufficient material stick-out (from the chuck) not only to allow for the full length of the finished work-piece (as required on the drawing), but also to provide an extra clearance distance to provide sufficient room for a parting tool of say 3 mm width, (necessary for the final parting off of the finished component), plus an additional safety clearance of at least another 2 mm required between the parting tool's left-hand side cutting edge and the jaws of the chuck.

Section 18/ 27
The special techniques required when using parting-off tools, and how various parting-tool problems can be dealt with. Also important information on the safety precautions required specifically when machining long workpieces.

It is important to note that the parting tool being used should possess a front cutting edge that has been ground to be as narrow as possible, consistent with it being robust enough in its shank area to support its cutting edge fully when under maximum load. The reason for this is that with a narrow front cutting edge, the parting tool will require considerably less power to cut into the material, and place less load on the tool and work-piece being parted off.

It should also be emphasized that it is important to always part off a component as close as possible to its supporting chuck's jaws, (or collet), in order to maintain the maximum rigidity of the set-up. (It follows that one should never attempt to part off a component further away from the chuck without the work-piece being positively supported by a fixed steady or, in difficult cases, by using a greased tailstock center's support to *partially* part off the component, by leaving (say) a 6 mm diameter diameter to be finally cut through with a hand-held hack saw, with the work-piece stationary. Also, one should never attempt to part off a component while it is being held solely between centers, as this would allow the work-piece to be completely unsupported in the latter stages of the parting-off operation (this fact will become obvious and abundantly clear as machining experience is gained).

When machining differing types of work-piece in the future, the actual component being machined will probably be much longer than our fig. 42 illustrated example; therefore, the machining of a longer work-piece will need to be given serious consideration regarding the correct method chosen for its support. The diameter of the gripped end of a very long component will have a definite advantage if it can pass through (with clearance), the hollow throat diameter of the lathe's spindle. The outer right-hand end of the work-piece requiring machining must be capable of being supported by either a greased dead center (secured in the tailstock), or a fixed steady positioned as close as possible to the point where the machining is actually taking place.

Care should be taken to ensure that only the minimum stick-out of material is allowed to protrude from the outer (left-hand) end of the machine's hollow spindle; if this must be the case, then protective guards must always be put in place to protect passers-by (for obvious safety reasons) to prevent them walking into a dangerously rotating piece of material.

The author mentions this safety problem particularly, because he once witnessed a dangerous material stick-out accident that occurred on the lathe immediately in front of the machine he was working on.

Having casually noticed that the machinist immediately in front of my machine was allowing a rather long stick-out of excess material to extend from the left-hand end of his lathe's spindle, and before a warning could be given, it was seen that he switched on his machine (which he had unfortunately left in a wrongly chosen high-speed range), causing the unsupported material, that immediately reached the lathe's maximum rotational top speed, to momentarily wobble, then alarmingly bend to about 45 degrees (due to the very high centrifugal forces involved), followed by proceeding to flail around dangerously, accompanied by a loud clattering noise as it reduced its protective guard to scrap.

Being in close proximity to this incident proved to the author, beyond any doubt, how dangerous it is to allow long unsupported stick-out lengths of material to project from the lathe's left-hand 'outer material entry end' of its spindle, particularly when the machinist is attempting to turn flimsy, easily distorted materials at high spindle speeds.

It became immediately obvious that the machinist should have used a much slower spindle speed, rather than the lathe's high-speed range to perform this task. The moral to be learned from this is, do not hang out an excessive length of structurally weak material beyond the outer extremities of the lathe's hollow spindle, and if this must be done, the machinist should use a *low* spindle speed and provide substantial visual guards to protect any passers-by.

If, for example, a very small-diameter material is being machined (of, say, 6 mm diameter), then the stick-out working length allowed from the front of the chuck's jaws must of course be proportionately very much shorter in order to maintain sufficient rigidity for the turning operation to take place; for example, it may be found virtually impossible to turn a piece of material of a smaller diameter than 6 mm with any accuracy, if it is allowed to possess a stick-out length of more than twice its diameter, without adequate support being given. This problem is mainly due to the probability of the work-piece material being pushed off by the cutting tool, particularly if the tool possesses a 'larger than is necessary' cutting radius. The tool for this type of work should ideally possess a very small cutting radius of 0.05 mm in order to reduce any push-off of the tool occurring].

The following paragraphs explain what the machinist should do if the work-piece drawing specifically states that it must not possess a drilled support center in its exposed outer end.

If the work-piece designer has indicated on his drawing that the work-piece's right-hand end should be left perfectly flat, (that is, by the omitting a drilled support center hole in its end), and we are required to machine say, a flat top hydraulic piston or a similar component requiring a completely flat and smooth end), we would in this case have to use, a female support center held in the tail-stock for its support; this would then require the need to machine a *male center* on the outer end of the work-piece to suit the female support center being held in the tail-stock.

Of course, after the work-piece has been turned, the now redundant male center will need to be machined off as it is now surplus to requirements, having provided the necessary tail-stock support for its previous turning operation.

Before a female support center can be used to support the work-piece (particularly when working on a long large-diameter workpiece that possesses a diameter that cannot be passed through the lathe spindle's hollow bore), the work-piece's outer diameter must in this case be supported by the use of a fixed steady (as shown in fig. 48).

The fixed steady must of course only be used in support of a *perfectly round* diameter. If the work-piece is of (say) black mild steel (which is not considered, in engineering parlance, to be actually round), then a suitable area of its outside diameter's surface must first be turned so that it is perfectly round in order to allow the fixed steady to be accurately used in this area to support the work-piece. (A support steady should always be used where there is considerable material overhang from the chuck.). This then provides the firm support necessary to enable a male center to be machined on its outer end face by the use of an appropriately oriented 30-degree angled chamfer tool.

[The dimensions and angle of the male center that needs to be machined on the work- piece must be identical to the basic male

60 degree included angled greased center normally used in the tail-stock for workpiece support. Its overall dimensions must therefore be machined to be perfectly compatible with the greased receiving female center held in the tailstock. However, on smaller diameters or shorter work-pieces when they are found to be capable of being entered into (and passed through) the spindle's hollow throat, the support of the chuck alone will usually be found to give sufficient support for this machining operation, provided the stick-out length from the chuck is kept to the absolute minimum. The male center' that is now to be machined onto the end of the work-piece must always be faced off on completion of the job by using a fixed steady to support the work-piece's major turned diameter during this facing-off operation. This operation is best performed by using a sharp right-hand knife tool.]

FIG. 42.

THE 'SET - UP' FOR EXTERNAL 'ANGLE APPROACH' SCREW - CUTTING ON THE CENTER LATHE.

USE THE FIG. 42a CALCULATION TO CONVERT THE 'STRAIGHT PLUNGE' DEPTH OF THREAD (1.84032), INTO THE REQUIRED 30 DEGREE 'ANGLED DEPTH OF THREAD'. WE THEN USE THE 'PROMPT' $\angle c, a = c$ DIVIDED BY COS. L, WE DO THIS BY ENTERING INTO THE CALCULATOR THE SEQUENCE 30, COS. (0.866025403), X - M; WE THEN ENTER 1.84032 DIVIDED BY RM = 2.1250 · · · · · THE ANGLED DEPTH OF THREAD. THIS USE OF A 'CONSTANT' METHOD IS SHOWN IN THE TEXT, AND ALSO IN FIG. 42b.
TO FIND THE 'STRAIGHT PLUNGE' DEPTH OF THREAD BY USING THE 'CONSTANT' NUMBER 31.526688, WE ENTER INTO THE CALCULATOR 31.526688, TAN, X 3, =, 1.84032 MM, THIS THEN GIVES US THE 'STRAIGHT PLUNGE' DEPTH OF THREAD FOR THE 'BOLT'. (NOTE THAT THE NUT FEATURED IN FIG. 44. WILL REQUIRE LESS DEPTH OF THREAD, THIS BEING 1.62381 MM + 0.108 MM = 1.73181, (ROUNDED UP TO 1.732 MM). (NOTE THAT IT WILL NOT BE FOUND NECESSARY TO USE MORE THAN TWO OF THE DIGITS AFTER THE DECIMAL POINT IN METRIC CALCULATIONS. TO CORRECTLY SET THE INDEX SCALES ON THE CENTER LATHE'S CROSS AND COMPOUND SLIDES.)

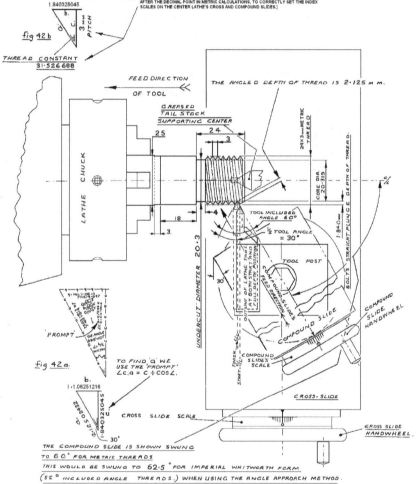

fig 42 b

THREAD CONSTANT
31·526688

THE ANGLED DEPTH OF THREAD IS 2·125 m m.

FEED DIRECTION
OF TOOL

GREASED
TAIL STOCK
SUPPORTING CENTER

LATHE CHUCK

24×3mm METRIC THREAD

CORE DIA. 20·319

1·84·0mm

BOLT'S STRAIGHT PLUNGE DEPTH OF THREAD.

UNDERCUT DIAMETER 20·3

TOOL INCLUDED ANGLE 60°

½ TOOL ANGLE = 30°

TOOL POST

COMPOUND SLIDE

COMPOUND SLIDE HANDWHEEL

COMPOUND SLIDE

PROMPT

fig 42 a.

TO FIND 'a' WE USE THE 'PROMPT' $\angle c, a = c ÷ COS \angle.$

CROSS SLIDE SCALE

COMPOUND SLIDE'S SCALE

CROSS-SLIDE

CROSS SLIDE HANDWHEEL

THE COMPOUND SLIDE IS SHOWN SWUNG
TO 60° FOR METRIC THREADS
THIS WOULD BE SWUNG TO 62·5° FOR IMPERIAL WHITWORTH FORM
(55° INCLUDED ANGLE THREADS.) WHEN USING THE ANGLE APPROACH METHOD.

ALL DIMENSIONS IN M.M.   DRAWN 20-5-14 G.N.R.

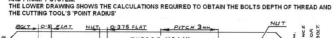

Fig 43.

FIG. 43. SHOWS A CROSS - SECTION OF A TYPICAL METRIC NUT AND BOLT. THIS EXAMPLE OF A 24 MM X 3 MM PITCH METRIC COARSE THREAD, SHOWS THE WORKING CLEARANCES REQUIRED FOR THE NUT AND THE BOLT. ALSO SHOWN IS A METHOD OF OBTAINING THE DEPTH OF THREAD FOR SCREW - CUTTING BY USING THE PROBE AND PROMPT SYSTEM.

THE LOWER DRAWING SHOWS THE CALCULATIONS REQUIRED TO OBTAIN THE BOLTS DEPTH OF THREAD AND THE CUTTING TOOL'S 'POINT RADIUS'

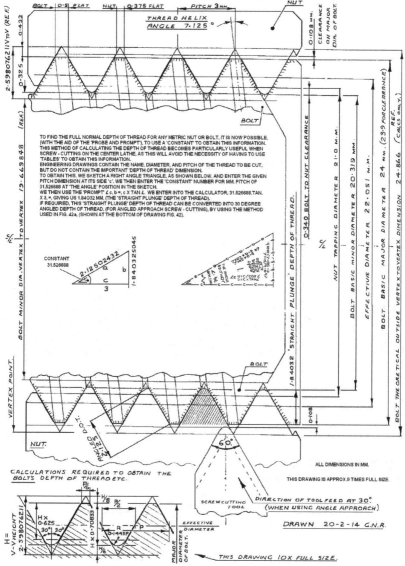

TO FIND THE FULL NORMAL DEPTH OF THREAD FOR ANY METRIC NUT OR BOLT, IT IS NOW POSSIBLE, (WITH THE AID OF THE 'PROBE AND PROMPT'), TO USE A 'CONSTANT' TO OBTAIN THIS INFORMATION. THIS METHOD OF CALCULATING THE DEPTH OF THREAD BECOMES PARTICULARLY USEFUL WHEN SCREW - CUTTING ON THE CENTER LATHE, AS THIS WILL AVOID THE NECESSITY OF HAVING TO USE 'TABLES' TO OBTAIN THIS INFORMATION.

ENGINEERING DRAWINGS CONTAIN THE NAME, DIAMETER, AND PITCH OF THE THREAD TO BE CUT, BUT DO NOT CONTAIN THE IMPORTANT 'DEPTH OF THREAD' DIMENSION.

TO OBTAIN THIS, WE SKETCH A RIGHT ANGLE TRIANGLE, AS SHOWN BELOW, AND ENTER THE GIVEN PITCH DIMENSION AT ITS SIDE 'c'. WE THEN ENTER THE 'CONSTANT' NUMBER FOR MM. PITCH OF 31.526688 AT 'THE ANGLE' POSITION IN THE SKETCH.

WE THEN USE THE 'PROMPT' ∠ c, b =, c X TAN ∠. WE ENTER INTO THE CALCULATOR, 31.526688,TAN. X 3, =, GIVING US 1.84032 MM. (THE 'STRAIGHT PLUNGE' DEPTH OF THREAD).

IF REQUIRED, THIS 'STRAIGHT PLUNGE' DEPTH OF THREAD CAN BE CONVERTED INTO 30 DEGREE ANGLED DEPTH OF THREAD, (FOR ANGLED APPROACH SCREW - CUTTING), BY USING THE METHOD USED IN FIG. 42a. (SHOWN AT THE BOTTOM OF DRAWING FIG. 42).

CONSTANT 31.526688

CALCULATIONS REQUIRED TO OBTAIN THE BOLTS DEPTH OF THREAD ETC.

ALL DIMENSIONS IN MM.

THIS DRAWING IS APPROX.9 TIMES FULL SIZE.

SCREWCUTTING TOOL

DIRECTION OF TOOL FEED AT 30° (WHEN USING ANGLE APPROACH)

DRAWN 20-2-14 G.N.R.

THIS DRAWING 10X FULL SIZE.

Section 18/28.
The necessary preparations required to screw-cut an external metric thread on a component by using the center lathe.

There follows a full description of how to set up and prepare the center lathe for a screw-cutting operation. How one deals with work-piece support problems, and how one uses the screw-cutting ring gauge to check the fit of the newly screw-cut thread.

The work-piece material shown in the fig. 42 screw-cutting example, will be seen to possess a stick-out length (from its chuck) of approximately 47 mm. It must be said that this would be considered to be a rather short stick-out needing the use of a 'greased dead center's support, but the author has deliberately chosen this particular length of stick-out to emphasize the fact that a workpiece of this or any longer stick-out length would definitely require this method of support being used, particularly when a screw-cutting operation needs to be performed on the work-piece. We must do this in order to ensure there is sufficient rigidity in the setup.

Every effort should be made to keep any unsupported stick-out lengths of material from the chuck as short as possible for reasons of safety and work rigidity. Ultimately, this will of course depend on the diameter and length of the material being worked on.

The machinist will require a considerable amount of expertise to prepare and set-up the center lathe in a satisfactorily condition that will allow a screw-cutting operation to be accurately carried out. The *initial basic* preparations required will be cleaning the main bed's slides and applying lubrication, cleaning the lead-screw's external threads using a brush, followed by lubricating the lead-screw and its outer end bearing, ensuring that the compound and cross-slides are free in their movement but are not at all loose, checking that the tail-stock is in true alignment, and ensuring that the fitted four-jaw chuck is running truly.

The work-piece material (shown in the fig. 42 example) will need to be turned down to its required screw-cutting diameter to receive its 24 mm × 3 mm pitch screw-cut thread, but before this is done, the lathe's controls must initially be checked to ensure they are in their correct positions for normal straight turning. Its screw-cutting lift-up lever must be placed out of gear and in its down position. The lathe's turning selector lever (push/pull selector knob, as is used on some lathe designs)' must be moved into its *straight turning* mode.

If the hollow bore of the lathe's spindle is found to possess a sufficiently large diameter to allow the work-piece to be passed through into its bore, then the preliminary stick-out of material from the chuck's now tightened front jaws should initially be approximately 25 mm; the outer end of the work-piece is then faced off smooth, using preferably a right-hand knife tool (shown in fig. 47 item 1), and this must be checked for its correct cutting height by using the lathe's height checking gauge. This is followed by center drilling the workpiece's end using preferably a number 3 center-drill and a high work-speed of 800 rpm, using plenty of coolant in order to prevent damaging its vulnerable cutting tip. The size of center drill used will of course depend on the finished diameter and the weight of the workpiece. The workpiece is then moved further out of the chuck and secured with a stick-out of approximately 47 mm with its end now being supported by using a greased dead center held in the tailstock. The existing 25 mm diameter of the work-piece is then turned down using a right-hand knife tool (fig. 47 number 1, to 23.9 mm diameter (the major diameter of its thread), to its required 24 mm length, followed by taking a very light skimming cut along the surface of the holding 25 mm diameter material sufficient to clean it up and run true to within 2 mm of its holding chuck, ensuring that the holding material is now running true. This is most important as there may be the need to remove the work-piece prematurely, and this will allow it to be replaced accurately back into the chuck (by using a DTI (clock gauge) to maintain its previous concentricity).

However, if the diameter of the work-piece material cannot be passed into the lathe spindle's bore, then the material will require an additional holding length in order to secure it into the chuck's jaws, plus the stick-out of at least the length of the finished work-piece (42 mm), plus the required additional (say) 5 mm to take account of the width of the (later to be used) 3 mm wide parting tool and the 2 mm safety clearance required between the parting tool and the chuck, adding up to an approximate 47 mm stick-out (plus the 25 mm chuck's holding requirement), giving us the 72 mm length required for cutting the material from the stock length of the store-issued material.

If not already done, the work-piece's end is then faced off at a medium spindle speed of 205 rpm and center drilled using a high work-speed of 800 rpm while using plenty of coolant to prevent damaging the drill's vulnerable cutting tip.

[A small-diameter workpiece would naturally only require a number one, or small, (say 3 m.m diameter) center-drill to be used. On the other hand, a very large diameter work-piece would require a number five, 8 mm diameter center-drill, which enables it to provide a much larger and substantially heavier support for the extra size and weight of the work-piece. In our particular case, we have chosen to use a number 3 center drill (of 6 mm diameter), to machine the required 60 degree internally tapered support in the end of the work-piece. This is then used to provide a positive support to the work-piece by using a greased supporting dead center held in the tailstock. (The author includes the word 'greased' to remind the student that a dead center support should always be lubricated with grease and should never be used dry.).

Note that (as previously stated), before screw-cutting an external thread on a work-piece, it is normal practice to reduce its drawing stated diameter of 24 mm, down to a clearance diameter that is smaller by 0.1 mm (0.004 inches) than the thread's major drawing stated diameter; this then allows sufficient clearance to suit its fitted nut. We therefore turn this work-piece down to 23.9 mm over a longitudinal distance of approximately 24 mm.

The exposed sharp end of the work-piece, should now be chamfered by using a chamfer tool (shown in fig. 47 tool number 8), with its cutting edges ground and set to an angle of 45 degrees. This chamfer is then machined over a nominal length (in this case) of approximately 2 mm; this will then ensure that the full chamfer actually reaches down to the core diameter of the thread about to be cut. This chamfer will then give a suitably unobstructed lead-in for the screw-cutting tool at the starting end of its screw-cut thread, and will also prevent the later need to chamfer the component after the screw-cutting operation has been completed, thereby preventing any possible burring occurring at the beginning of the thread should it be introduced after the screw-cutting operation.

[The accurate diameters required by the screw-cut male thread are shown in fig. 42, and in fig. 43, these being consistent with the screw-cut nut's thread shown in fig. 43a, and fig. 44. These drawings should now be very carefully studied, particularly with regard to the thread's clearance areas required.]

[*It should also be noted that in the fig. 42 drawing,* **(this being a student's screw-cutting test example work-piece)**, *that the thread length shown in this example is shorter than its diameter; this has been deliberately planned and shown (in this case) to point out to the student, that in general engineering terms, this short length of thread would be considered to be* **marginal** *in longitudinal strength. The generally accepted rule in engineering circles is that a screw-cut thread should always be at least* **as long as its diameter** *(for strength purposes); it can of course be much* **longer** *than its diameter, but as a general rule, it is only rarely machined with a shorter length than its diameter*].

The undercut tool clearance area (positioned immediately following the screw-cut area of the thread) is now to be machined. This clearance area will allow the screw-cutting tool to move into a safe (but rather short) area immediately following the threaded part of the component. This clearance area is very important, as it will be occupied for a short time by the screw-cutting tool's point radius on completion of its initial and subsequent threading cuts. This clearance area allows the tool and its radius point to enter into an area of free air. The dimensions of this small clearance area are considered by the designer of the component to be of sufficient width to allow the tool to be quickly *snatched out* in order to prevent it coming into contact with the following main body of the component.

This undercutting operation is performed in the work-piece by using a 4 mm wide, parting-type tool (with a greater-than-normal width but similar to fig. 47 tool number 5).

Full tool support must be provided by ensuring that the tool only possesses a very short overhang from its tool-post; the compound slide must also be rigorously checked to ensure that it is not overhanging its support casting. This precaution must be taken when machining any undercut or parting-off operation; in this case, it is doubly important because of the parting tool's greater-than-usual width. When using a very wide parting tool, its cutting edges can cause chatter to occur during the under-cutting operation; this is mainly due to the increased load being transmitted by the tool and the work-piece to its tool post.

To machine the undercut, the tool's left-hand side cutting edge is maneuvered into (in this case), a position that is 24 mm from the outer end of the workpiece. The saddle is locked in this position and coolant is applied to the tool and workpiece. The existing 23.9 mm outside diameter of the workpiece *about to be screw-cut* is then plunge turned by the tool down to a diameter of 20.3 mm by using the cross-slide's hand-wheel being turned very carefully clock-wise (note that this 20.3 mm diameter is also the work-piece's calculated minor diameter of the intended thread). The under-cut's diameter should always be considered to be important; this is due to it often being used in the latter stages of a screw-cutting operation to identify the final depth of thread that needs to be reached by the point of the screw-cutting tool where it often leaves a minute spiral witness mark on the under-cut's diameter indicating that the tool has reached its full calculated depth of thread.

The recommended tip radius honed onto the tool for screw-cutting this particular series of metric coarse male threads, is obtained by using the calculation 0.1433 × the pitch (as shown in the lower 10 times formula drawing in fig. 43); in this case the radius that needs to be honed onto the screw-cutting tool's tip will be 0.43299 mm (rounded down to 0.43 mm), (this being equivalent to a diameter of 0.865 mm (now rounded down to 0.86 mm) when using the flattened end of a twist drill or a section of wire of this diameter to check the radius practically, by the use of a 10 × magnifying glass.

As previously stated, this minute spiral mark now seen on the under-cut's diameter at the end of a screw-cutting operation, will serve as a helpful witness mark to the machinist, by providing a positive indication that the point radius of the tool has reached its full theoretical depth of thread.

It must be said that when an accurately machined under-cut diameter of 20.3 mm is produced, its existence becomes a very useful aid for checking the tool's exact and required depth of thread, particularly if the tool has needed to be replaced midway through a screw-cutting procedure, when nearing its final finishing cut.

For this component we are using the angle approach method of screw-cutting (as shown in drawing fig. 42, and explained in Section 18/43 Option 3 and Section 18/44 Option 4), the angled *compound slide* assembly must be accurately set to half the included angle of the cutting tool, before being finally secured. This enables the tool to approach the workpiece at half the included angle of the cutting tool and of the thread being cut (the center-line of the 60-degree included angle cutting tool must be aligned so that it is positioned at exactly 90 degrees relative to the work-piece's local surface).

The initial preparation required on the center-lathe when it is about to be used for the angle approach screw-cutting of a 60 degree included angle metric or American thread, is initially carried out by accurately installing a suitably ground and sharp, 60 degree included angle screw-cutting lathe tool into the tool-post (using cutting tool number 4 in fig. 47). It is important that this tool is located securely and robustly into a position into its tool-post that permits its supporting shank to possess the minimum of overhang (or stick-out) from the tool-post's base casting. This must be done in order to ensure that the maximum rigidity is being maintained throughout by the whole of the tool's support assembly.

A thread setting gauge of the square type (as shown in fig. 49), is positioned with its top edge aligned with the turned 23.9 mm diameter of the work-piece, it is then used to accurately align the screw-cutting tool's 60 degree included angled cutting edges accurately into the 60 degree precision ground cutout V in the gauge. (The tools cutting height must also be checked for accuracy by using the lathe's tool height setting gauge.)

A piece of white paper held under the tool will reflect sufficient light to assist in viewing the accurate alignment of the tool's cutting edges relative to the gauge's cutout V.

When correctly aligned, the tool-post's clamp screws can then be fully tightened.

As with the other modes of turning, screw-cutting, parting off, and boring, the tool's cutting edge and its point height is critical and must always be set at the height of the exact center-line of the work-piece, and of the lathe's spindle and chuck.

In the absence of a tool setting height gauge, a simple way of checking whether the tool's point is set at its correct cutting height (for external screw-cutting) can be obtained by carefully nipping a thin flat object between the tool's point and the near side of the turned diameter of the work-piece, such as a 6 inch (150 mm) steel rule held carefully between the tip of the tool and the surface of the perfectly round work-piece, and noting whether the rule remains vertical, (if it remains vertical, this will indicate that the tool is set to its correct height), or, if the top of the rule is found to tilt toward the operator, (this will indicate that the tool is set too low); alternatively, if the top of the rule tilts away from the operator, (this will indicate that the tool is set too high.). Most machine shop lathes will have of course a reliable *tool height setting gauge* for use in checking the tool's cutting height with accuracy and this should be used if it is available. (The dead center's point, secured in the tail-stock's quill, is also at the correct height for checking the cutting tool's height, and can be used for tool height setting purposes).

An experienced machinist will usually select from his tool-box his own screw-cutting tool for the job in hand; he will choose either an existing (previously ground, used, and verified as accurate), 60 degree included angle screw-cutting lathe tool (similar to fig. 47 tool number 4), or alternatively, he will need to spend a considerably longer time, in grinding a new tool from a standard-issue, solid high-speed steel tool blank, or possibly a basically shaped butt-welded high-speed steel / mild steel lathe tool (possibly a square-ended parting tool, similar to fig. 47 tool number 5) for modification into a 60 degree included angle V, by using off-hand grinding techniques to obtain the correct cutting and clearance angles required. However, if a cemented carbide screw-cutting tool that possesses the correct 60 degree included angle ground at its cutting edges is available, then this can be used, but it should be treated with respect, because carbide-tipped tools are considered more vulnerable to being chipped because of their extreme glass-like hardness. They are also more difficult to sharpen by using off-hand grinding techniques because of their extreme hardness and possibly the unavailability of the correct grade of grinding wheel needed to perform this task. (A diamond-impregnated wheel is of course considered best for grinding this particular tool.).

If a new screw-cutting tool (similar to tool number 4 in fig. 47), is not available from the firm's tool stores, then the nearest outline shape found most suitable for regrinding modification purposes into a screw-cutting tool will be the half-inch square butt-welded, left-hand offset parting-type tool (similar to number 5 tool in fig.47) as previously mentioned.

The required grinding modification to this tool will be performed manually by the tool-room machinist, using his considerable skill and the dexterity of his hand and eye, to carefully grind, by using off-hand grinding techniques, the required offset center line to its left, and its precision 60 degree included angle, together with its necessary (and very important) clearance and cutting angles; these will include in some cases, a one degree slope on the tool's top surface to provide an advantageous cutting angle relative to its left-hand side cutting edge. As previously mentioned, the center-line of the tool's two V-shaped cutting edge faces, can be ground, with advantage, so that it possesses a distinct offset to its left. This offsetting of the tool's center line then allows a more precise positioning of its radius point and its cutting edges relative to the work-piece and its tool post. It will also allow the workpiece to have increased clearance relative to the tool's holding shank.

This particular tool's cutting edges, by being ground to the left of the tool's center line, will prove to be a distinct advantage to the machinist when the tool is being used to screw-cut a thread in very close proximity to a work shoulder, or up to a large-diameter flange. This modification also reduces the possibility of the tool's main body accidentally contacting a large-diameter flange of a work-piece while at the extreme end of its passing cut.

Section 18/29
Continued screw-cutting advice including how one obtains the helix angle of a screw-cut thread in order to assist in producing the correct clearance angles required on the tool.

The helix angle of the thread being worked on must be established in order to obtain the screw-cutting tool's correct leading cutting edge's clearance angle. The helix angle of the thread is shown dimensioned at the top of the fig. 43, example drawing. This helix angle can be calculated (in our example's case of a 3 mm pitch thread), by entering into the calculator the following sequence:-, 3, (the pitch in mm) divided by 24 (the thread's nominal diameter in mm), =, (0.125); this is followed by entering the sequence INV. TAN, to give us 7.125 degrees. We must now add to the calculated helix angle, at least another four degrees, in order to provide the tool with *adequate* tool clearance in use, giving us approximately 11.125 degrees. This is generally considered the clearance angle that must be provided at the tool's leading angled cutting edge that is sufficient to allow the tool to fully clear the thread being cut.

Note that if a tool is ground with a too-large clearance angle (in the area below the tool's leading cutting edge), it will have the tendency to weaken the tool's cutting edge by providing less material support, resulting in the tool having less wear resistance and a shorter working life between the necessary regrinding of the tool finally becoming necessary.

The four extra clearance degrees, plus the 7.125 helix angle of the thread, will allow the lower part of the tool to fully clear the existing sloping angle of the thread being cut.

It should now be realized that the tool's right-hand side clearance, (its trailing side clearance when in use), produced on the screw-cutting tool when using the angle approach method, needs only a minimum side clearance to be provided; therefore in this case a one degree clearance angle will be sufficient for the trailing cutting edge of the screw-cutting tool. This minimal clearance angle is made possible by the helix angle of the thread actually assisting in the clearance of the tool, because it moves to its left on a cut and away from the surface of the material being cut.

Section 18/30
Describing the advantages gained by using the angle approach method of screw-cutting as shown in drawing fig. 42, and how one needs to calculate the 'straight plunge' depth of thread first, by using a thread constant and by using the Probe and Prompt.

Fig. 42
The fig. 42 drawn example of an external screw-cutting setup depicts a vertical view of the screw-cut thread and its tooling. The

tool's orientation and the direction and length of its travel are being indicated (in this case), by the use of dashed lines. The fig. 42a lower triangle drawing shows in detail the use of the Probe and Prompt method for calculating the dimensions required to obtain the angled depth of thread that is required for the angled approach method of screw-cutting. It also shows (as a matter of interest) the distance the tool will actually progress to its left along the work-piece on reaching its full depth of thread and will in so doing use up 1.06 mm of the under-cut's clearance width on completion of the thread, which will leave it considerably narrower than originally intended.

Also shown at the top of the fig. 42 drawing (in fig. 42b), is the method used to obtain the 'straight plunge' depth of thread, by using the thread constant (31.526688), and by using one's own hand sketched right-angle triangle and the Probe and Prompt system of triangle calculation to obtain it.

There now follows an in-depth explanation of how the angled approach screw-cutting method is used in practice, and the advantages gained over using the traditional straight plunge' method of screw-cutting, (which is often used for screw-cutting relatively small diameters that require a shallow depth of thread in their work-piece).

The fig. 42 drawing shows the necessary calculations required before screw-cutting the example of a 24 mm diameter by 3mm pitch metric coarse external thread.

It explains in minute detail a unique method for obtaining the elusive 'angled depth of thread' dimension required.

In fig. 42b, we first calculate the 'straight plunge' depth of thread by drawing our now familiar triangle (at the top of fig.42) and by now entering the author's new constant of 31.526688 at its The Angle position, using the prompt $\angle c$, $b = c \times$ TAN $\angle$, being entered in the sequence, 31.526688, TAN, (0.613441681) × 3, =, giving us 1.840325045 mm. We then move on and draw another familiar triangle at fig. 42a, (at the bottom of fig. 42), and enter the now calculated number into its side c, while also entering into its The Angle position the angle of 30 degrees (which is half the tool's angle); this now allows us to find the length of its side a (which will contain the angled depth of thread we require).

Because we now know both its The Angle and the length of its side *c*, we can now use the prompt ∠*c*, *a* = *c* ÷ COS ∠, by entering into the calculator the sequence 30, COS, (0.866025403) XM, followed by entering the sequence 1.840325045 ÷ RM, =, giving us 2.12502432 mm, the angled depth of thread we require.

Therefore when screw-cutting a thread using the angle approach method, it is always the case that the *cross-slide's scale* is set to read zero, which causes the whole screw-cutting operation, from start to the finish to be at the zero setting on its scale for each of the multiple passing cuts. The machinist must therefore always return the cross-slide and its scale assembly to its zero setting before taking its initial and for every passing cut.

It is the *compound' slide's scale* that is used to indicate to the machinist the depth of thread that is so far being reached, and it is this scale that is adjusted at the start of cutting the thread to be set at *its* zero setting, and for it to finish cutting the thread when its scale has reached (in this case), 2.12 mm, (the full calculated 'angle approach' depth of thread required).

To start the screw-cutting operation, the compound slide's hand-wheel is turned anticlockwise sufficient to ensure that its top slide is 2 mm short of the point of overhanging its support casting, this is followed by its hand-wheel being turned clockwise by a quarter of a turn to eliminate any back-lash in its lead screw and its scale set to read zero. The *cross-slide's* hand wheel is then turned clockwise by half a turn to eliminate *its* backlash and its scale is then set to read zero.

The screw-cutting tool's point is now positioned by using the *cross-slide's* hand wheel being rotated clockwise so that it touches the previously turned screw-cutting diameter of 23.9 mm, and its scale is then reset to read zero. The *cross-slide's* hand wheel is then turned anticlockwise by one revolution to allow the tool to clear the work-piece; the *saddle's* hand wheel is turned clockwise to reposition the tool's point to a position approximately 5 to 6 mm to the right of the work-piece. The cross-slide's hand wheel is then turned clockwise by one revolution down to its set scale's zero mark, making the tool ready for taking the first cut. The lathe is then switched on in its normal direction of rotation, at a work-speed of approximately 60 revs per minute. The *compound slide* is then indexed inward by turning its hand wheel clockwise by the minimum amount of 0.05 mm, in preparation for taking the first *exploratory cut* along the workpiece; its scale is then marked using pencil. The coolant supply is switched on to cover the tool and workpiece. We now study the lines and numbers situated on the *rotating screw-cutting dial*'s drop in scale situated on the front of the lathe's apron; we then select the number or line we have previously chosen and allow it to line up with the apron's fixed datum line. The *screw-cutting lift-up lever* is then engaged at this exact chosen point, and the first screw-cutting cut is then started. (At this point, a mental note should be made of the chosen line or number being used.).

On completing the first cut, the tool, reaches the undercut area and (by using the guidance of a carefully placed rotating warning marker situated on the chuck's periphery) the *quick method of tool withdrawal* is used (consisting of one quick and complete anti-clockwise rotation of the cross-slide's hand wheel), this is coupled with the *instant* operation of the *screw-cutting knockdown lever,* (which isolates the tool from its lead screw). The *saddle*'s hand wheel is then turned clockwise (with care) to return the saddle and tool to its start position of approximately of 5 to 6 mm to the right of the workpiece. The *cross-slide*'s hand wheel is then carefully turned one revolution clockwise to return the tool back to its originally set zero scale's position.

The angled *compound slide*'s hand wheel is then turned very carefully clockwise (by the minute amount of 0.05 mm) to index in the tool for the next cut to be taken, referring to and using its index scale for this precise 'depth of cut' setting. Its index scale should then be marked with a pencil (or a similar marking instrument) to retain (for the machinist) a lasting reminder of the depth of thread the tool has so far reached.

The rotating screw-cutting dial is studied again until one's chosen mark or number is again seen to line up with the apron's stationary datum line. The *lift-up screw-cutting lever* is then re-engaged using the same chosen mark (or number) on its dial. This accurate re-engagement (by relying on the lathe's *rotating screw-cutting dial* and its fixed datum mark) ensures that the tool will mesh accurately into the previously cut thread.

[It is important to note that failure to engage the *screw-cutting lever* at the identical mark each time on its *rotating screw-cutting dial's scale* for each and every passing cut will result in the tool wrongly engaging with the previously partially cut thread, and will in this case cause damage to the workpiece by this misalignment. Any damage caused to the thread in this way can cause the workpiece to suffer damage that may be beyond reclamation.

[A wrongly engaged tool (particularly in the latter stages of a screw-cutting operation) can also cause the tool's cutting edges to be damaged, which necessitates a tool replacement to be made, requiring a rather lengthy resetting up process to be undergone; it is therefore considered extremely wise *not* to allow this to happen. To reiterate, the student should note that when using the *angle approach* method of screw-cutting, it is the *cross-slide's* hand wheel assembly that is always returned to its set zero position at the start of each of its screw-cutting passes. This is followed (immediately after its passing cut has been completed), by using the *instant snatched method of tool withdrawal* and the instant *screw-cutting knockdown lever* routine, at the end of each passing cut while using the guidance of the chuck's carefully placed rotating warning marker.

The student should also note that because we are using the *angle approach method* of screw-cutting for this operation, it is the pencil marked *compound slide's* hand wheel's reference scale that is used to indicate to the machinist the gradual increase in depth of each and every succeeding cut needing to be taken by the tool during its following screw-cutting passes. The actual depth of cut needing to be indexed in by using the compound slide and its scale, for each cut, will normally be 0.05 mm for each cut (for steel), (or can be slightly deeper if machining softer materials such as aluminum or brass), until the full 2.12mm calculated angled depth of thread has been reached by the tool, is seen indicated on the compound slide's scale, showing that the full depth of thread has been reached.

The machinist will also find it helpful if he forms the habit of always marking the compound slide's index scale regularly with pencil (or a similar marking device), *just before* each cut is taken.

After taking a series of 0.05 mm depth passing cuts that are considered to be sufficient for the tool to be now approaching its full depth of thread, the *thread ring gauge* is used to check the fit of the screw-cut thread to establish whether it has the correct depth and size to fit the *thread ring gauge* now being used for this check.

If the ring gauge can be screwed onto the thread without exhibiting any tight-ness, loose-ness, wobble, or any end float being present, then this screw-cut thread can be considered to be in an acceptable condition for submission to the inspection department.

The component can now be parted off using tool number 5 (or 6) described in fig 47.

It will be found expedient (in this case) to use the existing screw – cutting tool's point to chamfer the work-piece at the site of the parting off operation, this being positioned exactly 42 mm from the outer end of the work – piece. After moving the tool to this position, the lathe is switched on, and we then index the tool's point 2 mm into the work-piece using the cross-slide's scale while using coolant, it is then withdrawn,and the lathe switched off. The screw – cutting tool is now removed from the tool-post and replaced by the selected parting tool, which should possess (for this particular job), a cutting width of 3mm. It should be set at its correct cutting height, oriented perfectly in line with the cross-slide's center-line and possess the minimum of stick out from its tool – post consistant with allowing sufficient clearance for the existing diameter of the work-piece. Its right-hand cutting edge is now positioned on the center-line of the previously machined V cut-out, the saddle is now locked, the lathe switched on at 135 rpm, the coolant is switched on, and the parting off process is begun by indexing the tool inward clock-wise positively and carefully (by an estimate of 0.05mm per revolution), until the component ceases to revolve indicating that the operation is complete. The component is then withdrawn from the lathe by slackening off the supporting greased dead center.

[There are four main methods generally used when externally screw-cutting a thread on the center-lathe:]

(1) *straight plunge method* used with a *permanently engaged* lead screw, and using just its *cross-slide* and its scale to index the tool inward for each cut.

(2) *straight plunge method* used with a *disengaged* lead screw, also using just its *cross-slide* and its scale to index the tool inward for each cut

(3) *angled approach method* used with a *permanently engaged* lead screw, using its *cross-slide scale* always set to zero prior to each cut being taken, and then by using its *compound slide's scale* to accurately index the tool inward, for each cut.

(4) *angle approach method* used with a *disengaged* lead screw, using the *cross-slide's scale* always being set to zero prior to each cut being taken, then using its *compound slide's scale* to index the tool inward for each cut.

The *straight plunge*, traditional method of screw-cutting (as explained in Section 18/41. Option one and Section 18/42 Option two), uses its compound slide and its tool post set in its (normally used) horizontal position that is in line with the center line of the lathe.

In the straight plunge case, the cutting tool will be cutting the thread into the work-piece by using (for steel), minute infeeds of, say. 0.05 mm (indicated on its *cross-slide's scale*), directly into the work-piece at 90 degrees relative to its local surface, the center line of the tool always being set at 90 degrees relative to the work-piece's local surface.

Section 18/31

Missing << image 48 >> Here

Extensive notes on the off-hand rinding of lathe tools, including the correct preparation of the grinding machine and its grinding wheels, sufficient to enable the production of an accurately ground cutting edge and the required clearance angles being ground on the lathe tool.

The finished 60-degree included angle screw-cutting tool must be accurately ground by hand; it is most important that the grinding operation includes all the necessary clearance angles relative to the tool's cutting edge relief areas. Note that safety glasses must always be worn when using a grinding machine, even if a transparent safety glass shield is provided on the machine, (it should be constantly remembered that we are only supplied with one pair of eyes, therefore extreme care should always be taken to prevent one's eyes being damaged by particles of grit and grinding dust etc. being ejected from the grinding machine). This precision off-hand grinding operation, is usually performed by hand, by using a pedestal-type grinding machine. The grinding wheels fitted to this machine should ideally have been prepared, mounted, and dressed by the management's appointed person (who would normally have attended and passed a grinding wheel safety course), which qualifies him to do this particular job). The chosen grinding wheel grit material for grinding high-speed steel tools is normally a material called aluminum oxide, and for grinding cemented carbide-tipped tools it is normally a material called green silicon carbide, (although diamond-impregnated wheels are used in the tool-room for the precision grinding of this material). The two types and grades of grinding wheel mentioned are normally available for use; these are fitted one at each end of the tool grinding machine. This allows the machine to be used for sharpening either types of tooling, but beware, this twin setup has not been provided to allow two persons to operate the machine at the same time, and this practice should definitely be disallowed for safety reasons. The 'green grit' wheel will (in practice) be found to be the slightly softer of the two wheels. As a general rule in grinding parlance, a soft wheel is normally used for grinding hard materials, and a hard wheel is generally used for grinding soft materials, although (in this particular case, high-speed steel, if compared to cemented carbide (which is very close to diamond in hardness), is considered to be the softer material of the two).

Some work-shops possess a type of off-hand tool grinding machine that differs in design to the two-wheel pedestal grinding machine. This machine is equipped with a recessed grinding wheel that possesses a special backing flange specifically designed to support the back face of the wheel on its driven spindle; it is designed to withstand and reinforce any sideways pressure being exerted onto the fragile grinding wheel, by the tool being ground. The recessed designed shape of this wheel also allows its securing clamp nut together with its front support flange to be recessed sufficiently to allow the machinist to have unimpeded access to both the front and side face of the grinding wheel. The advantage gained by using this type of grinding wheel for tool grinding is that it allows the machinist to produce a completely flat and angled cutting edge on the tool, complete with its accurately ground clearance angles. This grinding machine will also provide a superior support to the actual cutting edge of the tool that gives a much improved resistance to wear compared with a tool that has been ground on the front part of the wheel fitted to the two-wheeled pedestal grinding machine. The two-wheel pedestal grinding machine, suffers from the fact that virtually all the tool grinding is being carried out on the front curved surface of its grinding wheel, and this is known to produce a 'hollow ground' surface on the tool being ground, particularly in the area of clearance situated immediately below the tool's cutting edges. This hollow grinding of the clearance angles on the tool, is caused by the distinct curve of the wheel's periphery. The actual radius of this curvature is dependent of course on the diameter of the wheel in use. This hollow grinding of the tool, provides a cutting edge that, although sharp, provides marginally less support to its cutting edge, and as a result its cutting edge will suffer from being less resistant to wear when in use.

It is generally considered in the work-shop that it is bad practice to use the side of the wheel on a pedestal grinding machine for grinding the side clearances on the tool, because of the wheel's fragility and its liability to shatter if it is used in any way robustly in a sideways direction.

However, if it is discovered in your particular machine shop, that all of the tool grinding must be performed on just the two-wheel grinding machine, then it is grudgingly permitted (provided one uses extreme care and uses the absolute minimum of sideways pressure on the wheel), to allow the tool's previously ground clearance angles (that were produced on the two-wheel machine by using the front of its wheel to grind below the tool's cutting edges), for the tool to be allowed to momentarily touch the side face of the wheel while being firmly supported by hand, and by using a very gentle forward and backward reciprocating motion, while the tool is being held at its correct clearance angle, to produce on the tool a very narrow flat (of say half a millimeter wide), in the area of tool that is below and supporting its cutting edges, with the object of producing a well-supported and wear-resistant (non-hollow-ground) cutting edge on the tool.

However, it will always be found that the cutting edge of a tool ground by using the side of the wheel (on the supported wheel machine), will provide a far superior, flat, longer-lasting cutting edge, and, in consequence, will allow increased periods of time to elapse before there is the need to re-sharpen the tool.

Grinding wheels are not just hewn from a piece of abrasive rock, as is sometimes thought by the engineering uninitiated; they are of course man-made from multiple combinations of grain size, grade, and hardness of abrasive. Grains are molded into shape, bonded together by using varying strengths of selected bonding agents, followed by heat treatment to solidify them and finally produce the finished grinding wheel.

Prior to the grinding wheel being mounted onto the machine's spindle, it must always be provided with two soft cardboard washers, these are interposed between the wheel and its support flanges in order to take up any irregularities present in the slightly uneven wheel's side surfaces, thereby preventing possible cracking of the fragile wheel while it is being tightened. One should note that the wheel's securing nuts are always loosened in the direction of the wheel's rotation (should this be required), regardless of which end of the machine the wheel is fitted. It follows from this statement that the wheel's securing nuts are always tightened in the opposite direction to the wheel's rotation, therefore the wheel-nut on the left-hand side or end of the machine will always be fitted with a left-hand thread. The reason for this is that should the nut come into accidental contact with any object while it is rotating, it will always be automatically tightened and not loosened.

A grinding wheel is generally spoken of as being soft if its grit's bonding agent is relatively weak; therefore in this case it will have been specifically designed to allow its grains (or grits) to be shed as they become dulled by the grinding process, in order to expose new sharp cutting grains that will form a new sharp grinding surface. This automatic replacement of dulled cutting grains tends to extend the period before the wheel will need to be finally re-dressed. That is of course providing the wheel has received just normal tool grinding use, and has not been abused (illegally) by other visiting workshop personnel, attempting to grind banned materials on the grinding machine, such as brass, copper, or aluminum (even wood), thereby causing extreme glazing or loading of the wheel's grits, which makes the wheel incapable of cutting, with the requirement that it must now be re-dressed and balanced before it can be used for regrinding hardened lathe tools.

A hard grinding wheel may use a similar grade of sharp grit in its makeup, but its bonding agent is designed to hold the grains more firmly in place, making the wheel less able to shed its grit when becoming dulled; this type of wheel will usually require more frequent dressing to expose new layers of sharp grits. A hard wheel is therefore more prone to become glazed or loaded with metal particles between its individual grains; this condition is called a glazed wheel, and when in use, it will rapidly become apparent to the observer by it's shiny glazed and loaded appearance, and its inability to cut satisfactorily. A loaded wheel will also show a marked reluctance to cut or grind material from the tool being ground because it is actually rubbing and not cutting. This rubbing action will cause serious overheating to occur to the tool, because of the added amount of friction generated. The machinist must be very careful therefore not to allow the tool being ground to become over-heated in this way by using a glazed wheel; the tool should never be allowed to become so hot that its cutting edges are seen to change to either a straw or blue color, for if allowed to reach this high temperature range, (exceeding 300 degrees centigrade), its cutting edge will in all probability have become softened by having its temper withdrawn from its cutting edge area, (thereby making the tool less hard), this condition will be found to markedly reduce its ability to remain sharp over long periods, and allowing this to happen will have seriously reduced its working life between resharpening. A glazed wheel should always be re-dressed to expose a new clean sharp cutting surface of grits to allow a lathe tool to be re-ground without it becoming excessively hot in the process. The twin-wheeled pedestal-grinding machine should be serviced regularly to allow its grinding wheels to be balanced, and dressed periodically in order to produce a flat, open, and clean grinding wheel cutting surface. This is achieved by the careful use of a hand-held wheel dresser (called a Huntington dresser), a hand-held diamond dresser, or one of the other types of professional precision dressing tools that include the use of an industrial diamond embedded and brazed into a mild steel shank or block. This tool being accurately traversed across the face of the wheel to remove any irregularity in the wheel's roundness and across its cutting surface in order to produce an accurately balanced and sharp wheel. It is most important that the grinding wheel is dressed accurately so that on completion of its bout of servicing, it is handed over to the machinists and tool-makers in a condition that is found to possess extremely useful (to the machinist), square corners on its wheels. It is therefore most important to the machinist that the grinding wheel has not been left with large radii at its outer corners, which then makes the wheel much less suitable for tool grinding.

A perfectly balanced grinding wheel should run smoothly, and be vibration free. If it is found that the grinding machine vibrates, (this phenomenon is particularly noticeable as the machine is slowing down after being switched off), this will indicate that at least one of its wheels is out of balance and needs re-dressing. Where square corners have been conveniently left on the grinding wheel by the appointed person, it will be found very useful for use in flute thinning during the grinding and sharpening of twist drills, or for performing grinding modifications to the clearances and the cutting edges of milling cutters and slot drills, should a tool and cutter grinder not be immediately available for precision tool re-sharpening.

If the sharpening of cemented carbide lathe tools is required, then the most suitable wheel for performing this task (on the shop floor) is the 'green grit' silicon carbide grinding wheel (as previously stated). This wheel should be used dry, and the operator should always allow the resulting hot lathe tool to cool down naturally and should *not* plunge it immediately into cold water to cool it down. It should be noted that if a hot high-speed steel tool, or particularly a hot cemented carbide-tipped tool, is plunged into cooling water immediately after grinding, the temperature change shock sustained by the tool is likely to cause its tip to crack or its cutting edge to be seriously degraded. Of course when working in the tool-room and being employed working on precision grinding machines, it is the diamond-impregnated wheel that is used for grinding the precision angles on cemented carbide tooling.

One should also note that the grinding wheel's tool rest should always be set with a clearance gap of no greater than 1 mm (0.03937 inches), between the wheel's front grinding face and the tool rest; this narrow gap must be maintained not only for safety reasons, to prevent objects being jammed in the gap, but to comply with the published safety regulations.

After the wheel has been balanced and dressed by the appointed person, (the author should mention at this point that he has been elected as the appointed person in several of the engineering establishments he has worked in), this safety gap should continue to be maintained at this close level in order to prevent any object becoming jammed between the wheel and the tool rest. It is also recommended that while in the process of grinding a lathe tool by using the off-hand grinding technique, the tool's cutting edge should always be kept in motion by slowly moving the tool's cutting edge across the face of the wheel while using a slow reciprocating motion to smooth out and prevent any of the wheel's surface irregularities being transferred onto the tool's required flat cutting edge. This reciprocating motion given to the tool while it is being sharpened, also helps to maintain the grinding wheel's grinding surface in a flat and non-grooved condition, thereby allowing it to be accurately used for much longer periods.

The setting height and angle of the tool rest, of a pedestal grinder, (and for that matter, all off-hand tool grinders) is very important. The tool rest height is normally set (for convenience), so that it will allow the usually ground, and typical, half-inch (12.4 mm) square shank tool's cutting edge to be approximately level with the grinding machine spindle's center-line. The tool rest's backward tilt should be set to approximately 6 degrees. This set angle then makes the rest capable of rigidly supporting the tool's shank at an averagely correct clearance grinding angle, to provide accurate clearance angles for the majority of half-inch square lathe tools.

When faced with the problem of grinding the angles on a new screw-cutting tool from scratch, it will be found helpful to both the trainee and even the skilled engineer (particularly when the machine is being used for tool and drill re-sharpening purposes), if the top face of the tool rest is temporarily painted with marking blue to enable it to be accurately marked out visually on its top surface (we can do this by using a scriber together with an angle protractor), in order to produce two distinct scribed guide lines. These are placed adjacent to both sides of the wheel, and are scribed at 30 degrees (relative to the front face of the grinding wheel). These scribed lines can then be used to accurately position the shank of a 60-degree external screw-cutting tool to be accurately held (by eye) at its correct grinding angle.

The second line (scribed at 31 degrees), on the tool rest adjacent to the wheel's left-hand side, is used to simulate the approximate angle of a drill's shank while it is being handheld for sharpening its cutting edges. This accurate positioning of the drill's shank is achieved by aligning it by eye with the scribed line. While doing this, the drill should be securely and accurately held while being supported on its underside by the machinist's, gripped fingers which are in turn are supported by the tool rest with sufficient accuracy to allow its cutting edges to be ground to its recommended 59-degree angle of grind.

The 30-degree scribed lines positioned on the top face of the tool rest, will allow the 60-degree included angle *screw-cutting tool's shank*, to be viewed and aligned by eye (from above), thereby simulating the correct angle that needs to be ground on the screw-cutting tool's cutting edges. These guide lines and the 6-degree slope of the tool rest will then automatically establish the initial clearance angle required below the tool's cutting edges. However, the leading (or left-hand side) clearance angle given to a *right-hand external screw-cutting tool* must always be ground at an angle that will exceed the helix angle of the thread being cut by at least an extra 4 degrees (in our case this will be 7.125 + 4 = 11.125 degrees), in order to provide the sufficient amount of clearance for the tool during its screw-cutting operation.

When grinding internal screw-cutting tools (as shown in fig. 47 tool number 10), the hand ground clearance angle provided at its leading cutting edge must be considerably greater than the helix angle of the thread being cut, and in our case (in fig. 44), this clearance angle should be by at least an extra 10 degrees and possibly more, in the area of the tool's underside, in order to ensure that adequate clearance is being maintained between the tool's heel and the trailing area of the thread being cut.

The accuracy of the screw-cutting tool's angles must be carefully checked and verified as being correct during the off-hand grinding process, by the use of a screw-cutting grinding and setting gauge. One should always ensure when setting up the lathe for screw-cutting, that the actual finish ground tool's cutting edges exactly fit into the precision angle profile of the gauge. The accuracy of these ground angles is most important, particularly when the tool is being used for straight plunge screw-cutting. As previously mentioned, it is considered good practice, when grinding lathe tools, to always keep the tool's cutting edge moving slowly to and fro across the grinding wheel's grinding surface while holding it at the correct cutting and clearance angle, in order to maintain a straight cutting edge on the tool. Adopting this practice during the tool grinding process, also extends the period of time the grinding wheel's cutting surface can be maintained in a suitably flat condition for further use.

To reiterate, the line scribed at 31 degrees on the tool-rest (relative to the front face of the wheel and to the wheel's left-hand side) is used to accurately establish the correct angle (when viewed from above by eye) of the drill's shank while it is being firmly and securely handheld over the rest with its shank being carefully supported by using the underside surface of the machinist's gripping fingers being placed under its shank and on the rest to maintain positive contact; the outer end of its shank is then angled both downward and toward the operator to obtain the correctly ground drill clearance angle required. It will be found with experience, that this scribed line can be used as a very accurate aid to maintaining the correct sharpening angle of the drill during its re-sharpening process. It is most important that both of the drills' angled cutting edges are ground to identical angles, and that they are ground to possess the same angled cutting length relative to the drill's center-line, in order to maintain the concentricity required for drilling an accurately dimensioned hole. The identical cutting height of each of the drill's cutting edges can be accurately measured by using a vernier caliper gauge to measure the dimension existing between the relatively flat (gripped end) of the drill, and the junction of its angled cutting edge at the outside diameter of its land on the drill's major diameter. When this measurement is found to be identical for both of its cutting edges, and its center point has been accurately measured to be exactly centrally disposed, it then ensures that the drill will cut an accurately dimensioned hole.

However, the extreme accuracy that is normally required for the screw-cutting tool's 60-degree included angled cutting edges', can be considered (when in an emergency) to be slightly less critical if the machinist is using the angle approach screw-cutting method, provided the tool is being used with an accurately set angular setting on its compound slide. In this particular case, any slight minus (in angle) inaccuracy in the off-hand grinding of the tool's 60-degree included angle can be assisted to a certain extent, by using the accurate angled setting of the lathe's compound slide. This slide's accurate setting will then allow the lathe to cope with a more acute angle (that is, more pointed), inaccurately ground included angle on the tool. This will of course depend on the tool being correctly set up in the tool-post, by using the screw-cutting gauge's V-profiled shape relative to the tool's *left-hand (leading side) cutting edge*, which will then leave its right-hand side cutting edge to provide just additional tool clearance relative to the profile of the thread being cut.

A fine-grit sharpening stone should be used, worked by hand, to hone a small radius around the screw-cutting tool's tip. This operation must be performed very carefully in order to shape the tool's very vulnerable sharp point into a very small radius. The dimension of this radius will of course vary according to the diameter and pitch of the thread being cut, and can be calculated for our particular example of a 3 mm pitch ISO metric coarse external thread, (and for all other male external metric coarse threads), by entering into the calculator the sequence 0.1443 × the pitch =, so for our example thread this will be 0.4329, × 3 which will give us 0.4329 mm (0.017 inches), the actual radius that needs to be hand honed onto the tool's existing sharp tip. We must of course ensure that the sharpening stone is kept in continued contact with the tool's clearance angles during this process. The radius can be checked visually for accuracy by comparing the finally 'stoned radius', (by using a 10 × hand-held lens), and comparing it with the flat ground end of a twist drill's diameter, or its drill blank, or a short length of round wire gauge material, of approximately 0.86 mm (0.034 inch) diameter.

To reiterate, during the hand honing process (with the tool viewed in position from above), the sharpening stone (as previously stated), must be held horizontal relative to the tool, and must be kept in close contact at all times with the vertical clearance angles previously hand ground below the tool's two cutting edges; a sliding, reciprocating motion around its point must be very carefully carried out until the required very small radius of 0.43 mm is produced, taking extreme care to ensure that continuous contact is being maintained between the stone's honing face and the tool's clearance angles, to avoid any possibility of the stone slipping and dulling the tool's vitally sharp cutting edges.

For screw-cutting steel materials (and most other so-called tough materials), there have evolved in the industry over time, several 'schools of thought regarding the correct angle that should be ground onto the top surface of the screw-cutting tool. The majority vote among many workshop shop engineers is that the tool's top surface (its top rake) should have a one degree side rake and approximately one degree back rake when used for straight plunge screw-cutting;. Alternatively, if the tool is being used for the angle approach method of screw-cutting, it can (with advantage) be provided with a slightly more positive cutting angle at its leading left-hand side cutting edge; this is done by increasing its 'top top rake to one and a half degrees. In this case, the tool's top surface is angled away to its right and downward from the tool's left hand (leading) side cutting edge by approximately one and a half degrees. It is possible to further increase (very slightly) the tool's top rake angle in order to provide certain screw-cutting advantages, particularly when screw-cutting very soft materials such as aluminum, to approximately two degrees. However, if brass material is being screw-cut, then the tool's top surface should always be left completely flat, and should not be provided with any back or side rake.

Note that the important clearance angle ground below the leading (left-hand side) cutting edge, on both internal and external designs of screw-cutting tool, must always be ground to a considerably greater angle than the thread's existing helix angle (the amount of increase in the clearance angle required must be by at least four degrees, for external threads, in addition to the helix angle of the thread), and will need to be considerably more for the screw-cutting of internal threads in order to allow adequate clearance to be maintained between the material of the thread being cut and the leading cutting side face of the screw-cutting tool. It is therefore vitally important that this clearance is scrupulously maintained, for if the tool is ground with less clearance than this specified amount, it will be found to either rub (i.e. be reluctant to cut), or to produce a poor surface finish that will eventually lead to a tearing of the thread's surface, and can possibly also start tool chattering.

However, if the one to two degree top rake cutting angle has been produced on the top of the tool for use in the angled approach method of screw-cutting, the tool will in consequence be found to be less suitable for use in the straight plunge method of screw-cutting. The increased suitability for it to be used during the angled approach method of screw-cutting will be due mainly to its improved cutting angle now existing at the tool's leading left-hand side cutting edge. The fact that its trailing edge will now possess a negative rake will not (in this particular case) affect its screw-cutting performance, provided of course this tool is retained for use solely in the angled approach method of screw cutting. It should be noted that in this case, the tool's right-hand, trailing edge will just be brushing the thread's surface during the angled approach screw-cutting operation, and should not (in practice) be removing any significant amount of material from the trailing side of the work-piece thread during the screw-cutting procedure.

Section 18/32
The following notes explain the possibility (when one is in an emergency situation), of using an imperial 55-degree included angle screw-cutting tool, to cut a metric or American 60-degree included angle thread in the work-piece, if the machinist discovers that he is not in possession of the correct 60-degree tool to do the job.

If we study the foregoing screw-cutting tools angled approach method of use as a typical working example, then the student may now realize that by using just a little lateral thinking, he or she can now find it possible, after making the necessary 60-degree accurate angular setting to the lathe's compound slide (from its normally used horizontal position), and by orienting the tool's leading left-hand side cutting edge *into alignment with the left-hand side of the screw-cutting gauge's 60-degree cutout V,* to use a tool, that was originally ground for cutting Whitworth threads, (with its angles ground for cutting imperial, 55-degree included angle threads), to cut a 60-degree angle thread, if a 60-degree included angle screw-cutting tool is not available.

It will therefore be found perfectly in order and acceptable to use a 55-degree included angle screw-cutting tool to cut a 60-degree included angle metric (or American thread), provided the angled approach method of screw-cutting is used, and the tool is accurately positioned in its tool-post by using specifically, the screw-cutting gauge's left-hand side of its 60-degree V cutout, to exactly match the tool's left-hand side cutting edge to obtain the tool's accurate orientation in the tool-post. In use, it will be found that the 5-degree difference between the two tools' cutting angles will just provide additional clearance for the tool's right-hand side cutting edge during the screw-cutting process. The correct angled orientation given to the compound slide will ensure that the tool will actually cut the thread to the required 60-degree included angle when using just its left-hand side cutting edge and its point. It will also be discovered when using this particular setup for screw-cutting, that the surface finish of the screw-cut thread will not be degraded in any way.

Section 18/33
The disadvantages one can experience if using the straight plunge method of screw-cutting.

If the machinist has preferred to use the, (conventional) straight plunge method of screw-cutting, (where the tool, is fitted with its center-line in line with the cross-slide and therefore approaches and cuts into the workpiece at exactly 90 degrees to its local surface, and it is using both of its tools' angled cutting edges to remove all the material from the cut thread), it will be seen (in this case) that both of the tool's cutting edges and its radius point will be removing material from the work-piece.

The practical problem of using this particular method of screw-cutting on deeper and larger threads, is that the tool's cutting edges will now have the tendency to suffer from a considerably higher loading during the latter screw-cutting passes, because the tool gradually encounters an increasingly larger surface area of material to be cut, as the depth of thread increases. This gradually increasing depth of its cut combined with encountering increased loading, particularly when nearing its full depth of thread, can result in the tool being seriously overloaded, which causes chatter to occur, particularly when the machinist is working on large diameter deeply threaded work-pieces. This chatter problem will usually occur at the most awkward, crucial, and nerve-racking moment in the screw-cutting process.

This gradual overloading of the tool during this particular screw-cutting process, when accompanied by the possibility of eventual tool chatter occurring, can lead to the actual tearing or damaging of the surface finish of the almost completed thread.

This unfortunate situation is more likely to occur during the final stages of straight plunge screw-cutting when the tool is being used on large-diameter, deeply threaded work-pieces.

As previously stated, this potential problem can therefore be alleviated to a certain extent by adopting the angled approach method of screw-cutting, which uses just one side of its cutting tool and therefore suffers much less overloading in consequence, particularly when working on large-diameter deeply threaded work-pieces.

Section 18/34
An alternative method that can be used to obtain a 'not shown on the drawing' depth of thread, by the use of trigonometry calculations, the Probe and Prompt system, and the author's calculated numerical thread constants.

The author has included at the end of this section a list of alternative 'depth of thread' constants for use in obtaining the depth of thread for other metric and imperial thread pitches.

Most drawings issued by the engineering drawing office, contain important details such as the number and name of the work-piece to be screw-cut, including its diameter, and the pitch of the thread required to be cut, but they often fail to indicate the vital *depth of thread* dimension required by the machinist to enable him or her to accurately screw-cut the thread, (without having to thumb through various reference books to obtain this vital 'but not shown' information).

When screw-cutting a component on the lathe, it is therefore very necessary for the machinist to obtain an accurate depth of thread dimension before the screw-cutting operation can proceed.

Because of this drawing omission, the machinist is forced to use the various means at his disposal to discover this absolutely vital information. This will include reference to screw-cutting tables (when available) to obtain this information. It should be noted that the generally used metric threads are identified as being either the ISO metric coarse thread series, with graded pitches ranging from 0.035 mm pitch on 1.6mm diameter material through to 6 mm pitch on 68 mm diameter material, or the ISO metric fine thread series, with graded pitches, ranging from 1 mm pitch on 8 mm diameter material to 4 mm pitch on 64mm diameter material. The ISO metric coarse thread series are further identified (on issued drawings) by the prefix M before their stated diameter;.For example, M 24 will indicate a thread with a 24 mm major diameter in the coarse thread series, this being identical to our example shown in fig. 42, which has a pitch of 3 mm.

In practice, the coarse thread series gives a good resistance to stripping and is suitable for threaded fasteners where the wall thickness of the receiving female thread is robust enough to accommodate the thread's dimensions. This particular thread design will be found particularly advantageous for use on lower tensile strength materials, such as cast iron, mild steel, and the softer materials such as brass, aluminum, plastics, etc.

The ISO metric fine thread series is recommended for applications where a finer pitch is required, particularly where it is necessary to provide extra core strength for the male thread. This series is less resistant to stripping and to the effects of repeated tightening than that of the coarse series. This thread will also require a deeper length of engagement than the coarse series. This series is identified by the symbol M, its nominal thread diameter, followed by the 'x' sign, followed by the pitch in mm, e.g. M24 × 2, which will identify a thread with a 24 mm, major diameter and a pitch between its threads of 2 mm.

There does also exist, a 'constant pitch' series in the range 2.5 mm diameter × 0.35 pitch, up to 300 mm diameter × 6 mm pitch. These fine pitches make this series suitable for adjusting collars, retaining nuts, or thin nuts on shafts and sleeves, for use where the design needs to be compact.

The author's new constant method is now used for calculating the depth of thread for metric coarse threads (to be calculated in conjunction with using a sketched triangle) by using the constant number 31.526688.

[Note that a differing constant number will be used when calculating the depth of thread for metric fine threads (in conjunction with also using a sketched triangle), this being the number 31.52481224.]

If we now refer to fig. 42-b, (at the top left-hand side of drawing fig. 42), it will be seen that to discover the basic 'straight plunge' depth of thread for metric *coarse threads on bolts*, it is now only necessary for the machinist to use the known (drawing given) pitch of the thread, and the now-known thread constant, and to use these in two simple triangle calculations as follows.

We first make the freehand sketch of a right-angle triangle as shown in fig. 42-b, using our now-familiar right-angle triangle, and using its side *c* inscribed with the pitch of the thread, which in this case is entered as 3.

The new constant to use for Metric coarse threads is 31.526688. This is now entered into a position in the triangle that is normally occupied by The Angle. (Note that it will be side *b* of this triangle, after the calculation that will contain the straight plunge depth of thread required.).

The calculation is performed as follows:

The constant number is first entered into the calculator, as 31.526688, (this constant figure is unique for metric coarse threads, and would of course have a different numerical value if metric fine depth of thread or other thread pitches were being calculated). (A list of screw thread constants for other thread pitches is shown later on in the text.).

We then study the probe and select the prompt that is normally used for obtaining the length of side b in the right-angle triangle; in this case, this will be $\angle c$, $b = c \times$ TAN $\angle$.

We enter into the calculator the following sequence, (entering the numerical constant number first): 31.526688, TAN, (display shows 0.613441681), ×, 3, =, 1.840325045, this gives us the basic 'straight plunge' depth of thread required for screw cutting a metric 3 mm coarse pitch thread. This is then entered into the triangle at its side b.

If we now decide to use the angled approach method of screw-cutting to cut the thread in the work-piece, we will now need to find the elusive (but slightly longer) 'angled depth of thread' dimension that enables us to use the angled compound slide's accurate measuring index scale, to feed in the angled tool, (in increments of, say, 0.05 mm for each cut) into the work-piece prior to each cut being taken.

In order to complete this second calculation, we need to sketch a second triangle, shown as 42-a (at the bottom left of fig, 42).

We now re-use the previously obtained straight plunge 'depth of thread' figure of 1.840325045 mm (obtained in the fig. 42-b calculation), and use this figure again in the second triangle calculation, with this number now positioned (in this case) at the triangle's side c, to enable us to find the angled depth of thread required.

This calculation will then give us the actual dimension traversed by the tool when using the angled approach screw-cutting method, as will be shown in this case (after calculation) in the triangle's side a.

The side a of this new fig. 42a calculated triangle simulates the tool's angle of approach into the work-piece while it is being traversed by its compound slide. This angle is therefore entered into the triangle's The Angle position as now being 30 degrees.

Now, still referring to fig. 42-a, and using the Probe and Prompt system for the calculation, and the fact that we now actually know The Angle and c, and need to know the length of side a, we therefore select the appropriate prompt for this to find the length of side a in our triangle, by using the prompt calculation sequence $\angle c$, $a = c \div$ COS $\angle$.

We enter the sequence 30, COS, (display shows 0.866025403), XM, (transferring this displayed figure into the memory), followed by entering the sequence 1.840325045, ÷, R M, =, giving us 2.12502432, the full 'angled depth of thread' required for screw-cutting the thread while using the angled approach screw-cutting method (as shown in fig. 42).

If we should have the need to discover the actual amount the tool will have moved to its left at the end of its full depth angled approach, we can use the following sequence.

We know The Angle and c and wish to know b; we therefore use the prompt sequence $\angle$ c, $b = c \times$ TAN $\angle$. We therefore enter the sequence 30 TAN (display shows 0.577350269) × 1.840325045 =, giving us 1.06251216 the length of side b in this triangle calculation. This calculation informs us that the tool will move to its left by 1.06251216 mm on reaching its full depth of thread. This is valuable information, as this means that the original 4 mm wide undercut (previously machined in the work-piece for the screw-cutting tool's clearance immediately following its threaded section), will actually be reduced in width by this amount on the tool finally reaching its full depth of thread, thereby making the undercut now only approximately 2.93 mm wide (this being considerably narrower than was originally intended for the screw-cutting tool's run-out clearance), Therefore this narrower clearance width should be remembered when making the quick tool withdrawal at the end of each screw-cutting pass, because the tool now has less room to maneuver, coupled with the strong possibility of it now overrunning the undercut area and colliding with a work-piece's following existing shoulder.

This narrowing in width of the clearance undercut, will not of course effect the undercut's width if the machinist is using the *straight plunge* method of screw-cutting, for, in this case, the straight in-feed of the tool ensures that the tool enters the workpiece at exactly 90 degrees relative to its local surface and retains its full original 4 mm clearance width of undercut.

[If on the other hand we have decided to cut metric fine series of thread using, for example, the M24 × 2 mm pitch thread, we will first need to discover the straight plunge depth of thread for this thread. The constant we would use in this case is 31.52481224.

It is therefore now only necessary for the machinist to use the known pitch of the thread, and the known constant, and to use this in the now familiar two-triangle calculation series as follows:

We first make a freehand sketch using our now familiar right-angle triangle, but with its side c containing the pitch of the thread entered as 2. The constant used for fine threads is (31.52481224); this is now entered into the position in the triangle that is normally occupied by The Angle. It is the triangle's side b (after calculation) that will contain the required 'straight plunge' depth of thread. The constant 31.52481224 is now entered into the calculator. We then study the probe and select the prompt normally used for obtaining side b in the triangle; in this case, this will be $\angle c$, $b = c \times$ TAN $\angle$. We enter into the calculator the following sequence (by entering the numerical constant first), as follows: 31.52481224, TAN (display shows 0.613396624) $\times$ 2 = 1.226793249; this gives us the required 'straight plunge' depth of thread for this fine-pitch series of threads. This is now entered into the triangle at its side b.

If we were now going to use the angled approach method of screw-cutting for this particular fine-pitch series of threads, we now need to find the elusive (slightly longer) angled depth of thread that will enable us to use the angled compound slide's accurate measuring scale to feed the tool inward (in increments of, say, 0.05 mm) into the work-piece for each cut to be taken.

In order to complete this second calculation triangle (which in this case has not been shown in the text), we must now draw a second triangle, (similar in orientation to the original 42a triangle, with its line c vertical as used in fig.42a); we then re-use the figure of 1.226793249 in this second drawn triangle, again located in its side c. (Note that the fine pitch dimensions now being calculated are not shown in the original fig. 42a and 42b drawings as the dimensions shown in those triangles were for the coarse thread series.). We therefore simulate the method previously used in this newly drawn triangle to enable us to discover the 'angled depth of thread' required that will give us the actual dimension traversed by the tool during its angled approach, being simulated in its side a.

The Angle indication now shown in our new triangle simulation will show the angle of approach traversed by the compound slide together with its tool. This is now entered into the new triangle's The Angle position as 30 degrees.

We enter into the calculator the sequence 30, COS, (display shows 0.866025403) XM, we then enter the sequence 1.226793249, ÷ by RM, =, giving us 1.416578825 mm the angled approach depth of thread required for the 2 mm pitch thread.

There are of course other ways of finding the 'straight plunge' depth of thread for metric bolts and screws. This can be achieved by multiplying the calculated 'magic' figure 1.2269 by the pitch of the bolt or screw, then dividing by 2. For example, in the case of our 24 mm × 3 mm pitch bolt's thread, we multiply 1.2269 × 3 equals 3.6807, we then divide this figure by 2, to give us 1.84035 mm the depth of thread required.

To quote another example, (in the case of a metric fine 12 mm × 1.25 mm pitch screw thread), we can find the 'straight plunge' depth of thread for this screw thread by multiplying the 'magic' figure of 1.2269, by its pitch 1.25, then dividing it by 2 to give us 0.7668 mm, the depth of thread required.

Also, if we already know the 'straight plunge' depth of thread, we can obtain the angled depth of thread for metric bolts and screws by multiplying the straight plunge depth of thread by the figure 1.1547006, to obtain its angled depth of thread.

For example, for our metric 24 mm by 3 mm pitch bolt having a depth of thread of 1.840325045 mm, we can multiply this by the calculated 'magic' figure of 1.1547006 to give us 2.125024434, the approximate angled depth of thread; this being virtually identical to our original calculation used in our first method, the minute difference between the two dimensions can therefore be ignored.

[It should be noted that when physically referring to the lathe's cross-slide and compound slide's calibrated scales in order to 'put on' an accurately dimensioned cut, it will only be found possible to use the first two places after the decimal point of a metric calculation; we are therefore forced to ignore the value of any of the following decimal places, as these cannot be interpreted effectively on the lathe's measuring scales.].

Calculating the depth of thread for imperial Whitworth threads

If we need to calculate the depth of thread required for imperial (English) Whitworth threads, and for (English) BSF (British standard fine) threads, we can use the following method to obtain the depth of thread required.

The pitch of imperial Whitworth (BSW) and BSF (British standard fine) threads is always referred to in threads per inch or tpi.

If we take for example a 1 inch diameter Whitworth thread, it will have a pitch of 8 threads per inch.

To find its depth of thread, it is useful for calculation purposes, to use the calculated 'magic' number of 1280 as follows. We divide the figure of 1280 by the tpi (8), giving us 160.0. We now divide this number by 2 to give us 80.0. We must now mentally refer to this number (80) as if it were in thousandth parts of an inch (we do this by mentally moving its decimal point three places to the left) to give us 0.080 inches, the actual depth of thread required for a 1 inch diameter Whitworth external thread.

A similar method can be used for calculating the depth of thread for an American National Standard Unified external screw thread (including the English Unified screw thread) by using the calculated 'magic' number of 1226.

If we take for example a 1 inch diameter UNC thread having 8 threads per inch (tpi), we divide the 'magic' figure of 1226 by 8, giving us 153.25; we then divide this by 2 to give us 76.625. We must now mentally refer to this number as if it were in thousandth parts of an inch (we do this by moving its decimal point three places to the left), giving us 0.07625 inches, the actual depth of thread required.

The list of the following constants can now be used to choose an alternative way of calculating the depth of thread for other thread pitches. These constants have been specifically calculated to enable them to be used (with the assistance of the Probe and Prompt system), to obtain the depth of thread by using a similar method to that being used earlier in Section 18/34.

The method used, again consists of a sketched right-angle triangle, with the constant figure placed in the original The Angle position, with the pitch dimension being placed at side *c* of this triangle, followed by making the calculation using the Probe and Prompt calculation system.

The constants that can now be used in the following 'depth of thread' calculations are as follows:

For ISO metric fine, we can use the constant number 31.52481224.

For (imperial) British Standard Whitworth / British Standard Fine, we use the constant number 32.61111297.

For OBA, we can use the constant number 30.92096656.

For UNC, we can use the constant number 31.52628496.

For UNF, we can use the constant number 31.51662572.

FIG 43a.

THIS DRAWING SHOWS A CROSS - SECTION OF A TYPICAL 24 MM X 3 MM PITCH METRIC
NUT, WITH THE SCREW - CUTTING TOOLING BEING USED FOR EITHER THE 'STRAIGHT
PLUNGE' OR THE 'ANGLE APPROACH' METHOD OF SCREW - CUTTING.

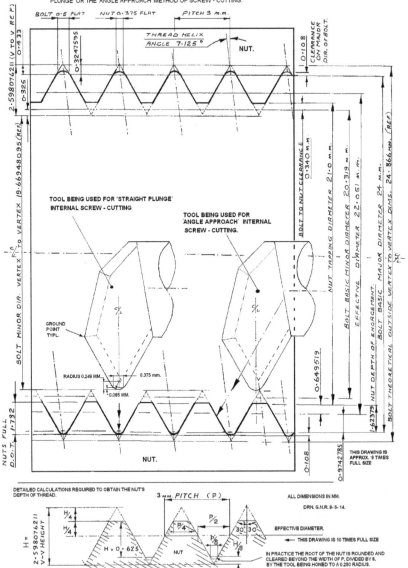

BOLT 0·5 FLAT   NUT 0·375 FLAT   PITCH 3 m.m.

THREAD HELIX
ANGLE 7·125°

NUT.

CLEARANCE
ON MAJOR
DIR. OF BOLT.

0·108

2·59807621I (V TO V REF.)
0·433
0·326
0·3247595

TOOL BEING USED FOR 'STRAIGHT PLUNGE'
INTERNAL SCREW - CUTTING

TOOL BEING USED FOR
'ANGLE APPROACH' INTERNAL
SCREW - CUTTING.

BOLT MINOR DIA. VERTEX TO VERTEX 19·66948095 (REF)

GROUND
POINT
TYPL.

RADIUS 0.249 MM.   0.375 mm.

0.065 MM.

NUT.

NUTS FULL
D.O.T. 1·732

BOLT TO NUT CLEARANCE 0·340 m.m.

NUT TAPPING DIAMETER 21·0 m.m.

BOLT BASIC MINOR DIAMETER 20·319 m.m.

EFFECTIVE DIAMETER 22·051 m.m.

NUT DEPTH OF ENGAGEMENT 1·62379

BOLT BASIC. MAJOR DIAMETER 24. m.m.

BOLT THEORETICAL OUTSIDE VERTEX TO VERTEX DIMS. 24·866 mm. (REF.)

0·649519

0·108

0·974278S

THIS DRAWING IS
APPROX. 9 TIMES
FULL SIZE

DETAILED CALCULATIONS REQUIRED TO OBTAIN THE NUT'S
DEPTH OF THREAD.

3 MM. PITCH (P)

ALL DIMENSIONS IN MM.

DRN. G.N.R. 8- 5- 14.

$H = 2·59807621I$
V-V HEIGHT

H/4
H/4

P/4
P/2
P/8

H × 0·625

NUT

30° 30°

H/8

EFFECTIVE DIAMETER.

THIS DRAWING IS 10 TIMES FULL SIZE

IN PRACTICE THE ROOT OF THE NUT IS ROUNDED AND
CLEARED BEYOND THE WIDTH OF P, DIVIDED BY 8,
BY THE TOOL BEING HONED TO A 0.250 RADIUS.

Section 18/35
There follows a full explanation of the fig. 43 drawing that gives full details of the 24 mm × 3 mm pitch metric coarse external thread.(This drawing also shows both the internal and the external thread's required clearances) in minute detail.

The fig. 42 drawing shows the tool's angled setup required when using the angle approach screw-cutting method. (This section also includes a method of dealing with engineering inspection problems.).

The fig. 43 drawing shows a very much enlarged (approximately nine times) cut-away view of a male, metric 24 mm × 3 mm pitch thread, together with its mating female fitted nut.

In order to fully enlighten the machinist with all the clearance requirements required for this particular screw-cut male thread and its mating female nut, the fig. 43 drawing shows a typical right hand screw-cut 24 mm diameter by 3 mm pitch male thread, together with its mating female nut. It will concentrate the mind on the importance of the clearance dimensions that must be maintained between the nut and its bolt. These clearances are extensively detailed in order to permit the machinist to study the interaction between the two components while they are in close contact with one another. The machinist must always be aware of the fact that these clearance gaps must always exist when both of the components are assembled together. These so-called clearance gaps are designed to prevent any unnecessary metal to metal contact being made between the major and minor diameters of the male thread, and the major and minor diameters of its mating nut; it therefore points out these important clearances that must be allowed for and taken into account when a machinist is screw-cutting a thread with a single point tool. The clearances to be maintained between the major diameter of the male bolt and the minor diameter of the female threaded screw-cut thread are of particular interest to the machinist, who will be depending to a large extent on knowing the correct dimension of the actual radius that must be honed onto the external screw-cutting tool in order to cut the thread in the bolt, and the internal screw-cutting tool's radius that will need to be honed onto the internal screw-cutting tool in order to screw-cut the nut accurately (as explained in fig. 43a). These externally honed tip radii on the tools are most important, and must be absolutely correct in order to allow each respective tool to obtain the exact depth of thread required for each component. The drawing also enlightens the machinist, later on, when the component is in service; there may be the necessity (due to rust contamination of the combined threads) to introduce easing agent or penetrating oil into the assembled nut and bolt to free a seizure. The aforementioned clearances provided by the machined clearance gaps will then allow the entry of penetrating oil, by permitting this oil to creep throughout the whole of the threaded part of the assembly, by using a process called capillary attraction.

The fig. 42 drawing specifically details the use of the angle approach method of screw-cutting, where the lathe's compound slide is swung through an angle of 60 degrees from its normally used horizontal position, followed by it being securely clamped to allow its screw-cutting tool to approach the work-piece at half the included angle of (in this case) the 60-degree included angle of the tool in use.

Therefore, for cutting metric and American threads (when using the angle approach method of external screw-cutting), the compound slide is swung through 60 degrees to allow the 60-degree included angle screw-cutting tool to produce a 60-degree included angle thread in the work-piece). This method actually uses just its left-hand side cutting edge and the tool's point radius to cut the thread in the work-piece.

Should the lathe be used later to cut imperial (English) threads, then in this case, the compound slide will need to be swung through 62.5 degrees from its horizontal position, in order to allow its 55-degree included angle screw-cutting tool to produce the 55-degree included angle thread in the work-piece and similarly by using just its left-hand side cutting edge and its point's radius to cut the thread in the work-piece.

Therefore, the use of the angled approach method of screw-cutting allows the cutting tool to cut the thread in the work-piece by using just its left-hand side cutting edge and its point's radius to produce the thread. This method has certain advantages over using the straight plunge method of screw-cutting, which by tradition retains its compound slide set to its normally used horizontal set position and allows its screw-cutting tool to cut using both sides of its two cutting edges and its point's radius.

In all cases however, the tool-post should be retained in a square horizontal position relative to the longitudinal center-line of the lathe. This then allows the center line of the screw-cutting tool to be secured always at 90 degrees relative to the center-line of the work-piece while using either of the two screw-cutting methods.

The fig. 43 drawing illustrates both the theoretical, (shown in the lower calculated part of the drawing) and the practical dimensions of the bolt's screw-cut thread, together with its fitted nut. As previously stated, it shows the theoretical air gap clearances that must be maintained between the two components to obtain the required correct fit when the screw-cutting operation has been completed.

It will be seen (when studying the fig. 43 enlarged thread example), that the outside diameter of the bolt actually retains a flat of 0.50 mm at the peak of its thread, thereby giving the nut a theoretical radial clearance of 0.108 mm relative to its major diameter. The nut also retains a theoretical flat of 0.375 mm relative to the bolt's major diameter. The clearance between the core diameter of the male screw thread and the nut's major diameter is also shown as a theoretical 0.340 mm.

The indicated effective diameter (considered by engineers to be a theoretical point located at exactly half the depth of thread), is the exact point where the male screw thread can be checked and measured for correct size by using a precision thread micrometer. The micrometer's anvils together with its central conical point can then be used to measure the effective screw-cutting diameter of the thread. This method is used as an alternative way of checking the thread's size rather than by using the thread ring gauge for this purpose.

During the final stages of screw-cutting a screw thread, a thread ring gauge (a hardened accurately ground and checked internally threaded gauge), is used to check the screw-cut component for correct fit.

In use, this gauge must be capable of being screwed freely on to the work-piece by hand, to ascertain that there is no rock, side-play, or end float, occurring between the component and the gauge. A perfect fit of the gauge relative to the component will then ensure that the work-piece will conform both to the inspection departments and to the drawing's requirements; the screw-cut component can then be deemed compatible with all standard-fitting production nuts. The machinist should bear in mind that an identical thread ring gauge will be used by the mechanical inspector, during his thorough checking of the work-piece, and the use of this gauge will highlight to him any slackness or tightness present in the machined, screw-cut thread.

One should also realize that should an inspector discover during his inspection of the thread that it has been machined too loosely, or that it possesses too much rock or end play, he may consider the component to be out of limits to the specification shown on the drawing, and therefore render the component to be scrap.

One should also be reminded that if the machinist finds himself in the unfortunate position where his work has been considered to be scrap, then the inspector's decision must be considered as final. Therefore when an inspector considers the work-piece to be scrap, then it is scrap, and one should not be tempted to argue this point.

It is always considered good practice for the work-shop staff to always do one's best to maintain a good working relationship with the inspection department, whatever the outcome of a suspected scrap enquiry. The machinist therefore, should resolve, on future occasions, always to use a thread ring gauge to check the fit of the work-piece, much sooner during the screw-cutting process, and always before the theoretical full calculated depth of thread has been reached by the tool.

While on the subject of dealing with engineering inspectors and their problems, I feel I must include the following occurrence that happened to the author, and how it was best resolved.

If, during one's daily machining work, a dimensional problem is discovered on a component that has been received in your own department from another part of the factory, while the machinist is setting up this particular piece to receive its second machining operation, a problem is found, it is always the best policy to inform the inspection department of the problem before carrying on with any further machining. Much time and expense can be saved by adopting this practice and by doing so will also promote continued good relations with the inspection department (this being a very important practice to adopt).

The author once experienced this situation, where, having been instructed to finish bore a large diameter heavy gear-wheel, previously machined all over its outside surfaces and having a finished gear-cutting operation already performed on its periphery, it was necessary to finally finish the component, by boring a precision 'tight limit' hole through its center to fit an existing 'close limit' shaft.

During the setting-up procedure involving the use of a crane, adjusting a four-jaw chuck to get the work-piece to run concentrically, using the assistance of a DTI (clock gauge) and a precision dowel for checking each of its individual tooth's contours to obtain work-piece concentricity, it was discovered while the workpiece was being rotated by hand in the lathe's chuck, tooth by tooth, to obtain an accurate 'clock gauge' setup, the gear cutting's overall dimensions were incorrect, in that the gear cutting operation previously performed on the component was decidedly oval in shape with regard to its gear-cut pitch circle's diameter. The inspector was called, and after his further careful checking (in this case, in situation while the component was still set up in the lathe's four-jaw chuck)' it was confirmed that the oval shape of its previous gear-cutting operation was indeed faulty, and the error now found in the gear-cutting's concentricity was sufficient for him to make the decision that the job was indeed scrap. He thanked me for my observation, adding that he would have been unable to check the inaccuracy of this particular gearwheel had he been forced to use his available checking equipment and his normal checking procedures that are used in his inspection room. This inability to accurately check this dimensional discrepancy in his department was mainly due to the exceptional unwieldiness and extremely heavy weight of the component. This experience proved to the author that the center lathe is actually very capable of being used as a precision inspection tool, (particularly in this case). The moral to this story is that one should always report inaccuracies that may have the potential of causing further problems down the production line regarding inaccuracy, to the inspection department for their perusal, before allowing them to be passed on as being OK, without their consultation.

Section 18/36
Further checks required to prepare the center lathe for screw-cutting, also the recommended precautions one should take before using a parting-off tool on the lathe in order to prevent possible accidents occurring.

The following checks will be needed when preparing a center-lathe for a screw-cutting operation; this advice applies whether one intends to use the straight plunge or the angled approach method of screw-cutting.

It is quite a usual occurrence for the lathe's compound slide, (during its normal daily use), to become gradually over extended beyond its normally used fully supported range of use, This situation often goes unnoticed by the operator; it is therefore very important for the machinist to realize that if the lathe's compound slide is used in this overextended condition (particularly on an older well-worn machine), then because of this overextension of its slide, it cannot then provide a sufficiently rigid platform from which to accurately perform either a straight turning, screw-cutting, or in particular, a parting-off operation. The compound slide must therefore always be returned to a fully supported position immediately before attempting to perform any machining work. It is also essential that the machinist makes a check of and adjusts, if necessary, any slackness found in the actual fit of the compound slide relative to its casting, or the cross slide's tapered adjustable jib strips relative to its male cross-slide (this adjustment can be achieved by making very fine adjustments to the cross-slide's adjusting screws) in order to remove any excessive sideways play or perceptible slackness throughout its total sliding movement. Any slackness discovered in the compound slide assembly should always be reduced to the minimum in order to provide the maximum rigidity of the set-up. Any overhang of the compound slide beyond its support casting must also be reduced by turning the compound slide's hand-wheel anticlockwise until all its overhang has been fully reduced beyond nil, followed by it being turned clock-wise by approximately 1 mm before any overhang is reached where its scale is then set to read zero. The cross-slide's hand wheel is then turned anti-clockwise by a full turn, followed by turning it clock-wise by half a turn to remove all the existing back-lash in *its* lead screw and its scale set to read zero.

(Note that the overhang adjustments we have just made to the compound slide's travel are quite often a neglected practice, and these adjustments should be rigorously adhered to in order to allow the lathe to perform at its absolute best.).

To reiterate, as a skilled machinist, one must ensure that at all times there is the minimum of overhang of the compound slide relative to its supporting base casting. This situation will apply particularly when attempting to perform a parting-off operation, because, when using the lathe for this operation, the maximum rigidity of the setup is *absolutely paramount*, in order to prevent tooling accidents occurring.

The parting tool should always be considered a fragile tool, and in all cases the tool's unsupported shank's stick-out (from its supporting tool-post), should only be allowed to possess the absolute minimum of overhang from its supporting tool post, in order to maintain the lathe's maximum rigidity during the *very critical* parting-off operation.

In other words, the tool's front cutting edge should only be permitted to have the minimum of protrusion from its tool-post base's support. The actual amount of stick-out allowed will of course ultimately depend on the diameter of the work-piece about to be parted off, and on the toughness of the material itself being parted off. It will usually be found to advantage if the leading front cutting edge of the parting tool is purposely ground so that it can possess a slight angled lead protruding forward at its front right-hand cutting edge. This then enables the right-hand corner of the tool to lead its left hand corner during the parting-off process. This then allows the tool's right-hand front cutting edge to cut into and part off the work-piece *before* its left hand cutting edge has actually reached the work-piece's center line during its parting-off cut. (The amount of this lead will of course depend on the width of the tool's front cutting edge.). By adopting this angled modification to the tool's front cutting edge, the work-piece can then be cleanly parted off before its left-hand cutting edge has reached the center-line of the holding work-piece.

By making this angled modification to the parting tool's front cutting edge, it will also be found that it leaves virtually *no residual pip* on the parted-off component's end surface when its left corner has reached the center-line of the holding material, thereby leaving the holding material's end face with a clean end. This modification to the tool will then avoid the extra work that would normally be required to face off the surplus pip from the end of the parted-off workpiece, that would require the use of a separate facing-off operation.

It is also considered a wise procedure to use the parting-tool with the narrowest cutting edge available, that is considered to be strong and robust enough to perform the task in hand. The parting tool must of course be extremely sharp, robust, and have sufficient cutting and clearance angles (provided by accurate off-hand grinding) to clear the sides and length of the slot that is about to be produced in the work-piece during the operation. The wider the parting tool's front cutting edge is, the greater the load that will be transferred to the unsupported part of the tool, its work-piece, and the lathe chuck's spindle support bearings.

It must be emphasized at this point that the machinist must always ensure that full support is being provided for all of the lathe's tooling; if this is not the case, it can result in a combination of poor screw-cutting performance, erratic and inaccurate turning operations, poor work surface finish, and very poor, possibly dangerous parting-off performance.

Therefore, before one uses a parting tool, it is imperative that the material being worked on is very securely held in the chuck, (preferably by using the four-jaw chuck if possible), for all parting-off operations.

Ideally, the held material should be fully supported by using the full length of the chuck's jaws to ensure the maximum rigidity of the setup is maintained. (It should be noted that the practice of insufficiently securing a work-piece in the lathe's chuck by gripping the work-piece material near the front of its jaws, is a recipe for disaster). As previously stated (provided one has the choice), it is always the best practice to use the four-jaw chuck for the parting-off operation. It will be found that the four-jaw chuck provides the most secure method of holding the work-piece, because it possesses a far superior gripping power compared to that of the three-jaw chuck when being used for parting off.

The machinist should always ensure (1) that the parting tool is perfectly sharp. Resharpen if there is any doubt at all regarding its sharpness, also check that its side clearances have been ground away sufficiently to provide clearance over the full length of its stick-out so that its cutting edge can cover the full depth of material being parted off; this clearance should allow the tool to clear the full width and depth of the work-piece's slot throughout the whole of the expected inward travel of the parting-off tool.). The parting tool shown in fig. 47 (tool number 6) has been modified by careful and meticulous grinding to increase its strength. This has been carried out to allow the tool to possess considerably stronger *parallel* sides which also provide the necessary side clearance, while leaving its front cutting edge of normal width, combined with a much shorter side clearance angle. This modification to the tool's normal design provides a much stronger butt-welded joint at the junction with its shank (its usual weak point). (2) that sufficient clearance area has been provided at the sides of the tool throughout the whole length of its unsupported stick-out area from its tool post; this clearance must be sufficient to allow its full depth to be completed without hindrance, on reaching its full parting-off depth. (3) that the front cutting edge of the tool is set at the exact (and correct) cutting height, and that the center-line of the tool's stick-out section is aligned exactly square and perfectly in line with the cross-slide's center-line; this then ensures that the tool's ground side clearances fully clear the sides of the machined slot that develops progressively in the work-piece prior to it reaching its final parting-off point. (4) that there is the minimum of overhang of the tool's cutting edge from its tool-post (also ensure that there is no overhang of the compoundslide relative to its support casting). (5) that the saddle is firmly locked to the lathe bed for *all* parting-off operations; this must be done in order to prevent any possible (dangerous) sideways movement of the tool during the operation. Also note that the parting tool should never be used to cut in a sideways direction, because it possesses a long unsupported stick-out from its tool-post, and is extremely weak if one attempts to machine the work-piece in a longitudinal direction.(5) that there is an adequate and *continuous* supply of coolant available and this is being used throughout the whole of the parting-off operation. (6) that the spindle speed selected is compatible with the diameter and toughness of the material being parted off. (Small diameters will require a much faster parting-off work speed. Large diameters will require a relatively slow parting-off work speed and will also require sturdier, more robust, and larger design of tool due to the greater amount of stick-out required. (Tough materials will generally require a slightly slower work speed than normal.). (7) that the whole parting-off operation should be carried out within 2 to 3 mm of the chuck's jaws in order to ensure the maximum rigidity of the set-up. (8) that one feeds the parting tool into the work-piece *very carefully and progressively*, and it is considered preferable to use one's left gripping hand located on the outside diameter of the cross-slide's hand-wheel (the machinist should **not** be using its handle, as this will only provide a jerky inaccurate rotary movement). Also, this operation is ideally performed by using just one's left hand, in order to provide a very progressive and accurate rotational in-feed movement to the tool (this can be assisted at times by using the right hand to allow a change of hands). The movement being given to the tool must be very *smooth, progressive, and have a continuous inward travel*. (Note that it is considered bad practice to allow the tool to pause or dwell at any time during the parting-off operation. (One must therefore refrain from allowing the tool to stop its movement forward during its progression inward), as any pause will cause its cutting edge to become dulled by the abrasive rubbing action caused by its non-cutting activity where it rubs against the diameter of the moving work-piece; in other words, there should be no intermittent feed given to the tool during its inward travel while performing the parting-off operation). One should also make every effort to acclimatize oneself into acquiring the necessary delicate feel that is needed for this particular operation. This can be achieved by keeping the in-feed hand wheel's inward movement as constant as possible and by using extreme care and diligence during the whole operation.

Experienced machinists will often listen attentively to the various noises that emanate from the parting tool during its progression into the revolving work-piece; they will also pay close attention to the various noises made by the lathe's drive motor itself during the parting-off operation. The student should be encouraged to become accustomed to these subtle changes in the sound of the motor and the pitch emitted from the tool during the operation, particularly when the tool is nearing its final parting-off point, where there is an audible rise in the tone of its pitch that provides a valuable warning to the operator of the approach of the exact and final parting-off point. The lathe itself will also warn the operator if too high a parting-off speed is being attempted (this is usually announced by the lathe emitting a high-pitched squeal), or in the case where the in-feed being given to the tool by the operator is too great, by the lathe emitting an audible grunting (or loud chattering noise indicating its disapproval). The ideal parting-off tool's feed of progression inward into the workpiece should ideally produce a neat coil of swarf that curls away from the top of the tool that is of consistent thickness, (a consistent thickness of swarf being produced by the tool will indicate to the machinist that the infeed's speed of progression', using the cross-slide's hand wheel, is correct); ideally of course this swarf should be ejected from the top of the tool's cutting edge in a shape that closely resembles a loosely wound clock spring, but this of course will only normally occur if the tool has been ground with its cutting edge perfectly square to its shank, that was purposely ground to reduce the frontal area of the cut being produced to its absolute minimum width. (This modification may be found necessary when there is the need to provide the absolute minimum of loading on the tool and the work-piece).

The ideal thickness of swarf produced in this coil, should always be aimed for, but may only rarely be achieved, particularly if the tool's front cutting edge has been purposely ground to produce a leading right-hand cutting edge designed to produce a neat parting-off surface on the cutoff workpiece.

It should also be remembered that to achieve a correct cross-slide/ infeed and rotational speed of feed during the parting-off operation, the whole procedure will require very delicate and the minimum force being applied to the tool throughout the whole 'parting off' operation.

If the tool is found to require a *higher-than-normal* infeed force to be applied to the cross-slide's hand-wheel to get the tool to cut, and is also accompanied by a marked reluctance for the tool to cut into the material, then this problem should be investigated immediately, and importantly, before attempting to apply any extra force to the infeed hand-wheel in order to force the tool to cut.

Failure to heed this warning can result in the tool (while under forced pressure from the machinist while being reluctant to cut), initially hesitating, followed by actually grabbing the work-piece, momentarily causing the tool to take an instant alarmingly deep cut (much deeper than was originally intended), which overloads the tool to such an extent that it may dig into the workpiece, causing the complete breakage (or snapping off) of the extended part of tool. (This break will usually occur at or near its butt-welded joint to its shank.). Therefore, if there is the tendency for the tool to be reluctant to cut, this should be *immediately identified*, and appropriate remedial action taken (this remedial action should include either re-sharpening, or lowering the tool very slightly), in order to perform a *temporary* cure. Initially, if the tool is found to be reluctant to cut but still found to be sharp, this problem can often be cured by lowering (very slightly) the tool's cutting height (in effect, this marginally increases the tool's top rake's cutting angle sufficiently to allow the tool to cut). If this does not cure the problem, then try re-sharpening the tool and increasing the angle of its top back rake by (say) at least an extra two to three degrees more than normal, then re-checking the tool's cutting height, then adjusting its cutting speed to either faster (in the case of parting off small diameters, or slower in the case of parting off large diameters). (Usually a slower chosen work speed when parting off large diameters will provide the required result.). It will also be found advantageous if the 'parting tool's front cutting edge stick-out and its cutting edge width is also reduced by the use of off-hand grinding. This modification to the setup will then reduce the load being transferred to both the tool and to the workpiece. The reduction of the parting tool's cutting edge width will be found particularly effective when parting off very small-diameter workpieces.

If one is working on very small diameter workpieces, the parting tool's width can be reduced substantially to a very narrow 1.5 mm, and in extreme cases, this can be reduced still further, to a 1 mm width (for exceptionally small workpieces); this modification will allow the parting tool to part off very small diameters even down to, say, 2 mm diameter, with less chance of damaging the work-piece.

When the correct infeed and work-piece speed have finally been established for parting off the work-piece while using a practical hand-operated procedure, then, provided an identical work speed and hand infeed speed can be selected on the lathe's infeed gearbox, an automatic in-feed can be used to part off all of the following batch of similar components. One must always remember to use in this case a plentiful and continuous supply of coolant on the work-piece. (However, this *automatic feed* parting-off operation should not be attempted by the student until he has first established an identical in-feed, work/speed, and coolant supply by a previous hand feed operation. A safe parting-off operation can only be relied upon if one is using a process of duplicating an initial practical hand-operated parting-off operation that has been proved beyond any doubt to be successful. Failure to observe this *duplicated parting-off* procedure will invariably result in a broken tool or a damaged work-piece, should the *automatic* parting-off method be attempted without first proving practically that all the lathe's spindle speeds, and tool feed speeds are working satisfactorily for the job in hand.

The advantage of using one's left hand to control the in-feed of the tool when parting off is that it allows an experienced machinist to use his right hand to support (and physically catch) small parted off components in the very last stages of the parting-off operation to prevent the work-piece falling into the drip tray of the lathe and becoming lost among the piles of metallic swarf in this area.

Trainees should also note that the lathe's drip tray should always be cleaned out at the beginning and end of each operation, particularly when there is a change of material being machined. If copper, aluminum, cast iron, or stainless steel materials are being machined, these components should be machined separately, and their swarf should always be collected and placed in separate bins for recycling.

Section 18/37
The dangers it is possible to experience when operating a badly worn or maladjusted lathe.

There follows an explanation of the variety of problems that can be found in the practical working functions of a badly worn lathe.

Any looseness found to be present in the cross-slide, compound slide, or the saddle, relative to its precision ground V ways, will be found to provide poor tooling support, and will often be the cause of tool chatter to occur during heavy turning or parting-off operations. In exceptional cases, there may also exist the major problem of the lathe having seriously worn spindle bearings. This particular problem will of course tend to occur on the older well-used machines. Badly worn main bearings will be incapable of supporting the main spindle and its work-piece with sufficient rigidity to allow the lathe to perform accurate (and safe) parting-off or turning operations. Excessive bearing wear in this area will permit the work-piece to be partially lifted by the cutting action of the tool during the machining operation. In practice this gives the machinist the impression that the workpiece is trying to momentarily lift and partially climb up the tool; this situation can occur particularly during parting-off operation, or when performing heavy straight turning cuts. This problem can become particularly alarming (and even dangerous), when attempting to part off a component on a lathe that is known to possess seriously worn spindle bearings that are providing insufficient support to the main spindle, and its chuck-supported workpiece.

When worn bearings are found to be giving this poor mechanical support to either the tool, the chuck spindle, or the work-piece, this situation can, in exceptional circumstances, cause the parting tool, while under maximum load, to be literally grabbed and snapped off at its weakest point. This breakage of the tool will be mainly caused by its heavily chattering vibration, causing an intermittent cut that enables the work-piece to momentarily lift and fall, thereby taking up the existing wear in the lathe's support bearings. Invariably the parting tool will break at a point that will occur in the area thinned down by excessive clearance grinding of the tool's now diminished cross-sectional area situated in front of and close to its butt-welded joint relative to its supporting shank. This weak point in the parting tool, is an inherent problem found evident in virtually every well-worn off-hand ground' parting tool. As stated, this problem is usually caused by excessive clearance grinding in the area close to the tool's butt-welded shank joint. This snapping off of a parting tool, due to worn bearings, combined with the existing tool thinning, (caused by excessive clearance grinding at its weld joint), and by using a possibly too heavy an infeed being given to the tool, will cause it to break without prior warning, and with alarming rapidity; it is usually accompanied by an embarrassingly loud crash and clatter, often in full view of other shop floor colleagues. The follow up this occurrence, it is often accompanied by a loud and embarrassing chorus of "'Go to be a farmer's boy' by your work-shop colleagues. This disaster should therefore be avoided at all costs, not only to avoid your embarrassment in the use of the lathe, but also in the obvious loss of face among your colleagues.

The broken tool's cutting edges will usually become embedded in the parting-off slot formed by the tool, and be submerged in the work-piece; its removal will prove difficult and almost impossible to achieve, as its extremely hard cutting edge will be found difficult to extricate without damaging the end face of the work-piece.

To resolve this problem, one can either move the work-piece assembly to a newer, more efficient lathe in order to part it off, while taking the necessary precaution of leaving sufficient excess material on the work-piece's length to allow for its inevitable facing-off operation, or attempt a second parting-off operation on the same machine, but with the work-piece relocated to a new position that is much closer to the chuck and by using a more substantial (stronger and more robust), very sharp, and very rigid parting tool (similar to the modified tool number 6 in fig. 47), while also ensuring that there is the minimum of overhang of the tool from its tool-post and the tool-post from its compound slide's support.

This parting-off operation is then followed by re-chucking the component to face off the excess material, while taking extreme care not to allow the new tool's cutting edges to come into contact the hard and embedded end of the broken off tool, as this new tools cutting edge will also be damaged if contact with it is made. By using extreme care, the broken end of the tool can often be removed during this process, and the work-piece again faced off carefully to enable it to be reclaimed.

If the machinist does suffer this total destruction of a parting tool, a thorough investigation should be made of the lathe's working parts and particularly checking the amount of wear present in the chuck's spindle bearings and in the compound slide's support guides.

To make this check, the lathe should be switched off at its main electrical switch to enable a safe thorough check of its bearing wear to be made. This can be undertaken by utilizing a 12 inch (300 mm) long × 20 mm diameter round test bar, securely mounted in the chuck. A dial test indicator is then positioned so that its pre-loaded probe is just touching the test bar with its scale set to read zero. A hefty tug on the bar (by using both hands), will spring the spindle over from its normally located center-line position; the extent of this movement will reveal as it does so the extent of wear present in the lathe's main bearings, by the machinist mentally noting the deviation of the clock gauge's needle relative to its scale.

A new machine in this situation would show virtually no deviation at all (except for the expected minimal spring in the test bar). A very small amount of deviation of up to 0.001 inch (0.025 mm) could be considered acceptable on a partially worn machine. A large deviation of 0.002 inch (0.05mm) or above will indicate that remedial action will need to be taken, probably by the maintenance department, replacing the lathe's bearings).

Most tapered-type center-lathe's spindle bearings can be adjusted for small amounts of wear by using its special spindle bearing adjusting nut. This tightens up the internally tapered support bearing sufficiently to take up the existing wear, (this job will normally be handed over to the maintenance department). Unfortunately, there are other designs of lathe main bearing spindles that do not have adjustable bearings; therefore, this particular design of lathe will need to have a complete set of new bearings fitted to its main spindle and its work-head. This work will involve a much longer time-consuming fitting job, normally performed as a major maintenance job by the maintenance department.

[As a matter of interest and to provide some useful information on parting-off methods, some of the very old lathe designs were equipped with extremely long cross-slides, that possessed very long cross-slide lead-screws; a lathe of this older design will often allow the machinist to position the tool-post well beyond, and to the rear of the work-piece, where there will now be found sufficient space for the parting tool to be installed in the tool-post in a position where it can now approach the work-piece from the rear. This tool must also be oriented into an upside-down attitude, which then allows the parting-off operation to be completed while the workpiece is *still being rotated in its normal direction of rotation*. It will often be found possible with this setup (when using this older type of lathe), to perform a relatively safe parting-off operation, even if the lathe's bearings are *slightly* worn.

[Using this rear upside-down tool method of parting off will be found to have the advantage that the tool will now be actually holding down the lathe's spindle and the work-piece during the parting-off operation and by doing so will allow a safer parting-off operation to take place. This particular setup will be found particularly useful when using an older type lathe when there has previously been the need for parting off a component with suspected worn main bearings, (providing of course the lathe used possesses the required 'long' cross-slide and its long lead-screw assembly).]

This method of parting off (by using an upside down parting tool in a rear mounted tool post) is often performed on capstan lathes. These particular lathes normally possess two tool posts, one in front of the work-piece, used for normal lathe turning tools, and one behind the work-piece used mainly for the inverted parting-off tool. This versatility makes the capstan lathe capable of turning and parting off a component without the need to change the tooling between each operation.]

Section 18/38
A general description of straight plunge external screw-cutting, where the lathe is using a combination of a disengaged lead-screw, a quick method of tool withdrawal, and using the thread-cutting knockdown lever to isolate the tool from the lead-screw

Having now prepared the lathe and work-piece for external screw-cutting, as described in section 18/28, the compound slide assembly and its tool-post must now be secured so that it is square to the center-line of the machine. The lead-screw should now be in its disengaged position; this is done by ensuring that the screw-cutting knockdown lever is in its down position, which in turn leaves the work-piece disengaged from the lead-screw's drive.

With the lathe now switched off and with the work-piece already turned down to its required 23.9 mm diameter in the area about to be screw-cut (preferably by using a sharp right-hand knife tool for the operation), and with the screw-cutting tool now fitted and correctly oriented into the tool-post by using the square-type screw-cutting gauge, the compound slide's hand-wheel is now turned anticlockwise sufficient to eliminate any overhang being present in its slide relative to its support casting, its hand-wheel is then turned clockwise by (say) a quarter of a turn to remove all the backlash in its lead-screw, followed by its scale being reset to read zero. The cross-slide hand wheel is then turned clockwise to position the newly fitted screw-cutting tool's point radius so that it contacts the 23.9 mm screw-cutting diameter; its measuring scale is then set to read zero.

It should be noted that when setting up the lathe for either the straight plunge or angled approach screw-cutting method, it is considered to be a distinct advantage if the machinist adjusts the position of the cross-slide's hand wheel's operating knob (its handle) so that it now occupies its topmost (twelve o'clock') position while the tool's point is in contact with the turned 23.9 mm diameter previously turned for screw-cutting.

This positioning of the cross-slide hand-wheel's knob is not just a frivolous move; it will be found in practice that it becomes very important, for it allows a much speedier hand access to allow the machinist to prepare for using the snatched, or (quick, method of tool withdrawal') from the work-piece, which must be performed exactly as the tool reaches the under-cut area, situated at the extreme end of the screw-cut portion of the newly screw-cut thread.

At this exact extraction point, it is also good practice to mark the periphery of the chuck with a very distinct warning marker. This enables an accurately timed tool withdrawal to be made from the thread. We mark this by using a strip of clearly visible adhesive masking tape or similar material, positioned so that it can be seen to pass through the machinist's peripheral vision while his main attention is being concentrated on close scrutiny of the tool's position as it nears the end of its screw-cutting pass. By using this marker, the machinist is given instant warning of the exact moment the tool will need to be quickly withdrawn, together with instantly operating the knockdown screw-cutting lever to disengage the tool from the lead-screw. This quick tool withdrawal is comprised of one quick revolution of the cross-slide's hand- wheel in an anticlockwise direction) to allow the tool to clear the work-piece with rapidity and safety. The timing of this quick tool withdrawal snatch is most important, for, with the cross-slide's handle now positioned in its uppermost position and within easily accessible reach, the withdrawal of the tool can now be performed with much greater speed and accuracy. The machinist will now be able to make just one complete and quick anti-clockwise revolution of the cross-slide's hand-wheel at the end of each screw-cutting pass, immediately as the tool is seen to have reached the under-cut area of the thread, and instantly operating the knockdown screw-cutting lever.

As previously stated, by using the guidance of this introduced warning marker, the screw-cutting knockdown lever is then operated at exactly the correct time and place needed to disengage the tool from the lead-screw, at the end of each passing cut.

The tool is now returned to its start position by turning the saddle's hand-wheel in a clockwise direction sufficient for the tool now to occupy its required start position of approximately 5 to 6 mm to the right of the work-piece.

[It should be noted that any premature use of the knockdown lever (even if only slightly before the tool has actually reached the undercut area on a passing cut), will result in the screw-cut thread being damaged (possibly seriously) by what will now be a stationary cutting tool; therefore, any premature movement of this lever should not be attempted during the pass.]

The cross-slide's hand-wheel is now turned clock-wise, by one revolution down to its zero setting, followed by a 0.05 mm increase of depth ready for its next passing cut by using its index scale for reference. This scale is then marked with pencil (as a constant reminder of the tool's depth). The lathe is now switched on in its normal direction of rotation using a low work speed of approximately 60 rpm, and the coolant supply to the tool and work-piece is now switched on.

A close scrutiny in now made of the 'rotating screw-cutting dial' situated on the lathe's front apron, until ones chosen number or line is seen to line up with the apron's fixed datum line, whereupon the *lift-up screw-cutting engaging lever* is operated to engage both the saddle and tool with the lead-screw, initiating its passing cut. On the tool being seen to reach the thread's under-cut area, and by relying on the chuck's rotating warning marker (as it is seen to pass through his peripheral vision), the *snatched method of tool withdrawal* (of one revolution anti-clockwise is again given to the cross-slide's hand-wheel) together with instantly operating the screw-cutting knockdown lever at exactly the same time. The tool is then returned to its starting point of 5 to 6 mm to the right of the work-piece by turning the saddle's hand-wheel in a clockwise direction. The cross-slide's hand-wheel is then turned clockwise by one revolution down to its scale's newly indicated pencil mark, followed by it being further indexed inward clock-wise by 0.05 mm, ready for the next cut to be taken; this new position is again marked with pencil.

This is followed by an identical series of screw-cutting passes with the depth of cut increased by 0.05 mm for each passing cut, until 1.84 mm is indicated on the cross-slide's scale, (showing that the full theoretical depth of thread has been reached). A thread ring gauge is then used to check the accuracy of the newly screw-cut thread.

We return now for a full explanation of all the minute details required during the initial setting-up procedure for screw-cutting this particular thread.

Your attention is drawn to the necessity of making a visual check of the whole screw-cutting 'setup' in order to ensure that adequate longitudinal clearance has been established between the vulnerable cutting edges of the tool, and its possible closeness to the dead hard (stationary) supporting dead center during the following screw-cutting operation; this must be done in order to prevent a possible clash occurring between these two pieces of tooling.

Should a clash occur, it is likely to cause the tool's cutting edges to be blunted or 'chipped (this will be considered a minor disaster at this fully set-up stage of the operation). We must also visually check that a 'safety gap of approximately 10 mm is rigorously maintained between the saddle's moving casting, and the tailstock's stationary base casting at the starting end of the operation, and monitored on the tool's return to its starting position during the operation. This gap should never be less than 10 mm in order to allow adequate clearance for a possible tool or saddle overrun in the tailstock's direction.

This over-run can occur in practice (particularly if one is using the 'fixed lead-screw method of screw-cutting), and can occur just before the saddle has finally come to rest during either a manual or an automatic return traverse prior to a next cut being taken.

A further observation should be made of the *rotating thread cutting dial's indications* (seen situated on the lathe's front apron), and a mental note should be made of the chosen number relative to its datum line, that one intends to use (or is currently using) throughout this whole screw-cutting operation. This chosen point should be memorized and later used for accurately *lifting-up screw-cutting engaging lever* into its fully engaged (i.e. fully up position) prior to making a new cut.

It is also very important that this lever is not accidentally left in a *partially* engaged position, for if left partially engaged, the screw-cutting operation will be unreliable, and will result in the transmission gear failing to engage fully into mesh with the lead-screw; this will in turn cause the screw-cutting tool's action to be partially delayed, allowing it to lag behind, resulting in the tool being slightly out of step with the previously cut thread on this, or on any later partially engaged screw-cutting passing cut.

It is important to note that a carelessly engaged *lift-up screw-cutting engaging lever* can also cause serious damage to the tool because it is temporarily out of step with the thread being cut),) and, in consequence, can be seriously overloaded on contacting a full 'out of out-of-step revolving thread). The work-piece's threaded portion itself can also be damaged by being out of sequence with the screw-cutting tool. The possible damage sustained by the workpiece, due to being out of step could be sufficient to prevent it being reworked or even salvaged (resulting in the work-piece being possibly scrapped).

During the whole of the screw-cutting operation, the machinist must therefore rely on accurately using his chosen line or number found on the *rotating screw-cutting drop-in threading dial* fitted to the saddle's front apron. Its numbered and positioned lines (engraved on the threading dials rotating scale) are specifically designed to indicate to the operator, the exact point where the *thread cutting lever* must be lifted up, or engaged, with the lead-screw, and this action must co-incide when the chosen number or mark is seen to line up with the fixed datum line engraved on the front apron.

With the lathe now switched on, the work-piece rotating, and with the tool positioned 5 to 6 mm to the right of the work-piece, the operator must now wait until the previously chosen rotating dial's position line (or number) is seen to line up *exactly* with the engraved stationary datum line located on the lathe's front apron, before lifting up the thread-cutting lever to start the cut. Correct alignment of these two datum points is *crucial*, and it is only by using accurate synchronization that the tool can engage accurately with the previously partially cut screw-cut thread.

[Experienced lathe machinists also regularly use a method that involves the quick tool withdrawal while relying on the visual warning marker fitted on the chuck, combined with using the *screw-cutting knockdown lever* to disengage the lead-screw from the moving saddle at the exact point where the tool reaches the undercut area at the end of each screw-cutting pass. This is followed by using the saddle's hand-wheel to return the saddle assembly, including the tool and its tool-post, manually to its start position.

[This method is often used when screw-cutting long work-pieces, as it will be found much quicker for the machinist to return the tool assembly to its start position manually, rather than waiting for the whole assembly to return slowly under its own power when using the permanently engaged lead-screw method of tool return.]

However, the use of the preferred disengaged lead-screw method of screw-cutting, will avoid the machinist having to wait patiently for the saddle and its tool assembly to return back slowly under its own power to its start position, particularly when using an extremely low work speed.

The practice of using the *screw-cutting knockdown lever* accompanied by using the *quick tool withdrawal* (while leaving the workpiece still in rotating mode), can of course be employed when using either the straight plunge or the angled approach screw-cutting method. In either case, the machine's spindle, chuck, and work-piece, are left rotating at the end of each pass, the tool is then quickly snatched out on reaching the under-cut area (by using the guidance of the chuck's rotating warning marker), followed by instantly using one quick and complete anti-clockwise rotation of the cross-slide's hand-wheel at the instant the tool has entered the clearance undercut area; the *thread cutting lever* is also instantly *knocked down* (to disengage the saddle, cross-slide, and tool from the leadscrew drive's influence).

When using the angle approach method of screw-cutting, the tool is then returned manually back to its start position by using a clockwise rotation of the saddle's hand-wheel, followed by using a full clock-wise rotation of the cross-slide's hand wheel to return the tool to its originally set zero scale's position.

Note that when using the straight plunge method of screw-cutting, the cross-slide is returned clockwise to the scale's pencil-marked *depth of thread so far reached* indication on its scale.

[One should note that if one is using the angle approach method of screw-cutting, it is the compound slide's scale that will be actually indicating to the machinist the depth of cut so far reached by the tool; its cross-slide with its zero set scale is then always returned to its normally set zero position before starting the next cut.]

Also note that the actual 'depth of thread so far reached', when using the straight plunge method of screw-cutting, is indicated solely by referring to the pencil-marked cross-slide's scale.

Section 18/39
To imprint on the student's memory, there follows a general description of how one uses the external angle approach screw-cutting method, while using the quick method of tool withdrawal (leaving the work-piece still rotating), followed by using the thread-cutting knockdown lever to isolate the tool from the lead-screw at the end of each pass.

With the machinist now using the angle approach method of external screw-cutting (as shown in fig. 42), the compound slide's clamping nuts should first be released and its assembly rotated clock-wise through 60 degrees (relative to the lathe bed) followed by locking the nuts to reposition this slide into its new position. (The compound slide assembly will now be occupying a position that is angled at 30 degrees relative to the cross-slide's center-line.). The compound slide's tool-post is then repositioned to enable it to be aligned perfectly square with the lathe's bed and secured. The compound slide's overhang from its support casting must then be reduced fully by turning its hand-wheel in an anticlockwise direction sufficient to leave no overhang relative to its support casting, its hand-wheel is then rotated clockwise by (say) a quarter of a turn, (to eliminate all its back-lash in its lead screw; its scale is then reset to read zero.

To ensure that all of the back-lash that exists in the cross-slide's lead screw is taken up in its correct clock-wise (working) direction, prior to using the lathe for external turning or external screw-cutting, we must now initially turn its hand-wheel in an anticlockwise direction by one complete turn, followed by turning it back clock-wise by half a turn, then resetting its scale to read zero. We now select a sharp 60-degree included angle screw-cutting tool, and fit this into the tool-post while ensuring that it is also set at its correct cutting height relative to the vertical center line of the lathe. It should also only possess a minimum of overhang from its tool-post. While being lightly nipped, the tool can be oriented into its correct angular position by using the precision 60-degree cutouts provided in the square-type screw-cutting gauge. This gauge is oriented by holding its top datum edge in alignment with the previously turned 23.9 mm screw-cutting diameter of the work piece, while the tool's cutting edges are being aligned accurately into the gauge's precision 60-degree cutout. A piece of white paper that reflects the light can be held underneath the tool to assist in viewing the tool's accurate alignment. The tool is then finally tightened in this position.

The screw-cutting tool's point is then repositioned (by using the cross-slide's hand- wheel being rotated clockwise sufficient to allow it to touch the previously turned screw-cutting diameter of 23.9 mm, and its scale is then reset to read zero. The cross-slide's hand-wheel is then turned anticlockwise by one revolution to allow the tool to clear the work-piece, and the saddle's hand-wheel is turned clockwise to reposition the tool's point at a position approximately 5 to 6 mm to the right of the work-piece. The cross-slide's hand-wheel is then turned back clockwise by one revolution and down to its set scale's zero mark, making the tool ready for taking its first cut. The lathe is then switched on in its normal direction of rotation, at a work-speed of approximately 60 revs per minute. The compound slide is then indexed inward clockwise by the minute amount of 0.05 mm using its hand-wheel, in preparation for taking its first *exploratory cut* along the workpiece; its scale is then marked using pencil. The coolant supply is now switched on to cover the tool and work-piece. We now study the lines and numbers situated on the *rotating screw-cutting dial's drop-in scale* situated on the front of the lathe's apron, we then select the number or line we have previously chosen to line up with the apron's fixed datum line. The *screw-cutting lift-up lever* is then engaged at this exact chosen point, and the first screw-cutting cut is then started. (At this point, a mental note should be made of the chosen line or number being used.)

On completing the first cut, the tool reaches the undercut area (and with the guidance of the carefully placed *rotating warning marker* situated on the chuck's periphery), the *quick method of tool withdrawal* is used (consisting of one quick and complete anticlockwise revolution of the cross-slide's hand-wheel), coupled with the instant operation of the *screw-cutting knockdown lever* (isolating the tool from the lead-screw). The saddle's hand-wheel is then turned clockwise (with care) to return the saddle and tool to its start position of approximately 5 to 6 mm to the right of the work-piece. The *cross-slide's* hand wheel is then turned carefully and accurately one revolution clockwise to return the tool to its originally set zero scale's position. The angled *compound slide's* hand-wheel then is turned very carefully clockwise (by the minute amount of 0.05 mm) to index in the tool for the next cut to be taken, referring to and using its index scale for this precise 'depth of cut' setting. Its scale is then marked with pencil (or a similar marking instrument) to retain (for the machinist) a lasting reminder of the depth of thread the tool has so far reached.

The *rotating screw-cutting dial* is now again studied until one's chosen mark or number is again seen to line up with the apron's stationary datum line. The *lift-up screw-cutting lever* is then re-engaged by again using the chosen mark (or number) on the dial. This accurate re-engagement (by relying on the lathe's rotating scale and its fixed datum mark) will ensure that the tool will mesh accurately into the previously cut thread.

[It is important to note that failure to engage the screw-cutting lever at an identical mark indicated on its *rotating screw-cutting dial's indicating scale* for each and every passing cut will result in the tool engaging wrongly with its previously partially cut thread, and will (if positioned wrongly) cause damage to the work-piece by this misalignment. The damage caused to the thread in this way can cause an almost finished workpiece to be damaged beyond reclamation.

A wrongly engaged tool (particularly in the latter stages of the operation) can also cause the tool's cutting edges to be damaged, necessitating an unwelcome tool replacement being made (this change will require a rather lengthy re-setting process to be implemented, so it is wise not to allow this to happen).

Note that when using the *angle approach* screw-cutting method, it is the *cross-slide*'s hand-wheel assembly with its *accurately zeroed index scale* that is always returned to its *set zero position* at the startof each *screw-cutting pass*. This is followed (immediately after the passing cut' has been completed), by using the *instant snatched method of tool withdrawal* and the *screw-cutting knockdown lever* routine, at the end of each passing cut by also using the guidance of the chuck's carefully placed rotating warning marker.

Also note that because we are using the angle approach method of screw-cutting, it is the pencil-marked *compound slide*'s hand wheel's reference scale that is used to indicate to the machinist the gradual increases in depth of each and every succeeding cut that needs to be taken by the tool during its following screw-cutting passes. The actual depth of cut needing to be indexed in by using the *compound slide* and its scale, for each cut, will normally be 0.05 mm for each cut (for steel), (or slightly deeper if machining softer materials such as aluminum or brass), until the full 2.12mm finished angled depth of thread, has been reached, as seen indicated on its scale.

The machinist will also find it helpful if he forms the habit of always marking the compound slide's index scale regularly with pencil (or a similar marking device), just before each cut is taken.

After taking a following series of 0.05 mm depth passing cuts, of sufficient number for the tool to be approaching its full depth of thread, the *thread ring gauge* is then used to check the fit of the screw-cut thread to establish whether it has been machined to its correct depth and size to fit this *thread ring gauge.*

Section 18/40
Describing the advantages of marking the cross-slide's scale when using straight plunge screw-cutting to indicate the depth of thread the tool has so far reached, also pointing out the advantages of marking the compound slide's scale when using the external angled approach method of screw-cutting.

There is also a distinct advantage in marking the chuck's outer periphery to give warning of the exact moment where the snatched method of tool withdrawal is to be used. This warning marker, carefully situated on the periphery of the lathe's chuck, can of course be used during either method of screw-cutting.

The student should note that when using the alternative *straight plunge* method of screw-cutting, it is the *cross–slide*'s index scale that is regularly marked with pencil prior to taking the next cut; this is done in order to retain a record on its index scale of the depth of thread so far reached by the tool during the operation.

However, when using the angle approach method of screw cutting (as used in drawing fig. 42 and also in Section 18/39), it is the *compound slide*'s hand wheel that is turned clockwise by 0.05 mm for each cut by using its index scale. This must be regularly marked with pencil before taking the next cut, to indicate the new and all the previous depth of cuts taken by the tool. This marking of the slide's index scales should be encouraged among both students and skilled machinists, as it will be found extremely helpful as a memory aid and should be encouraged among machinists so that it will eventually become a routine practice.

As previously mentioned, it will also be found most helpful, when performing either of the screw-cutting methods described, if the periphery of the chuck is also marked with a *warning marker* to indicate the exact point where the *snatched method of tool withdrawal* is to be used at the end of each screw-cutting passing cut. This can be done by securing a horizontal strip of highly visible drafting tape, (or similar visible adhesive material), in an accurately positioned place on the chuck's periphery which positively identifies the exact point on the chuck and on the work-piece's rotation where the screw-cutting tool *must be snatched out* quickly from the work-piece, by using (for external screw-cutting), one quick anti-clockwise revolution of the cross-slide's hand-wheel at the end of each screw-cutting pass. This personally introduced *helpful warning marker*, located strategically in its correct position on the surface of the chuck's outside diameter, will allow the machinist's eye to catch a momentary glimpse of it as it passes through his peripheral vision, indicating the exact position where the tool must be instantly extracted. This then allows the machinist to co-ordinate his tool withdrawal actions to occur at exactly the correct time and place at the end of each screw-cutting passing cut. This quick action avoids the possibility of an accident occurring should the tool be extracted either too early, where the thread can be possibly damaged by the now stationary tool, or operated too late, where the tool has overrun the undercut area and collides with its left-hand side face, where it can cause considerable damage to the work-piece and to the tool's cutting edge, by it *digging into* the work-piece and damaging the undercut's left-hand side face.

For the benefit of the less experienced student, and for trainees who still consider themselves to be still on steep learning curve, it will be found advisable, for training purposes, for them to initially adopt the simpler but slightly more time-consuming screw-cutting practice of using the *straight plunge* method of screw-cutting, where the leadscrew is left permanently engaged with the saddle and its tool for simplicity and ease of use.

This method of screw-cutting is carried out by leaving the saddle and tool assembly permanently engaged with its lead-screw, followed by using the *quick method of tool withdrawal* (of one revolution anticlockwise at the end of each passing cut), and by immediately stopping the lathe, reversing the motor, and allowing the screw-cutting tool and its saddle assembly to travel back under its own power to its starting point of 5 to 6 mm to the right of the workpiece where the motor is then stopped. This is followed by turning the cross-slide's hand-wheel clockwise down to its scale's pencil marked position, followed by indexing inward in a clockwise direction the next cut of, say, 0.05 mm seen indicated on its scale, and marking this with pencil, followed by switching the lathe back on in its normal direction of rotation and taking the next cut. We follow this by completing a full series of screw-cutting passes as found necessary, by using 0.05 mm depth of cut being indexed inward using the *cross-slide*'s scale for each of these passes until the calculated full depth of thread (*while using straight plunge screw-cutting*), of 1.84mm has been reached on its scale, indicating that the tool has now reached its full and final depth of thread. A thread ring gauge is then used to check the thread for fit, in compliance with the drawing's dimensional requirements.

It is strongly recommended that this method of screw-cutting is adopted by the student until sufficient practical experience has been gained in the art of screw-cutting. This training should be completed satisfactorily before being allowed to adopt the slightly more complicated *angled approach* method of screw-cutting, which uses the angled compound slide with the additional option of using either a *disengaged lead screw* or a *permanently engaged lead screw* for the screw-cutting operation.

The student should note that the *screw-cutting lever* is often referred to by using several differing names relating to its actual use at the time. At the start of a threading cut, it is often called the *lift-up screw-cutting lever* (or the *drop-in threading lever*). At the end of a cut, it is often referred to as the *knockdown lever* or the *lead screw disengaging lever*. All these names refer to the same lever, but its descriptive names are often changed by the operator when being used in its various alternative modes.

Section 18/41
**Because of the present need to explain all of the varying methods that can be used to screw-cut our example of a 24 mm diameter ×
3mm coarse pitch external male screw thread, it is now advisable, for clarity reasons, to explain in detail each of the possible options that can be used, in the following text.**

Option one.
The straight plunge method of external screw cutting, while leaving the screw-cutting tool engaged with the lead-screw and

using the quick method of tool withdrawal, followed by reversing the motor to reposition the tool back to its start position, under power, ready for the next cut to be taken.

With the selected diameter of the work-piece material now turned down to its screw-cutting diameter of 23.9 mm as shown in the fig. 42 drawing, (using preferably a right-hand knife tool for the operation), and the work-piece's clearance undercut already plunge turned to produce a screw-cutting tool clearance diameter of 20. 3 mm, (identical to the screw-cutting tool's intended depth of thread), to a width of 4 mm by using a slightly modified parting tool hand honed to possess two minute corner radii at its front cutting edge, the compound slide and its tool-post are now checked to establish (for straight plunge screw-cutting) that they are secured in their normal horizontal position relative to the center-line of the machine. The screw-cutting tool is then carefully installed in the tool-post by using a screw-cutting gauge to ensure the tool's two cutting edges and its point have been accurately oriented relative to the work-piece, and the tool's cutting height has been checked by using the lathe's tool height checking gauge. The compound slide's handwheel is then turned anticlockwise by a sufficient amount to eliminate any possible overhang of its slide relative to its support casting; it is then turned back clockwise by, say, a quarter of a turn (to take up any residual backlash in its lead-screw to provide its lead screw with sufficient solid back-up, while being used in its following clockwise direction); its scale is then set to read zero.

The honed point of the screw-cutting tool is then repositioned by rotating the cross-slide's hand-wheel clockwise by a sufficient amount to allow its point to just touch the screw-cutting diameter of 23.9 mm, (previously turned on the piece); the cross-slide's scale is then set to read zero. It will be found helpful if the machinist also contrives during this part of the setting-up procedure, to position the cross-slide hand-wheel's knob (its handle) into its topmost or twelve o'clock position; this is done for the convenience of easy hand access. If this positioning of the handle has not been carried out prior to the tool's orientation and the work-piece touching procedure, this process will need to be repeated and the cross-slide's scales re-zeroed.

The cross-slide's hand-wheel is then turned anticlockwise by one complete revolution to allow the tool to clear the workpiece. The tool is then re-positioned back at its start point of 5 to 6 mm to the right of the workpiece by using the saddle's hand-wheel. The cross-slide's hand-wheel is then turned clockwise by one complete revolution, plus an additional minute inward clock-wise movement of 0.05 mm (indicated on its scale), to put on the first cut; this scale is then marked with pencil to indicate the depth of thread reached on this passing cut. The coolant pump is now switched on and coolant now applied to the tool and the work-piece.

The lathe is now switched on in its normal direction of rotation at a low spindle speed of (approximately 60 rpm) and the screw-cutting lift-up lever is engaged at the exact point where the machinist's chosen line or number (situated on the front apron's rotating thread-cutting dial) is seen to line up with the apron's fixed datum line, (it should be noted that in this case we are using a permanently engaged lead-screw and that the lift-up screw-cutting lever will be left in this fully engaged position throughout the whole of the screw-cutting procedure). The lathe is then allowed to take its first passing cut. At the end of this passing cut, where the tool has just reached the under-cut area of the thread, and the machinist has glimpsed the chuck's warning 'tool extraction' marker (as it passes through his peripheral vision), the motor is instantly stopped by using the lathe's emergency stop lever (its emergency stop foot pedal, or its clutch if fitted); the cross-slide's hand-wheel is then instantly turned anticlockwise by one revolution, to make the necessary *snatched method of tool withdrawal*. The motor is then reversed; the machinist then waits until the tool assembly has reversed back fully to its original start position of 5 to 6 mm to the right of the work-piece, where the motor is then stopped by using the hand operated stop lever (or the lathe's clutch lever if fitted). The cross-slide's handwheel is then turned clockwise by one revolution to return the tool to its previously pencil-marked cross-slide scale's depth position; the cross-slide is then indexed minimally inward clock-wise by a further 0.05 mm (by using its scale as a reference). This position is again marked with pencil, in preparation for the next cut to be taken; the motor is then switched back on in its normal direction of rotation, and the next cut is taken.

Similar cuts of 0.05 mm depth of cut are then taken on each of the tool's following passing cuts, until the cross-slide's scale is seen to have reached 1.84 mm, indicating that the full depth of thread has now been reached. A thread ring gauge is then used to check the screw-cut thread for a correct fit.

Note that extreme care should be taken when choosing the exact moment to operate the stop and the reversing lever at the end of the forward screw-cutting pass, as this must be instantly carried out in order to avoid the tool's cutting edges overriding the under-cut and colliding with its left side). Care should also be taken at the end of the tool's return pass on its journey back to its start position, to avoid the tool and its saddle assembly overrunning its original starting point (as it is still under power and in reverse), in order to prevent the saddle making accidental contact with the tailstock's casting (which in this case is locked and supposedly immovable). Any accidental contact between the moving saddle's casting and the tail-stock's stationary casting on this powered return pass can be of sufficient magnitude to displace the (fixed) tailstock, by moving it along longitudinally to its right sufficiently to force the work-piece's supporting center to be withdrawn from the work-piece, thereby leaving it unsupported and in a dangerous condition; therefore, the possibility of this clash of tooling occurring should be closely monitored in order to avoid this potential problem occurring. Note also that the lathe can be switched off by using either its hand operated stop lever, or its foot-actuated stop pedal. On some lathe designs, this pedal or lever not only switches off the power (thereby stopping the motor), but also brings into use a powerful braking system capable of dead stopping the spindle, chuck, tool, and workpiece (this facility will be found useful when screw-cutting a thread up to a workpiece's shoulder). This braking system can be used during either normal working conditions or an emergency situation.

As previously stated, some designs of lathe have a clutch lever fitted and this provides an ability to slow the work-piece down when approaching a shoulder. This feature will prove to be particularly helpful when the tool is approaching the under-cut area, and will therefore prove to be a very useful addition to the lathes controls. As previously stated, the foot pedal can also be used to dead stop the lathe, on the tool reaching the,undercut area, or if the tool is seen to be about to cut into a work shoulder or the left-hand side of the undercut at the extreme end of its screw-cutting pass. Note that after using the 'dead stop' pedal, the on/off lever must first be returned to its off position before the lathe can be restarted.

The full sequences of screw-cutting passes are then carried out until the cross-slide's scale is seen to read 1.84 mm, indicating that the full depth of thread has been reached. A thread ring gauge is then used to check the thread for a correct fit.

Section 18/42

Option two

Straight plunge external screw-cutting, while using the 'disengaged lead screw' technique, leaving the work-piece in its still-rotating mode, using the quick tool withdrawal, then returning the tool manually to its start position, followed by using the guidance of the apron's rotating drop-in scale to allow the correct timing for the thread-cutting lift-up lever to be operated and the next cut to be taken.

With the diameter of the work-piece now turned to its 23.9 mm screw-cutting diameter (using preferably a right-hand knife tool), and with the relief under-cut also machined to provide a clearance diameter for the point of the screw-cutting tool of 20.3 mm diameter × 4 mm wide, (this diameter being identical to the full depth of thread to be cut, as shown in fig. 42), and by also using in this case a modified parting tool that possesses two minutely honed corner radii at each end of its front cutting edge.

The screw-cutting tool is then carefully installed in the tool-post by using the square-type thread cutting gauge's top face being held in alignment with the turned 23.9 diameter of the work-piece to establish that its cutting edges have accurate orientation in the precision cutouts of the gauge. A piece of white paper can be held under the tool to reflect the available light and assist this accurate installation; its cutting height must then be carefully checked by using the lathe tool height setting gauge. The cross-slide's hand-wheel is then turned clockwise by approximately half a revolution to eliminate any backlash

in *its* lead screw, and its scale is reset to read zero. Any overhang of
the compound slide from its support casting should also be reduced
to nil by turning its hand-wheel firstly in an anticlockwise direction
sufficiently to allow its top slide to be supported by its base casting
by a minimum of 1 mm. The inherent back-lash in this particular
slide's lead-screw is then removed by turning its hand-wheel clockwise
by, say, a quarter of a turn and resetting its scale to read zero (this
movement has now backed up and fully supported the position of the
tool-post and its slide; this prevents the tool being pushed back on
starting to cut the thread). The machinist should also contrive during
this setting-up procedure to position the *cross-slide*'s hand wheel and
its operating knob (its handle), to occupy its uppermost twelve o'clock
position at the same time as the screw-cutting tool's point has been
maneuvered to just touch the 23.9 mm diameter about to be screw-
cut; its scale is then reset to read zero. The position of this knob (or
handle) then allows an accurate location and easy hand access during
the following screw-cutting procedure.

The cross-slide's hand-wheel is then turned anticlockwise by one
complete revolution to allow the tool to clear the workpiece. The
tool is then repositioned 5 to 6 mm to the right of the work-piece
by turning the saddle's hand-wheel clockwise. The *cross-slide*'s hand
wheel is then indexed inward clockwise by one complete revolution,
followed by an additional 0.05 mm inward clockwise movement being
indicated on its scale, in order to 'put on a cut and prepare for taking
its first cut. Its scale's depth should now be marked with pencil. The
lathe is now switched on in its normal direction of rotation at a low
speed of approximately 60 revs per minute, and coolant is applied to
the tool and to the work-piece. A close watch is now kept on the lathe's
rotating screw-cutting dial, (situated on its front apron), until one's
chosen number (or line) is seen to line up with the apron's stationary
datum line, (a mental note should also be made of its exact chosen
position for future use). At this exact moment in time, the thread-
cutting operating lever is lifted up into its engaged position, (thereby
closing the split nut into mesh with the lead-screw and allowing the
lathe's drive to be taken up allowing the tool to make its first screw-
cutting passing cut.

The tool, on completing this screw-cutting pass and on it just reaching the under-cut area, (while allowing the machinist to catch a brief glimpse of the rotating warning marker previously adhered to the chuck's rotating outer periphery as a *warning signal,* to disengage the lead-screw from the tool, by operating the screw-cutting knockdown lever. This is combined with using the instant snatched method of tool withdrawal of one complete anticlockwise revolution of the cross-slide's hand-wheel at the end of the passing cut to allow the tool to clear the work-piece. The tool is then returned manually back to its start position of 5 to 6 mm to the right of the work-piece, using the saddle's hand wheel. The cross-slide's hand-wheel is then rotated clockwise by one complete revolution down to the previously made pencil mark of 0.05 mm on its scale. The next cut, of 0.05 mm depth, is now indexed further inward in a clockwise direction using the cross-slide's scale as a reference; the tool's new depth position is again marked on its scale using pencil. The operator then waits until the chosen number or line indicated on the apron's rotating thread cutting dial is seen to line up with the apron's stationary datum line. On reaching this exact point, the lead screw is re-engaged with the tool by operating the *lift-up thread-cutting lever* enabling the lathe to take the next passing cut. On the tool reaching the under-cut area and also again catching a brief glimpse of the chuck's warning marker, the snatched method of tool withdrawal is used, coupled with instantly using the *screw-cutting knockdown lever* to disengage the tool from the lead screw. The tool is then returned manually back to its start position of 5 to 6 mm to the right of the work-piece by using the saddle's hand-wheel.

This whole screw-cutting procedure is then repeated by completing a series of 0.05 mm depth of passing cuts being made using the cross-slide's scale, by using pencil marks, until the full depth of 1.84 mm is reached on its scale, indicating that the full depth of thread has been reached. The component's thread should then be checked for correct size and fit by using a thread ring gauge.

To establish a correct fit, the ring gauge should be capable of being screwed easily on to the component by hand, and the thread should not be found to possess any slackness or end play being noticeable relative to the gauge.

Note that the lathe can be stopped or switched off by using either the hand-operated stop lever or the foot-actuated stop pedal, should an emergency arise requiring it to be used.

On some models, this pedal not only switches off the power, (thereby stopping the motor), but also brings into use a powerful braking system, capable of dead stopping the spindle, chuck, tool, and work-piece (this braking action is often found to be very useful when screw-cutting up to a work-piece's shoulder). This particular braking system can therefore be used either during normal working conditions or in an emergency situation. The foot pedal can also be used to dead stop the lathe, if the tool, on reaching the under-cut area, is about to cut into a work shoulder (or into the left side of the under-cut's clearance provided at the end of each screw-cutting pass); this pedal can of course also be used in an emergency situation. It should be noted that some designs of lathe have a clutch mechanism operated by a lever; this allows the machinist to slow the lathe spindle and work-piece down prior to reaching the under-cut area, and this option will be found very useful (if fitted).

(Note that after one uses the 'dead stop' pedal, the on/off lever must be returned to its 'off' position before the lathe can be re-started.). Also note that if one has the need to stop the lathe in mid-cut, then the cross-slide hand-wheel must be used quickly to perform an emergency snatched method of tool withdrawal immediately, before the lathe has stopped, we must do this in order to avoid possible damage being caused to the tool's cutting edges should it be re-started while it is still on a deep cut.

It is also good practice to always leave a clearance gap of (say) 5 to 6 mm between the screw-cutting tool and the right-hand end of the work-piece at the start of the threaded portion. This gap then enables the lead-screw drive's back-lash to be taken up before the tool actually starts its move to its left to cut the thread. The machinist must always be aware that this catch-up area must always be allowed to exist, in order to allow the tool to catch up with its lead-screw. Failure to realize that this area exists can be the cause of damage occurring to the start of the screw-cut thread by the tool being slightly out of step for its initial movement due to the inherent back-lash existing in the lead-screw's drive mechanism.

Note that a thread ring gauge must always be used in the latter stages of screw-cutting the thread. This check should be made preferably several cuts before the theoretical (calculated) full depth of thread has been reached. This then enables the operator to get a rough estimate of the number of finishing cuts required for the tool to reach its full depth of thread. The newly finished thread must be capable of complying with both the drawing's stated dimensions and the inspection departments test requirements of fit, within the drawing's stated limits.

Section 18/43

Option three.
The angle approach external screw-cutting method being used, while leaving the lead screw engaged, using the quick tool withdrawal method, stopping and reversing the motor to return the tool to its starting position, making the lathe ready for the next cut to be taken

With the diameter of the component to be screw-cut previously turned to the required 23.9 mm, by using preferably a right-hand knife tool for this, and the tool relief under-cut having been machined to 4 mm wide and to a minor diameter of 20.3 mm using a modified parting tool honed with small radii at its corners, the compound slide is then loosened and rotated clockwise through 60 degrees relative to its originally set horizontal position followed by it being secured. The tool-post is then set square to the center-line of the lathe, the screw-cutting tool is then set at its correct cutting height and at its correct orientation in the tool post by using the screw-cutting square-type tool setting gauge, using its top datum edge held in close contact with the 23.9 mm diameter about to be screw-cut. Note that the tool's 60-degree included angle cutting edges must be maneuvered into accurate alignment with the gauge's 60-degree precision cutout prior to the tool being finally clamped; this tool alignment can be assisted by using a piece of white paper held under the tool to reflect the available light. When found to be correct, the tool is securely clamped in the tool-post. The back-lash normally existing in both the cross-slide and the compound slide's lead – screws are then eliminated by turning both hand-wheels in a clockwise direction by (say) a half turn, followed by resetting their scales to read zero. Note that the lathe's compound slide must be fully supported on its casting with no overhang being present.

The *cross-slide*'s hand wheel is then turned clockwise sufficient to allow the point of the tool to just touch the 23.9 mm diameter previously turned for screw-cutting. Prior to actually doing this, it will be found very helpful if the handle (or knob) of the cross-slide's hand wheel is positioned at its topmost, twelve o'clock position. This re-adjustment of its position will allow a more familiar and accurate hand access for the later operation. Note that after this adjustment to the handle is made, the tool may require its orientation to be reset and the tool's point again made to touch the previously turned 23.9 mm diameter, followed by its scale being reset to read zero. The cross slide's hand wheel is then turned anticlockwise by one complete revolution to allow the tool to clear the work-piece. The *saddle*'s hand wheel is then turned clockwise to position the tool back to its start position of 5 to 6 mm to the right of the work-piece. The cross-slide's hand wheel is then turned one revolution clock-wise until the set zero position is reached on its scale. The *compound slide*'s hand wheel is then turned clockwise by the minimal amount of 0.05 mm to index the tool inward in preparation for taking the first cut. The machine is now switched on in its normal direction of rotation, preferably at the low work speed of approximately 60 revs per minute; coolant is then applied to the tool and to the work-piece. A close study is then made of the rotating drop-in screw-cutting dial situated on the lathe's front apron, until the chosen number or line is seen to line up with the apron's stationary datum line (this exact chosen point should of course be memorized and noted down for future reference). At this point, the thread-cutting lift-up lever is lifted up to engage the tool with its lead screw to commence the tool's first cut.

At the end of the ensuing screw-cutting pass, and as the tool has just reached the undercut area, (confirmed by the revolving warning marker, previously fitted to the chuck's outer diameter as it is seen to pass through the machinist's peripheral vision), the *snatched method of tool withdrawal* (of one anticlockwise revolution of the cross-slide's hand wheel) is then instantly used to allow the tool to clear the work-piece. The motor is then instantly stopped by using the on/off lever (or the on/off button switch). (Note that in this particular case the screw-cutting lever is now being left permanently engaged with the lead screw). The motor is now switched back on in its reverse direction of rotation, and, during the saddle's return to its starting point of approximately 5 to 6 mm to the right of the work-piece, the next cut of, say, 0.05 mm is indexed inward clockwise on the *compound slide*'s scale used as a reference; this new scale's position is then marked with pencil. The motor is then stopped. The cross-slide's hand wheel is then turned clockwise by one complete revolution, until its index scale is again seen to read *exactly* zero again; the motor is then switched back on in its normal direction of rotation to permit the lathe to take the next cut. On the tool reaching the under-cut area and with the warning marker again being used as a guide, the snatched method of tool withdrawal is used (of one complete anticlockwise rotation) of the cross-slide's hand-wheel. The motor is then again instantly stopped followed by it being reversed to allow the tool to return to its start position of 5 to 6 mm to the right of the work-piece.

This whole screw-cutting procedure is then repeated by using successive cuts of 0.05 mm in depth indicated on its compound slide's scale, until the machinist is aware that he is approaching the full angled depth of thread of 2.12 mm by the indications shown on the *compound slide*'s scale, whereupon a check of the work-piece's newly cut thread should be made by using the appropriate thread ring gauge. If the gauge is found to be still a rather tight fit, then it will be found necessary to take several more 0.05 mm deep passing cuts until the gauge is again used, and felt to be a perfect fit on the thread without any slackness or end float being detectable. The screw-cut thread will then be considered fit for passing on to the inspection department for their close scrutiny regarding its finished size and accuracy of fit.

Section 18/44

Option four
  The angle approach external screw-cutting method being used, while leaving the work-piece still rotating, disengaging the lead-screw, using the quick tool withdrawal, followed by

returning the tool manually to its start position, ready for the next cut to be taken.

With the screw-cutting diameter now already turned to its required 23.9 mm diameter, and with the tool relief under-cut also machined to its 4 mm width and to its required 20.3 mm diameter (this diameter being identical to the screw-cutting tool's full depth of thread), the lathe is then switched off. The compound slide securing nuts are now released to allow this assembly to be swung through 60 degrees relative to its normal horizontal position and secured; its tool-post is then reset square to the center line of the lathe and secured.

A sharp 60-degree included angle screw-cutting tool is then fitted into the tool-post possessing the minimum of overhang from its compound slide and is oriented correctly by using the square-type screw-cutting gauge's top datum face being held in contact with the 23.9 mm diameter of the previously machined work-piece. Its cutting edges are now aligned to fit into the 60-degree precision cutout machined in the gauge. A piece of white paper is then held below the tool to reflect the available light to allow the tool's cutting edges to be to be visually aligned correctly with the contours of the gauge, followed by the tool being finally secured in the tool-post. The compound slide's movement is then adjusted by turning its hand wheel anticlockwise sufficiently to ensure that its slide is not overhanging beyond its support casting. Its hand-wheel is then turned back clockwise by a quarter of a turn to eliminate any backlash existing in its lead-screw; its scale is then set to read zero. The *cross-slide*'s hand wheel is then turned clockwise sufficient to allow the point of the screw-cutting tool to just touch the 23.9 mm diameter of the work-piece, previously turned for screw-cutting, and its scale is set to read zero. It will be found advantageous if, during this setting-up procedure, the handle or knob of the cross-slide's hand-wheel is positioned to occupy its top most, twelve o'clock position while the tool is actually contacting the 23.9 diameter of the work-piece; this is advisable in order to allow a familiar and easy hand access during the following screw-cutting procedure. (If at this stage, the repositioning of the cross-slide's handle becomes necessary, it is most important that the tool's exact orientation and the setting- procedure is repeated, so that once again the point of the tool is allowed to touch the turned 23.9 diameter, followed by its scales being reset to read zero.

The cross-slide's hand-wheel is then turned anticlockwise by one complete revolution to allow the tool to clear the workpiece. The *saddle*'s hand wheel is then turned clockwise to position the tool approximately 5 to 6 mm to the right of the work-piece. The cross-slide's hand-wheel is then turned clockwise by one complete revolution back to the originally set zero position. The *compound slide*'s hand wheel is then turned minutely clockwise by 0.05 mm (using its scale as a reference), to prepare the tool for taking it's first cut. The lathe is then switched on in its normal direction of rotation at a low work speed of approximately 60 revs per minute; the tool and work-piece should now be supplied with a continuous supply of coolant. A close study is then made of the rotating *drop-in screw-cutting rotating dial* situated on the lathe's front apron. When the previously chosen line or number on the rotating screw cutting dial is seen to line up with the fixed datum line of the apron, the *lift-up screw-cutting lever* is engaged with the lead screw to allow the lathe to take its first cut. The exact position of this chosen line (or number) must now be mentally noted (or written down) at this time to retain it for future use.

At the end of this screw cutting pass, and on the tool reaching the under-cut area, the machinist, aided by catching a fleeting glimpse of the *tool extraction warning marker* (previously fitted on the outside diameter of the chuck for guidance), disengages the lead screw by instantly operating *screw-cutting knockdown lever*, together with instantly operating the *snatched method of tool withdrawal* (of one quick and complete anticlockwise revolution of the cross-slide's hand wheel). This instant quick tool withdrawal and the instant operation of the *knockdown screw-cutting lever* must be made in unison to ensure that the tool fully clears the work-piece while leaving it in its still rotating mode. The saddle's hand wheel is then used to return the tool assembly to its starting point of 5 to 6 mm to the right of the work-piece. During the tool's hand return, the next cut of 0.05 mm is indexed inward clockwise on the *compound slide*'s scale, and marked with pencil. The cross-slide's hand-wheel is then turned one complete revolution clockwise until its index scale is again seen to read exactly zero (this scale's zero position must be precise on this and all its following screw-cutting passes; it is also important that it is performed before taking the next passing cut).

The *thread lift-up engaging lever* is then operated at the exact number (or line) previously selected on the apron's rotating dial and at the exact point where the fixed datum line is seen to become accurately aligned. The next cut is then taken.

This whole screw-cutting procedure is then repeated by taking similar cuts of 0.05 mm in depth for each passing cut, by using the *compound slide* and its scale's reading, until its scale is seen to be approaching 2.12 mm (the calculated full depth of thread required when using the angled approach method of screw-cutting), whereupon the motor is then stopped, the tool is moved away from the work-piece and a thread ring gauge is used to check the thread's fit. If the gauge is found to be rather a tight fit, then extra passing cuts of 0.05 mm will need to be taken, by using the compound slide's scale as a reference until the gauge is found to screw onto the work-piece smoothly without exhibiting any shake or end float being present.

THIS DRAWING SHOWS A CROSS - SECTION OF A TYPICAL 24 MM X 3 MM PITCH METRIC
NUT, WITH THE SCREW - CUTTING TOOLING BEING USED FOR EITHER THE 'STRAIGHT
PLUNGE' OR THE 'ANGLE APPROACH' METHOD OF SCREW - CUTTING.

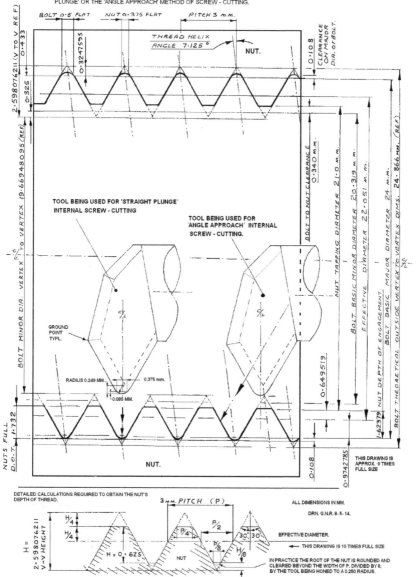

DETAILED CALCULATIONS REQUIRED TO OBTAIN THE NUT'S
DEPTH OF THREAD.

ALL DIMENSIONS IN MM.

DRN. G.N.R. 8- 5- 14.

EFFECTIVE DIAMETER.

THIS DRAWING IS 10 TIMES FULL SIZE

IN PRACTICE THE ROOT OF THE NUT IS ROUNDED AND
CLEARED BEYOND THE WIDTH OF P, DIVIDED BY 8,
BY THE TOOL BEING HONED TO A 0.260 RADIUS.

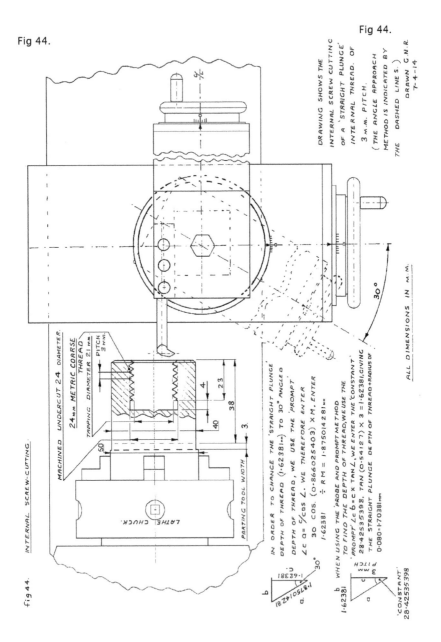

Fig 44.

Fig 44.

DRAWING SHOWS THE
INTERNAL SCREW CUTTING
OF A 'STRAIGHT PLUNGE'
INTERNAL THREAD. OF
3 m.m. PITCH.
(THE ANGLE APPROACH
METHOD IS INDICATED BY
THE DASHED LINES.)
DRAWN G.N.R.
7-4-14.

30°

ALL DIMENSIONS IN M.M.

MACHINED UNDERCUT 24 DIAMETER.

24mm METRIC COARSE
THREAD

TAPPING DIAMETER 21 mm.

PITCH
3 mm.

23
4
38
40
50
3.

PARTING TOOL WIDTH

LATHE CHUCK.

INTERNAL SCREW-CUTTING.

fig 44.

IN ORDER TO CHANGE THE 'STRAIGHT PLUNGE
DEPTH OF THREAD (1·62381mm) TO 30° ANGLED
DEPTH OF THREAD, WE USE THE 'PROMPT'
Lc a = S/cos L. WE THEREFORE ENTER
30 COS. (0·8660254503) X M. ENTER
1·62381   ÷ R.M = 1·87501 4 281mm

30°
b
1·62381
a.
1·8750142281

WHEN USING THE PROBE AND PROMPT METHOD
TO FIND THE DEPTH OF THREAD, WE USE THE
PROMPT Lc b= c x TAN L, WE ENTER THE 'CONSTANT'
28·42535398. TAN (0·54127) X 3 =1·62381,GIVING
THE STRAIGHT PLUNGE DEPTH OF THREAD+RADIUS OF
0·080=1·70381mm

b
1·62381
c
3
m.m
PITCH
a

'CONSTANT'
28·42535398

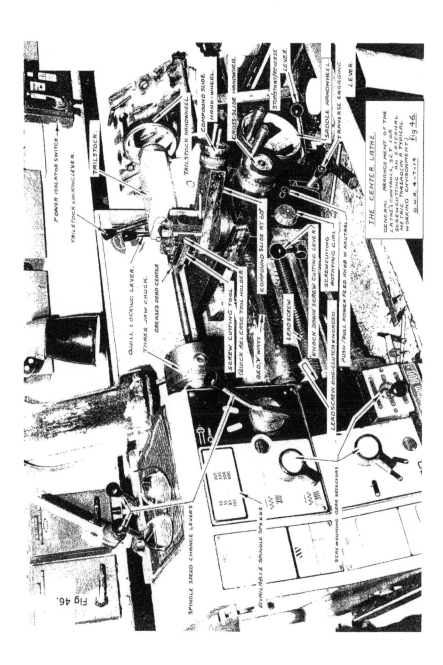

Fig 46.

POWER ISOLATOR SWITCH

TAILSTOCK LOCKING LEVER.

TAILSTOCK.

TAILSTOCK HANDWHEEL.

COMPOUND SLIDE HAND-WHEEL.

CROSS-SLIDE HANDWHEEL.

STOP/START/REVERSE LEVER.

SADDLE HANDWHEEL.

TRAVERSE ENGAGING LEVER.

QUILL LOCKING LEVER.

THREE JAW CHUCK.

GREASED DEAD CENTER.

SCREW CUTTING TOOL.

QUICK RELEASE TOOL HOLDER.

COMPOUND SLIDE AT 60°

BED/ V WAYS.

LEADSCREW.

KNOCK DOWN SCREW CUTTING LEVER.

SCREWCUTTING ROTATING DIAL.

LEADSCREW DOG-CLUTCH ENGAGED.

PUSH/PULL POWER FEED KNOB IN NEUTRAL.

SPINDLE SPEED CHANGE LEVERS

205
335
500
800

33
55
85
135

AVAILABLE SPINDLE SPEEDS

SCREWCUTTING GEAR SELECTORS

THE CENTER LATHE.

GENERAL ARRANGEMENT OF THE
LATHE'S CONTROLS, SET FOR
SCREWCUTTING AN EXTERNAL
METRIC THREAD (IN A TYPICAL
WORKING ENVIRONMENT.)
G.N.R. 4-7-14

fig 46.

Fig 48.

Fig.48.

SUPPORTED BORING ON THE
CENTER LATHE.
USING MY HOME MADE FIXED
STEADY, TO SUPPORT A BORING
OPERATION ON A STEERING
SHAFT.

fig 48

fig 48

fig 49.

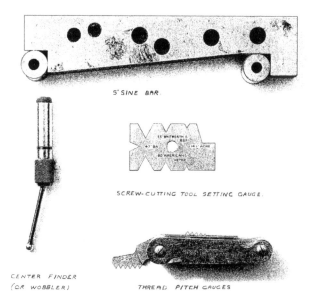

5° SINE BAR.

SCREW-CUTTING TOOL SETTING GAUGE.

CENTER FINDER
(OR WOBBLER)

THREAD PITCH GAUGES

Fig 49.

Fig 49.

Aid for trigonometry © copyright 2010 G. N. Reed.

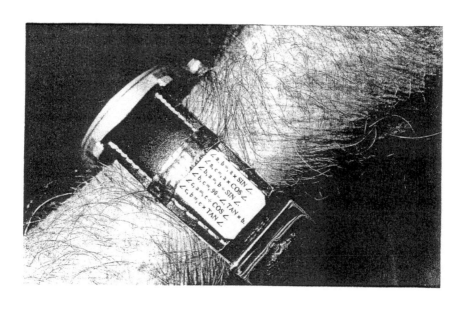

∠ a, b =, a × SIN ∠.
∠ a, c =, a × COS ∠.
∠ b, a =, b / SIN ∠.
∠ b, c =, 90 - ∠, TAN × b.
∠ c, a =, c / COS ∠.
∠ c, b =, c × TAN ∠.

∠ a, b =, a × SIN ∠.
∠ a, c =, a × COS ∠.
∠ b, a =, b / SIN ∠.
∠ b, c =, 90 - ∠, TAN × b.
∠ c, a =, c / COS ∠.
∠ c, b =, c × TAN ∠.

∠ a, b =, a × SIN ∠.
∠ a, c =, a × COS ∠.
∠ b, a =, b / SIN ∠.
∠ b, c =, 90 - ∠, TAN × b.
∠ c, a =, c / COS ∠.
∠ c, b =, c × TAN ∠.

∠ a, b =, a × SIN ∠.
∠ a, c =, a × COS ∠.
∠ b, a =, b / SIN ∠.
∠ b, c =, 90 - ∠, TAN × b.
∠ c, a =, c / COS ∠.
∠ c, b =, c × TAN ∠.

∠ a, b =, a × SIN ∠.
∠ a, c =, a × COS ∠.
∠ b, a =, b / SIN ∠.
∠ b, c =, 90 - ∠, TAN × b.
∠ c, a =, c / COS ∠.
∠ c, b =, c × TAN ∠.

∠ a, b =, a × SIN ∠.
∠ a, c =, a × COS ∠.
∠ b, a =, b / SIN ∠.
∠ b, c =, 90 - ∠, TAN × b.
∠ c, a =, c / COS ∠.
∠ c, b =, c × TAN ∠.

∠ a, b =, a × SIN ∠.
∠ a, c =, a × COS ∠.
∠ b, a =, b / SIN ∠.
∠ b, c =, 90 - ∠, TAN × b.
∠ c, a =, c / COS ∠.
∠ c, b =, c × TAN ∠.

∠ a, b =, a × SIN ∠.
∠ a, c =, a × COS ∠.
∠ b, a =, b / SIN ∠.
∠ b, c =, 90 - ∠, TAN × b.
∠ c, a =, c / COS ∠.
∠ c, b =, c × TAN ∠.

∠ a, b =, a × SIN ∠.
∠ a, c =, a × COS ∠.
∠ b, a =, b / SIN ∠.
∠ b, c =, 90 - ∠, TAN × b.
∠ c, a =, c / COS ∠.
∠ c, b =, c × TAN ∠.

∠ a, b =, a × SIN ∠.
∠ a, c =, a × COS ∠.
∠ b, a =, b / SIN ∠.
∠ b, c =, 90 - ∠, TAN × b.
∠ c, a =, c / COS ∠.
∠ c, b =, c × TAN ∠.

∠ a, b =, a × SIN ∠.
∠ a, c =, a × COS ∠.
∠ b, a =, b / SIN ∠.
∠ b, c =, 90 - ∠, TAN × b.
∠ c, a =, c / COS ∠.
∠ c, b =, c × TAN ∠.

∠ a, b =, a × SIN ∠.
∠ a, c =, a × COS ∠.
∠ b, a =, b / SIN ∠.
∠ b, c =, 90 - ∠, TAN × b.
∠ c, a =, c / COS ∠.
∠ c, b =, c × TAN ∠.

∠ a, b =, a × SIN ∠.
∠ a, c =, a × COS ∠.
∠ b, a =, b / SIN ∠.
∠ b, c =, 90 - ∠, TAN × b.
∠ c, a =, c / COS ∠.
∠ c, b =, c × TAN ∠.

∠ a, b =, a × SIN ∠.
∠ a, c =, a × COS ∠.
∠ b, a =, b / SIN ∠.
∠ b, c =, 90 - ∠, TAN × b.
∠ c, a =, c / COS ∠.
∠ c, b =, c × TAN ∠.

∠ a, b =, a × SIN ∠.
∠ a, c =, a × COS ∠.
∠ b, a =, b / SIN ∠.
∠ b, c =, 90 - ∠, TAN × b.
∠ c, a =, c / COS ∠.
∠ c, b =, c × TAN ∠.

∠ a, b =, a × SIN ∠.
∠ a, c =, a × COS ∠.
∠ b, a =, b / SIN ∠.
∠ b, c =, 90 - ∠, TAN × b.
∠ c, a =, c / COS ∠.
∠ c, b =, c × TAN ∠.

∠ a, b =, a × SIN ∠.
∠ a, c =, a × COS ∠.
∠ b, a =, b / SIN ∠.
∠ b, c =, 90 - ∠, TAN × b.
∠ c, a =, c / COS ∠.
∠ c, b =, c × TAN ∠.

∠ a, b =, a × SIN ∠.
∠ a, c =, a × COS ∠.
∠ b, a =, b / SIN ∠.
∠ b, c =, 90 - ∠, TAN × b.
∠ c, a =, c / COS ∠.
∠ c, b =, c × TAN ∠.

∠ a, b =, a × SIN ∠.
∠ a, c =, a × COS ∠.
∠ b, a =, b / SIN ∠.
∠ b, c =, 90 - ∠, TAN × b.
∠ c, a =, c / COS ∠.
∠ c, b =, c × TAN ∠.

∠ a, b =, a × SIN ∠.
∠ a, c =, a × COS ∠.
∠ b, a =, b / SIN ∠.
∠ b, c =, 90 - ∠, TAN × b.
∠ c, a =, c / COS ∠.
∠ c, b =, c × TAN ∠.

Chapter 19

Section 19/1

The fig. 44 drawing shows the lathe required to machine a straight plunge *internal* screw-cut 24 mm × 3 mm pitch metric coarse thread in the component. The internally threaded section of this component contains a design of thread that is referred to in the engineering industry, as being of a *blind nut* design. During the machining of this particular thread in the work-piece, the trainee machinist must be extremely diligent during this procedure due to the *minimum* clearance available for the tool *beyond* the internally screw-cut part of the thread.

The fig. 43a drawing shows a nine times enlargement of the nut's internal thread in minute detail, and includes its very specific and accurate internal dimensions. The lower drawing in fig. 43a shows the actual formula used to obtain the *very* important depth of thread required. This internally threaded component is designed to fit the previously screw-cut and similarly pitched, male screw-cut thread produced while using the fig. 42 drawings dimensions.

Section 19/1 refers to the fig.43, drawing, and explains (in part one) the '*initial* preparation of the nut's steel material, followed by the *basic* work-piece preparations required before the machinist can progrees on to internally screw-cut the work-piece with the required 24 × 3 mm pitch internal metric coarse thread.

The following guidance notes explain in minute detail the methods used, and the sequences of operations required that will enable the engineering trainee or student, to screw-cut this internal metric coarse thread into the component's tapping size internal bore.

There follows a very long and extremely detailed explanation of the complete turning and screw-cutting operation required for the manufacture of this component to the specifications set out in the fig. 44 and 43a drawings. These explanations and instructions are written specifically for the trainee and student machinist who, while using the center lathe, should now be capable of manufacturing this component with the minimum of further instruction being received from the supervision or from the works management on the subject.

There are certain safety problems that will arise during the machining of this component, these problems are mainly due to this component containing a blind nut in its design. The example shown in fig. 43a drawing depicts the internal screw-cutting dimensions required for this particular design of thread, but does not show that it is in fact a 'blind' nut. This design has been chosen by the author *specifically* to make the student machinist fully aware of the tooling difficulties that can be experienced when machining this particular design of internally threaded component.

Because of the complexity of the following explanations, coupled with the rather involved setting up procedure required for screw-cutting this internally threaded component, it has been decided to divide the full explanation into two main parts.

Part one includes the important details required for the basic setting-up procedure, followed by a complete outline of the actions the machinist is required to take in order to accurately complete the whole screw-cutting operation.

Part two includes (in meticulous detail), a partial repeat of the final setting-up procedures, followed by a full explanation of the whole internal screw-cutting sequences required to accurately manufacture this component that contains a *blind nut* in its design.

Part one.

An explanation (in general terms) of the actions the machinist must take to set up the center lathe for the manufacture of this internally threaded component described in fig. 43a and fig. 44.

It is not generally realized by the trainee machinist, that the total preparation time required to set up this particular type of internal screw-cutting operation will probably take more than half of the total time allowed to complete the finished component.

If on the other hand, a whole batch of identical components were needing to be screw-cut, then the very long initial setting-up time would be spread over the whole batch, resulting in this 'setting-up time becoming more acceptable.

If a skilled engineer or toolmaker is given this particular component to manufacture, it is quite likely that he or she will possess the required skills and most of the lathe tools needed to complete the job, allowing them to use their own tools taken from their own toolboxes. However if a student is given this particular job to complete, (particularly if it is being machined completely from scratch), it will be discovered that the whole operation will require a much longer preparation time in order to prepare the required tools to their precision accuracy sufficient to allow the accurate production of this internally screw-cut component.

The work-piece is initially machined using several separate operations in order to prepare it for receiving it's internally screw-cut thread. To do this, the machinist must perform the following series of machining operations:- Power sawing the steel billet to length; Center drilling its end for support; Externally turning its diameter; Facing-off and chamfering its exposed end to clean it up; Internally drilling its bore to its 'roughing-out diameter and required depth; Internal boring its tapping diameter to its correct depth and length; under-cutting its bore to provide sufficient end clearance for the screw – cutting tool; Internal chamfering to improve the thread near its under-cut area; screw-cutting its internal thread to its correct dimensions; parting off the finished component;. All these operations are required to allow the work-piece to receive its internal screw-cut thread while using a 'single point tool to produce an internal 24 mm × 3 mm pitch metric coarse thread in its machined bore.

Following these operations, the thread is finally checked for correct fit by using a thread plug gauge, followed by the component being finally parted off using a 3mm wide parting tool (shown in fig. 47 tool number 5).

The vital depth of thread required for this component must be obtained by using either tables or by calculation, before one attempts to screw-cut the thread.

Details of how to obtain the required internal depth of thread for this component are shown in the sectioned, enlarged, and fully dimensioned fig. 43a drawing. The fig. 43drawing shows the screw-cutting dimensions required for its closely fitting *bolt*'s thread used for comparison. The fig. 43a drawing therefore specifically shows the depth of thread required for the *nut*. The formula used to calculate this 'nut's specific depth of thread is shown in the lower sectioned part of the fig.43a drawing.

The internal screw-cutting operation required to produce this internal blind nut thread in the work-piece is generally considered by engineers to be rather more difficult to perform than if one were screw-cutting an external thread of similar pitch. The difficulties one can experience during the machining of this thread are largely due to the internal screw-cutting tools cutting edges being completely out of sight of the machinist for considerable periods of time during the thread's cutting and its extraction phase during its return to its starting point of approximately 5 to 6 mm to the right of the work-piece. In exceptional cases, this out-of-sight tool problem can be the cause of a possible tool overrun into the components 'blind end' that can be the cause of an accident occurring should the correct screw-cutting procedure not be fully understood or properly carried out.

It therefore requires considerable skill on the part of the machinist to perform this internal screw-cutting operation accurately safely, and successfully, mainly because of the drawing's requirement for the work-piece to possess an internally threaded blind nut in its design.

The difficulties one can encounter during this screw-cutting procedure are fully explained in minute detail in the following series of helpful operation notes. These explanations will point out most of the snags and difficulties likely to occur during this internal screw-cutting operation.

Section 19/2.

The fig. 44 drawing shows in detail the setup required to internally screw-cut a 3 mm pitch thread in a component using the

center lathe. The component's depth of thread, in this case, can be calculated by using a 'thread constant, combined with using the Probe and Prompt method of triangle calculation. (The 'depth of thread' dimension required before screw-cutting this thread can also be obtained by using the ten- times enlarged formula, shown in the lower part of the fig. 43a drawing, while using (in this case) a slightly differing method of 'depth of thread' calculation.)

If we now refer to the fig. 44 drawing of the component being screw-cut, it shows the setup required while using the straight plunge method of internal screw-cutting while using a center lathe for the operation. We can observe from the figures shown in the fig. 44 and the fig. 43a drawings that the basic depth of thread required for screw-cutting this internal 24 × 3 mm pitch metric coarse thread, can be obtained by using calculations in fig. 44, and the 'Reed constant' number 28.42535398. This constant has been specifically calculated for use in obtaining the depth of thread for *all* metric coarse *internally* threaded components).

This new constant number 28.42535398 is then multiplied by the pitch dimension of the thread about to be cut. Therefore in this case, the depth of thread required for this 24 mm × 3 mm pitch internal coarse thread is obtained by entering into the calculator the sequence 28.42535398, TAN, (0.54127) ×, 3, =, 1.62381 mm. This calculation gives us the theoretical depth of engagement of the nut relative to the bolt's screw-cut thread.

This information is shown in the thread's theoretical depth of thread dimension (shown in fig. 43a) of 1.62379 mm, which will differ slightly in practice by an insignificant amount of 0.00002 mm (0.000000787 inches).

In practice however, the actual depth of thread required for this component can be measured by the machinist from the point where the internal screw-cutting tools' cutting cutting tip's radius actually contacts the internal surface of the 21 mm diameter 'tapping size' bored hole. The student is also advised that this dimension will only provide us with the basic calculated depth of engagement of the nut's thread relative to the thread of its mating bolt.

To obtain a totally accurate internal depth of thread dimension (by calculation) for this thread, we must take into account the fig 43a, lower drawing which states the P/8 formulas' dimension, (P divided by 8). This particular calculation will of course vary according to the pitch of the thread being cut, but in our case, the calculation will be 3, (the pitch of the nut), divided by 8, giving us 0.375 mm, its theoretical flat, (this dimension is also shown at the top of the fig. 43a drawing of the internal screw-cut thread.). In practice, the root of the nut is then rounded and cleared beyond the width of P/8, (shown minutely in the lower drawing as a radius). This excess material in the nut is surplus to the nut's requirements, and is therefore removed by the screw-cutting tools *tip radius* during the screw-cutting operation. It is therefore very important that this tool's tip radius is hand honed accurately in order to produce a calculated and correct radius of 0.25 mm (0.0098425 inches). The relieved clearance area produced in the nut by this tool's honed radius then provides the necessary clearance gap to comply with the thread's accurate drawing; it will also provide the necessary clearance area that is considered sufficient to clear the male thread's overall diameter when it is later assembled into the nut's thread. The fig. 43 and 43a drawings show this clearance area to be 0.108 mm (0.004 inches) in depth, thereby proving that this clearance gap must exist between the core diameter of the nut and the 0.5 mm (0.0197 inch) flat that *should* exist on the correctly machined screw-cut major diameter of its mating bolt.

The total depth of thread required for screw-cutting this internal thread, must therefore include this minute area of clearance required for the bolt's major diameter. This clearance dimension must therefore be included in the final calculated dimension needed to obtain the nut's required depth of thread. We therefore add the 0.108 mm figure to the original calculated figure of 1.62381 mm, to give us a total depth of thread for the nut of 1.73181, (now deliberately rounded down to 1.70 mm), (to make this particular dimension compatible with the lathe's cross-slide's index scale), to allow it to accurately interpret this dimension in practice). This then gives us the full 'straight plunge' depth of thread required for screw-cutting this particular internal thread in a component that contains a blind nut in its design.

To reiterate, the design form of this particular nut's thread (in its maximum material condition) is depicted in the lower sectioned drawing of fig. 43a. In practice, and in order to avoid sharp corners being left at the root of the nut thread's major diameter, the root is rounded, as shown. This is also detailed in the drawing of the tool's point, (in this case by the use of minutely curved dotted lines); this very small radius area will therefore be cleared beyond the width of P/8 by the accurately dimensioned tip radius being honed onto the point of the internal screw-cutting tool.

It has been found (through experience and also through calculation), that the actual screw-cutting tool's cutting point (in this case), should be hand honed to a radius of 0.25 mm (0.0098425 inches). This radius can be checked practically for accuracy on the tool by using a 10 × hand-held lens, and by comparing it practically with the flattened end of a 0.5 mm diameter twist drill or alternatively using the flattened end of a short length of 0.50 mm diameter wire. It is very important that this particular radius has not been honed larger than the 0.25 mm radius required, because in practice, this will cause the flanks (or cheeks) of the screw-cut internal thread to be cut wider (and effectively looser) by the tool than is required for a correct fit on its corresponding bolt. This wider cutting action of the tool's side cutting edges (that occurs if the radius has been honed too large) will be seen (and discovered), on the tool reaching its final calculated depth of thread, which in practice would cause the thread to be cut too loose. Therefore, to be prudent (or in other words, to be on the safe side), it is considered best practice to always hone the tool's point to either the exactly correct calculated radius of 0.25 mm, *or*, to hone the tool's point to a very slightly smaller radius than the calculated dimension, in order to minimize the possibility of a too-loose thread being produced by the tool when finally reaching its full calculated depth of thread.

By accurately adopting this practice, the thread can now be screw-cut to its full calculated depth of thread without fear of the tool cutting the thread too loosely. A loose thread would, in all probability be beyond the range of the 'limits of fit required, coupled with the strong possibility that the component would be scrapped by the inspection department, because of this inaccuracy. This careful attention to detail by honing the tool's tip to its correct tip radius will be found to pay dividends by producing a tool that is now fully capable of producing a perfectly machined fit relative to the inspection department's thread plug gauge's requirements and the requirements of the drawing.

To reiterate, by adopting this meticulous 'tip-honing procedure, the thread produced in the nut by the tool, will now be machined so that it possesses an identical fit to its checking thread plug gauge, thereby providing the correct and required quality of fit that will allow it to accurately screw onto its mating bolt without any shake or end float being present. As detailed in fig 43.

Section 19/3.

A test piece being used to check initially that the lathe is set up correctly to produce the required 3 mm pitch thread. How one sets up the material correctly in the lathe's chuck. The importance of taking a light skim along the outside surface of the material to provide true concentricity of the whole workpiece as an aid to its accurate replacement (should this become necessary).

How one drills and bores the correct tapping diameter in the workpiece material.

When the machinist is initially engaged in setting up the lathe for screw-cutting this internal thread, it will be found best practice to first check that pitch of the thread about to be cut is correct by using possibly a spare piece of scrap, approximately 12 mm diameter round mild steel material, mounted in the lathe's chuck to be used as a test piece. The position of the lathe's screw-cutting gearbox levers must be checked to ensure they are positioned in their correct locations as indicated on the lathe's engraved information plate to produce the required pitch of 3 mm. This will then allow its lead-screw, through its gearbox, to rotate the lathe's spindle and chuck correctly to produce a 3 mm pitch between each revolution of the work-piece.

A basic pointed lathe tool (it need not be a screw-cutting tool at this stage), is then fitted temporarily into the tool-post and set at its correct cutting height, this tool is then indexed inward clockwise using the cross-slide's hand-wheel, to just touch the test piece's outside diameter; its scale is then set to read zero. The tool is then moved to a clearance area to the right of the test piece; the cross-slide's hand-wheel is then turned minutely clockwise by a further 0.05 mm

(0.002 inch) and its scale re-zeroed. The lathe is then switched on in its normal direction of rotation at a spindle speed of approximately 60 rpm and the *lift-up screw-cutting lever* is operated to allow the tool to make a dummy run along the test piece's outside surface forming a spiral groove over a nominal distance of approximately 15 mm. The lathe is then stopped and the distance between each of the machined grooves is then carefully measured to ensure that the required gap of 3 mm between each of the spirally grooved pitches is correct. To further check this pitch dimension, a thread pitch gauge (shown in fig. 49), a vernier caliper, or a precision metric rule can be used. When the pitch is found to be correct, a mental note should be made of the lever's actual positions for future reference.

Having now established that the distance between each of the screw-cut pitches is correct at these levers' settings, the tool and its test piece are now removed, and replaced by the chosen billet of steel material to use for the work-piece; this is then re-checked to ensure that it possesses sufficient overall length and diameter to complete the component's dimensions. This is then secured into the lathe's four-jaw chuck and checked that it is running truly in the chuck by using a dial test indicator.

Note that we should always allow an extra length of material to be added to the drawing's dimensions, to establish the (required) sawn cutoff length of the steel billet to allow for (1) the holding length required by the chuck's jaws, (2) facing off its outer end face, (3) the extra 2 mm allowance for the tool clearance required between the chuck's front face and the parting tool, and (4) the extra 3 mm to be allowed for the width of the parting tool required to part off the finished component.

The mounted work-piece is then re-checked for true running in the chuck, by using a clock gauge (or DTI) to ensure that the work-piece material is running truly.

The existing work-piece's outside diameter of 50 mm is now turned down to its 40 mm (drawing required) outside diameter by using a right-hand knife tool (shown in fig. 47 tool number 1), using coolant, followed by being faced off to its required length, (which must of course include the previously mentioned extra lengths). (Note that in practice it is always recommended that the machinist cleans up the outside diameter of any workpiece being turned, by taking a light skim along its entire visible length by using a right-hand knife tool, to ensure that the workpiece will now possess a diameter that is running perfectly truly relative to the center-line of the lathe. This true running of the workpiece's outside diameter becomes very important should the workpiece need to be removed from the chuck part way through a machining operation (such as a change in the *priorities of work* decided by the supervisor or by the management). This will then enable the workpiece to be re set accurately later in the chuck by the using a dial test indicator or clock gauge for its centralization.

[The machinist should now concentrate 100 per cent of his or her attention to the machining requirements of the internally threaded part of the component, this being the prime subject matter being dealt with in this chapter.]

Note that during the facing-off operation, the point of the right-hand cutting tool should only be allowed to progress inward up to the work-piece's center-line; the machinist should not allow the tool's point to travel beyond this center point because this action causes the tool's cutting edge to be blunted by the surface of the rotating work-piece (in this area) now rotating in an upward direction that will cause the tool to rub and not cut. (The student should therefore note that the work-piece material beyond the material's center-line will be travelling in an upward direction and will therefore not be capable of being machined by the tool).

The external sharp corner that now exists at the outer end of the workpiece at its junction with its outside diameter, must now be removed by using a chamfering tool (shown in fig. 47, tool number 8) to produce a conventional chamfer (this machined chamfer is universally undertaken for safety reasons and to improve the component's general appearance); this chamfered area is machined to approximately 1 mm in length at an angle of 45 degrees. (The now smooth faced-off end of the workpiece will later be used as a datum face to allow the square-type thread-cutting setting gauge to be used in close contact, to orient the screw-cutting tool's cutting edge accurately so that it fits into the gauge's 60-degree precision cutout. (This setting gauge is shown in fig. 49.).

The workpiece's outer end is then center drilled to a nominal depth of approximately 4 mm overall, using (preferably) a number 3 center drill held in the Jacobs drill chuck; this in turn is secured in the lathe's tail-stock. Note that when center drilling steel on the center lathe, we should always use a plentiful supply of coolant and a high work speed, the spindle speed selected must be at least 800 rpm, in order to protect and not to overload the drill's vulnerable extended and weak central cutting point, in order to prevent it being either damaged or actually broken off by the action of being jammed by an accumulation of dry swarf. The use of this center drill then aids the (following) larger-sized drill to centralize itself in the drilled hole produced. (The following larger-diameter drill will in this case normally possess, at its held end, a Morse tapered fixation to secure it into the tail-stock quill's taper.). This secure design of drill fixation then allows the drill's taper to absorb the extra load produced by the drilling operation, and will also provide the necessary rigid security for the drilling operation.

We now need to finally establish the exact core diameter (the so-called tapping diameter) required for the internal screw cut thread that is about to be cut. This tapping diameter is shown in the fig. 43a drawing, specified as 21 mm diameter.

When obtaining this tapping drill from the tool stores, it should first be checked for sharpness, and that its cutting edges are of equal angle and of equal length; this attention to detail will then ensure concentricity of the hole and a true diameter being produced in its drilled hole. Should this correctly sized drill be found to be serviceable, it can then be used to drill the hole to its required depth of 27 mm. (Note that it will be found an advantage (to the machinist), if this particular drill's point is allowed to reach the slightly deeper depth of 28 mm; to allow its pointed cutting tip to produce (a small conical clearance area, located at the extreme center at the end of the hole), which will then provide a run-out clearance for the following boring tool's cutting tip while being used to face off the inner end of the work-piece's bore to its *required* depth of 27 mm).

However, if the 21 mm diameter drill is not available, (which can be the case), then the next size smaller, say, 20 mm, can be used to provisionally drill the hole while using coolant and a work-speed of approximately 135 rpm (this drill must also be checked for sharpness etc. in the same way as the 21 mm drill originally required. This is then used to drill the hole while still enabling its *point* to reach a depth of 28 mm to provide the previously mentioned conical clearance area at the end of the finished bore. This operation is then followed by using the boring tool (number 9 in fig. 47) fitted horizontally into the tool post and set to its correct cutting height, in order to finish bore the hole to its required 21 mm diameter, by using a number of passing cuts of .05 mm deep, and by finally using this tool's relieved front cutting edge's tip, to face off the internal bore to produce a squared-off and correct overall depth of 27 mm.

Section 19/4.

Selecting a suitable internal under-cut tool for the job in hand, followed by setting up the lathe (as shown in fig. 44), to allow the tool to produce a correctly dimensioned and positioned under-cut area at the inner end of the tapping sized bore of the component.

(The student should also note that there is an alternative calculation method that can be used to obtain the vital tapping diameter required for this internally threaded component.)

We now select an *internal undercut tool* similar to (number 11 in fig. 47). This design of tool is normally obtained as an issued item from the work's tool stores, or, if a suitable tool is already available in the machinist's possession, then this is selected from his or her own toolbox. This specific tool is now used to machine the under-cut clearance area at the extreme inner end of the work-piece's bore. The width of the available tool's front cutting edge should now be measured by using a micrometer, a vernier, or a precision rule, and this dimension should be noted down for future reference. The tool is then fitted horizontally into the tool-post in a similar position and orientation to a boring tool, with its front cutting edge set to its correct cutting height, (this must be checked by using the lathe's tool height setting gauge). Its front cutting edge must also be set so that it is perfectly square with the bore of the component with its side clearances visually equally aligned relative to the cross-slide's center-line in order to provide equal and sufficient side clearance throughout its full depth of cut.

[It should be noted that the normal standard-sized undercut tool's cutting depth (as shown in fig. 47 tool number 11), may be found rather too long for this particular application; it will therefore need to be shortened substantially by using the machinist's off-hand grinding techniques, to make it more suitable for the undercutting of this and any later *smaller diameter* screw-cut threads. It will be found desirable therefore for the machinist to eventually possess both a shallow undercut tool (for small diameter threads), and an additional deep undercut tool (for larger diameter threads), for use later when undercutting future internal deep screw-cut threads.

It is important that the compound slide's hand-wheel is initially turned anticlockwise by a sufficient amount to ensure that its moving slide is fully supported by its casting. This action removes any existing back-lash present in its lead-screw, and prepares the tool for its expected move to its right later in order to extend the under-cut's width, without encountering any backlash problem. The tool is then maneuvered by using the saddle and cross-slide's hand-wheels, to enter into the work-piece's bore, where it is now positioned with its front cutting edge just touching the bore's diameter; its cross-slide's scale is now set to read plus 1.7mm; this is followed by using the saddle's hand-wheel being turned very carefully in an anticlockwise direction, to allow the tool to reach a touching point with the end of the bore. The saddle is then locked in this position.

[Because of the component's requirement for a 4 mm wide under'cut, a calculation is now made to take into account of the previously measured 3 mm width of the existing tool's cutting edge, while also realizing that the under-cut's required width is 4 mm.]

The lathe is then switched on at approximately 135 rpm with coolant applied to the tool and the internal bore of the workpiece. The *cross-slide* hand wheel is then turned very slowly and carefully *anti-clockwise* to allow the tool to enter into the bore of the workpiece and begin machining the undercut; this movement is continued very carefully until the zero mark is seen to have been reached on its scale, signifying that the tool will have now reached the full depth of the undercut required.

Because the cutting edge width of the undercut tool was previously measured at 3 mm, we must now extend the undercut's width to its required 4 mm width by now making a second plunge cut to complete the undercut to its full width and depth. To do this we must first initially turn the cross-slide's hand wheel in a *clockwise* direction by approximately half a revolution to allow the tool's cutting edge to fully clear the internal workpiece's bore, followed by turning the *compound slide's* hand wheel in an anticlockwise direction by approximately *1* mm (using its scale as a reference). The *cross-slide's* hand wheel is then turned anticlockwise very carefully to allow the tool to make its second plunge cut down to the same set zero mark (as before) on its scale; this then completes the undercut's 4 mm machined width to its full required depth.

The cross-slide's hand wheel is then turned *clockwise* by half a revolution to allow the tool to clear the workpiece's bore fully. The lathe is then switched off, the saddle is unlocked, and the tool assembly is very carefully extracted from the workpiece by turning the saddle's hand wheel carefully in a clockwise direction until the tool is now fully clear of the workpiece.

It is now very important that the student makes a full study of the fig. 43, and fig. 43a drawings that show in minute enlarged detail both the external and the internal thread. It is important that this study of the thread is thoroughly undertaken in order to allow the student to obtain a full awareness of the theoretical and the practical clearance dimensions that must be maintained between the nut and its bolt.

It should also be realized (after this full study of the thread's details has been made, that by adopting the correct screw-cutting procedure, the screw-cutting tool will be able to create all of the necessary metalto-metal clearances required between the two components, and that this situation can only occur if the fig. 43a drawing's stated dimensions are being strictly adhered to,and are being scrupulously maintained throughout the whole of the following internal screw-cutting procedure.

These thread clearances are designed to allow a precision fit to be obtained with the checking thread plug gauge, used on this thread's completion. This internal thread measuring instrument is to be finally screwed into the newly screw-cut nut, in order to ensure that the necessary clearance areas have been maintained between the nut and its bolt's thread for the final assembly made with the male threaded component detailed in fig.43.

[Note that the tapping diameter required for this screw-cut nut can also be obtained by using a calculation utilizing the bolts basic minor diameter (shown in the fig. 43a drawing as 20.319 mm), plus twice the bolt to nut clearance of 0.340 mm (also shown in the fig. 43a drawing), to provide us with the total internal diameter required for the screw-cutting (or tapping) diameter, of 20.999 (i.e. 21 mm), the tapping diameter required for this internally screw-cut thread.

Section 19/5.

Calculating the important depth of thread required for the internally screw-cut nut.

How one ensures that the finished screw-cut thread will fit the checking thread plug gauge.

An important note, warning of the danger of the screw-cutting tool being allowed to over-run the thread's undercut area, which can result in a collision occurring between the tool and the blind end of the bored hole.

Helpful notes that explain the reasons why the tool's clearance areas must be accurately provided when screw-cutting an internal thread. The physical marking of the top of the tool's extended shank in order to indicate to the machinist the exact position of the hidden undercut area, while the tool's cutting edge is completely out of sight of the machinist

The student should note that it is possible to use the detailed dimensions and formulas shown in the lower cross-sectioned drawing (shown below the fig. 43a 'nut' main drawing), to obtain the required depth of thread for the nut. We do this by using a series of calculations utilizing (from the lower drawing), the calculated vertical height of the theoretical 'vertex to vertex' dimension (the 'V to V' dimension) of 2.598076211 mm, shown at H. This dimension is also shown at the extreme top left-hand corner of the main fig. 43a drawing (for your perusal).

Utilizing this dimension in calculations, allows us to calculate the theoretical internal depth of thread required for the nut by using the sequence, (taken from the lower cross-sectioned drawing), H dimension (in this case 2.598076211) multiplied by 0.625, giving us 1.62381 mm; to this figure we must now add the 0.108 mm (0.004 inch) dimension (also shown in the fig. 43a main drawing). This minute clearance dimension is produced in practice by the tool's cutting point radius (also detailed in the main fig. 43a drawing), that now provides us with the full and final internal theoretical depth of thread required for this straight plunge internal 24 mm × 3mm metric coarse pitch thread, of 1.70381 mm, (rounded down to 1.7 mm to allow the lathe's measuring scales to reproduce this dimension).

Note that this 0.108 mm (0.004 inch) dimension added to the theoretical 1.662381 mm calculated depth of thread, now includes the internal screw-cutting tool's point radius; this is designed to remove this exact amount of material from the core of the nut to provide the necessary clearance area required between the major diameter of the nut and the major diameter of the bolt.

When the machinist is nearing the end of the internal screw-cutting operation, with the tool now approaching its full calculated depth of thread, there is now the need to check the current fit of the screw-cut thread during the last few passing cuts. We do this by first disconnecting the drive from its lead-screw by operating the thread-cutting knockdown lever then moving the tool well away from the component to allow sufficient space to allow the offering in of the thread plug gauge carefully into the thread to check its fit; it is important that this gauge is very carefully screwed into the thread in order to check if there is any end float, wobble, or looseness being present. The accuracy of the work-piece can only be considered correct if this final check allows the plug gauge to be fully entered into the thread without any undue force being applied, thereby proving that the screw-cutting operation has been satisfactorily completed. The whole screw-cutting operation may therefore require several extra passing cuts to be taken by the tool before a correct fit can be obtained to suit the fit of the thread plug gauge.

An area of uncertainty can be experienced by the machinist at this time, when it is realized that the depth of thread figures have so far been produced solely by calculation. The practical use of these figures will of course be affected if there has been any inaccuracy in the precision hand-working of the tool's honed radius. It is very important that this radius has been hand honed correctly, and has been checked to possess the required calculated radius of 0.25 mm. The actual sequence of operations required when screw-cutting this internal thread, will be basically similar to those previously described for screw-cutting the *external* thread of similar pitch and diameter, (detailed and explained in the previous chapter 18, Section 23 and onward); it will therefore be to the student's advantage if he (or she) reads through and inwardly digests all the contents of chapter 18 in order to allow the necessary comparative judgments and mental notes to be made regarding the main differences to be found between the two screw-cutting methods employed.

Internal and external screw-cutting operations will therefore require a basically similar screw-cutting technique to be adopted.

The main difference found between screw-cutting an internal thread, when compared to screw-cutting an external thread,is that the internally threaded component's back-lash, that is normally existing in its cross-slide's lead-screw must be eliminated (in this case) by turning the cross-slide's hand-wheel in an *anticlockwise* direction by (say) half a turn in order to ensure that the existing back-lash has been totally removed. Its scale must then be re-set to read zero immediately before using the tool for the internal screw-cutting operation. The machinist must also be aware that when internally screw-cutting a component, the *snatched method of tool withdrawal* (made at the end of each passing cut), must be performed (in this case) by turning the cross-slide's hand wheel in a *clockwise* direction to remove the tool from the thread. The amount of rotation given to the hand wheel at this time will depend on the *amount of clearance available* in the thread's tapping-sized bore, during and at the end of each passing cut. This clearance area must be of sufficient diameter to allow the tool's point radius to clear the core diameter of the internal thread fully, without the tool's back face coming into contact with the far side of the internal thread during the tool's withdrawal. This rapid extraction of the tool from the thread must of course be coordinated with instantly using the 'dead stop' lever to stop the lathe's motor and its work-piece, while the machinist also catches a fleeting glimpse of the rotating chuck's warning marker, followed by instantly applying the brake to stop any over-run of the tool past its undercut area.

To reiterate, the total clock-wise movement given to the cross-slides hand-wheel in order to obtain a full tool clearance, must be sufficient to allow the tool's tip radius to fully clear the thread by a minimum of (say) one millimeter, (preferably two millimeters for safety). This must be done in order to ensure that full clearance of the tool from its screw-cut thread has been achieved *before* the machinist allows the tool to be extracted from the bore and being returned back to its starting point.

With regard to machine clearances, it should also be noted that it is very important to maintain a sufficient longitudinal safety clearance area of at least 10 mm between the saddle's moving casting and the tail-stock's stationary casting, during any screw-cutting procedure, this will apply particularly at the entry end of the internal thread.

It is therefore very important that this clearance area is maintained though out the whole of the tool's withdrawal procedure.

Section 19/6.

This helpful note describes some of the clearance problems that will be encountered when internally screw-cutting relatively small diameter threads, including an explanation of the dangers that can be experienced during the internal screw-cutting operation, particularly if the tool's front cutting edge has been allowed to approach the work-piece's blind end face too closely. This potential problem can be overcome by scribing two warning marks on the top of the screw-cutting tool's extended shank, (using ideally a waterproof fine-point pen). These warning marks are to be accurately placed and dimensioned relative to the outside end of the workpiece in order to identify exactly where the very short length of the hidden undercut area is positioned; we do this to prevent any accidental overrun of the tool causing extensive damage to the tool, and to prevent a possibly dangerous accident occurring.

[This *back of the tool clearance problem* can also become critical when the machinist is internally screw-cutting very small-diameter threads, which of necessity possess only very small clearance areas between the back of the extended tool's shank and the far side of thread's bore. This problem occurs particularly during the tool's snatched method of withdrawal (in a clock-wise direction' from the thread prior to the tool being moved back to its starting position).

The tool's return journey to its starting position must in this case be very closely monitored in order to avoid the possibility of the back of the tool coming into contact with the opposite side of the internal thread being cut. If the amount of this clearance is insufficient, then the tool's shank can cause severe damage or burring to occur to the core of the screw-cut thread by its abrasive rubbing action.

For this reason, the total amount of tool extraction movement given (in a clock-wise direction), to the cross-slide's hand-wheel, prior to the tool's extraction phase, must be limited to possibly half a revolution *clock-wise* (when screw-cutting large diameter threads having plenty of clearance available), and considerably less extraction movement when screw-cutting small diameter threads, where the back of the tool's clearance will be very limited. In this case, the cross-slide's scale should always be marked (for safety reasons), with two clearly defined water-proof pen marks that indicate both the maximum and minimum limits of clearance space available during the tools extraction phase. Therefore, before one attempts to internally screw-cut small diameter threads, the back of the tool's extended shank should be very carefully examined to make sure that it will have sufficient clearance relative to the core diameter of the screw-cut thread during use. If necessary, this may entail its extended shank being very carefully off-hand ground along its *back face* and on its *underside edge* to provide sufficient working clearance between the core of the screw-cut thread and the back and under-side of the tool during its extraction phase. This clearance must of course always be maintained in both the tool's working position and during its extraction phase.

To reiterate:

The machinist should be made fully aware, and be on constant guard of the possibility of the end of the tool assembly contacting the end of the bore's blind end face, particularly when the tool is *approaching the extreme end of its screw-cutting passing cut, where the end face of the bore is of necessity being approached very closely by the tool while the lathe is in gear and under power,* very quick action will need to be taken by the machinist in order to prevent an accident occurring; this must be done by instantly stopping the lathe's motor and by instantly applying its spindle's brake.

Therefore, when internally screw-cutting a component that possesses a blind nut in its design, the tool's front cutting edge must *on no account* be allowed to over-run the width dimension of the clearance area provided in the relief undercut area, as this would allow the front cutting edge of the tool to collide with the solid end of the tapping diameter's bored blind hole, causing not only damage to the tool and the work-piece, but would also cause a possible screw-cutting disaster.

To avoid this problem occurring, the tool's extended shank must be very carefully and accurately marked with (preferably) a water-proof marking pen (on the top of its extended shank's surface); these marks must clearly indicate very accurately to the machinist the exact points where the undercut begins, and particularly where it ends, during the screw-cutting passing cut. To actually make these warning marks, we must first ensure that the lathe is switched off, (is out of gear, and the work-piece is stationary). It will be found an advantage if the top of the tool's extended shank is painted with typist's white correction fluid (if available), prior to marking the required position marks. The screw-cutting tool is then entered into the tapping-size hole by using the saddle's hand-wheel being turned in an anti-clockwise direction sufficient to allow the leading edge of the tool to just touch the end of the blind hole; we then mark this position, at this exact point relative to the datum end of the workpiece, on the top of the tool's extended shank. While again using the outer end of the workpiece as a length datum, we then mark out a second line, 4 mm to the left of the first line on the tool's shank, to indicate the exact width of the under-cut and the exact extraction point of the tool, relative to the available undercut's limited clearance area.

To reiterate, for safety reasons the clearly marked out extremities of the hidden undercut area in the workpiece must on no account be exceeded by the tool while it is under power and on a screw-cutting cut.

The exact positions and the width of the under-cut marked on the tool's shank can now be used as an accurate guide to enable the machinist to synchronize his use of the stop button (or lever), the snatched method of tool withdrawal, [and, if we were using the disengaged lead – screw technique, of operating the knockdown screw-cutting lever, which would disengage the lead screw, at this exact moment before the tool has reached the absolute end of the blind hole]. These very accurate moves must be very carefully synchronized by using the assistance of the aforesaid warning marker previously adhered to the chuck's outer periphery, which has been placed to provide a visual warning point for the tool's extraction. This warning marker will have been previously positioned on the chuck's outside diameter, (by the use of a highly visible material such as drafting tape (or a similar visible adhesive material). This material, when provided with a (preferably) inked waterproof datum mark, must be positioned very accurately to indicate to the machinist the exact point where, at the end of the screw-cutting cut, the tool is instantly extracted in a *clockwise direction* and the motor is *instantly stopped*, by using either the on/off switch/ or the on/off lever, (or (if fitted) the handle of the clutch's 'slowing down and stopping' lever).

These inked marks made on the top of the tool must be very closely monitored during the internal screw-cutting procedure, paying particular attention to the mark indicating the beginning of the undercut, while the machinist is also catching a glimpse of the chuck's rotating warning marker as it is seen to pass through his peripheral vision, indicating that the tool has reached the end of the screw-cut thread, and is now entering the beginning of the very short under-cut area.

[Of course if the machinist were screw-cutting a conventional nut, specifically designed with a 'straight through' clearance area situated beyond the thread being screw-cut (this would be the case if a conventional 'straight through' nut' were being screw-cut), then the tool's exact withdrawal point in this case would not be so vital because of the existing large clearance area available following the screw-cut part of the nut.]

Section 19/7

The importance of selecting just one *specific* and *chosen* revolution of the workpiece at the end of each screw-cutting pass, in order to accurately and instantly stop the lathe's motor within the very short clearance area provided by the narrow 4 mm width of the under-cut

The trainee machinist should now realize that the relief undercut, machined for the tool's clearance at the end of each screw-cutting passing cut, will (in our example's case), only be 4 mm wide, and as the present thread we are cutting only possesses a 3 mm pitch, this means that in practice it will only require *one* specific and *chosen* revolution of the chuck and its workpiece at the end of each screw-cutting passing cut, for the tool to have traversed virtually all of the safety area provided by the undercut's width. This fact makes it vitally important that the work speed initially chosen is as slow as possible, and that the *last vital revolution* of the chuck is very carefully chosen by the machinist at the end of each screw-cutting passing cut in order to allow the stopping of the motor instantly and accurately. This action must be very precise in order to allow the tool to be safely extracted and the motor instantly stopped *within* this *very short* 4 mm safety area.]

With the student now having read through and inwardly digested all of the foregoing information and warnings regarding the potential dangers that can occur when performing internal screw-cutting operations, we now carry on with the final setting up procedure required to actually screw-cut the thread in the component.

Section 19/8

Part two. The final setting up procedure required prior to performing the internal screw-cutting operation in the work-piece.

Choosing the correct design shape of the internal screw-cutting lathe tool, prior to screw-cutting the internal thread in the component.

Explaining the importance of the extreme accuracy required when setting up the tool in its tool-post, including the advantages of machining a 30-degree chamfer on the inner end of the tapping size bore prior to the material being screw-cut.

The importance of eliminating all the back-lash that exists in both of the slide's lead-screws prior to screw-cutting the thread, followed by confirming that the dimension of the cutting radius honed onto the tool's cutting point is exactly correct.

This section explains in great detail the final preparations and the tooling required to internally screw-cut the metric coarse 3 mm pitch internal thread into the work-piece's bore.

It is quite possible that the employer's tool store will stock and possibly issue an internal screw-cutting tool considered by the stores department to be suitable for screw-cutting the thread in this particular component; however, on receipt of this as-issued tool, it will, in all probability be found that it does not contain all of the required accurately ground cutting angles, clearance angles, and tip radii that are required for screw-cutting of this particular component. It should be pointed out that some companies will expect the machinist (by using his extensive 'off-hand grinding techniques), to off-hand grind the required tool's dimensions from an issued piece of solid high-speed tool steel that is normally stored and issued to the work-shop machinists from stock, thereby allowing them to grind their own screw-cutting tools to the required specifications from scratch in order to accurately screw-cut the component.

It will therefore be considered a distinct advantage, if the student has at least some experience of grinding his or her own lathe tools and in using their own skills in off-hand grinding techniques to allow the grinding of the new screw-cutting tool to the required accuracy.

Alternatively, the machinist may be required to finish grind an as-issued internal screw-cutting tool to the exact shape and angles required. This tool must be ground to contain its required 60-degree included angled cutting edges, complete with the clearance angles that are generally considered to be correct for use on a 3 mm pitch internal screw-cut thread to screw-cut the job in hand.

Of course if an internal screw-cutting tool, previously ground by the machinist, is found to possess the correct shape complete with all the required clearance angles is available, then this of course will be selected from one's own tool-box to screw-cut the thread in this particular component. This tool must possess (as previously stated), cutting edges that are ground to a very accurate 60-degree included angle (by using the guidance of the precision V cutout, in the thread cutting gauge). The tool must also possess (particularly for this job) a hand-honed radius at its cutting point's tip of 0.25 mm (as calculated using the lower drawing in fig. 43a), this radius being perfectly compatible with the 3 mm pitch of the thread being cut. The tool must also possess all the necessary accurately ground clearance angles that are sufficient to fully clear the cut surfaces of the thread being cut.

As previously stated the tool's tip radius can be checked for its correct dimensions and contour after its hand-honing process is completed, by using the flattened end of the diameter of a short length of 0.5 mm (0.019685 inch), wire, or alternatively the flattened end of a twist drill of 0.5 mm diameter, by using a ten-times magnifying glass for the comparison.

Note that when a tool is being used for internal screw-cutting, the clearance angle provided below its leading cutting edge becomes a very important criteria, because of the need for it to cut the thread in the very confined space available in the work-piece's bore. It must therefore be clearance ground sufficient to provide the minimum of 7 degrees greater than the helix angle of the thread about to be cut; this must be done in order to provide adequate working clearance between the tool's front cutting edge and the trailing cut surface of the thread during the tools screw-cutting traverse along the work-piece's tapping-sized bore. The machinist must also ensure that the *heel* of the tool is ground away sufficiently to allow its surface to fully clear the trailing surface area of the thread being cut. The amount to be ground away will of course depend on the diameter of the thread being cut, (small diameters will require considerably more clearance to be provided in order to ensure that its heel does not rub the surface of the thread.

The selected tool must be carefully fitted into the tool post and provided with just sufficient stick-out from this supporting tool-post that will allow its front cutting edge to reach the full longitudinal depth of the tapping-sized bored hole, *plus* an additional safety clearance of 6 mm to allow for its required *longitudinal safety clearance* relative to its tool post. Its precision cutting edges must also be set to their correct cutting height by the use of the tool height setting gauge. The tool's cutting edges should be oriented very accurately in its tool post to enable the center-line of its cutting edges to be set perfectly square to the center line of the cross-slide of the lathe, by using the square-type, thread-cutting gauges' assistance. This is accomplished by using (in the internal thread's case), the gauge's left hand side face (as shown in fig. 49), being held in close contact with the newly machined flat right hand end face of the workpiece. A piece of white paper held under the tool during this orientation process will be found to reflect sufficient light to allow the tool's honed point and its precision angled cutting edges to be accurately aligned into the precision 60-degree cut out V of the gauge, followed by the tool being finally tightened and secured into this exact position in the tool-post. The cross-slide hand-wheel and the saddle's hand wheel are then turned sufficiently to maneuver the tool into its approximate starting position of 5 to 6 mm to the right of the workpiece with its cutting tip positioned approximately 5 mm away to the right of the bore's diameter, and temporarily at the approximate longitudinal center line of the lathe.

The cross-slide's hand-wheel is then turned half a revolution anti-clockwise to 'take out any existing backlash' from its lead-screw; its scale is then set to read zero.

All the foregoing advice and setting information is required in order to allow a safe and accurate internal screw-cutting operation to take place.

[It should be noted that students are advised when screw-cutting this particular design of component, to adopt the *straight plunge* method of screw-cutting, and to use the permanently engaged lead-screw technique. The use of this method includes the *snatched method of tool withdrawal*, coupled with the instant stopping of the lathe's motor chuck and work-piece, followed by the motor being re-started in its reverse direction of rotation, to return the tool back to its start position under its own power at the end of each cut.

Having now fitted and oriented the screw-cutting tool accurately into the tool-post and checked the position of the marks previously inscribed on the top of the tool's shank that give warning of the exact position of the under-cut and the extremely close proximity of the blind end of the bored hole relative to the leading cutting edge of the screw-cutting tool, we now apply a thin smear of micrometer blue to the inner surfaces of the tapping-sized) bore, followed by switching on the lathe's motor at a chosen minimum work-speed of approximately 60 rpm (this relatively slow work-speed of 60 rpm, is the preferred option in this case). (The actual work-speed selected would normally depend on the spindle speeds available on the lathe, and on the diameter of the thread about to be cut.).

With the tools point now to the right of the component's bore, we now enter the tool's point longitudinally into this previously bored hole by about (2 mm), using the *saddle*'s hand wheel being rotated in an anticlockwise direction. The cross-slide's hand wheel is then turned very carefully anticlockwise until the tip of the tool is seen to be just touching and removing a minute quantity of blue and a minute sliver of metal from the inside diameter of the tapping-sized 21 mm diameter bored hole. The motor is then stopped and the cross-slide's scale is then re-set to read the *total depth of cut* required for this particular thread (in this case this will be 1.70 mm); this fine adjustment to the scale is made by grasping the outside diameter of the cross-slide's hand wheel securely between the thumb and forefinger of the left hand (to prevent any rotational movement), while the right hand grasps and turns its index scale to read 1.70 mm (in practice this setting will then enable the full calculated depth of thread to be reached by the tool when its scale is eventually seen to read zero).

We then turn the *cross-slide*'s hand wheel clockwise just sufficient to remove the tool's tip from contact with the workpiece's bore.

Having now gone through the process of setting up the lathe for screw-cutting the thread, the machinist should now realize at this stage that there is a distinct advantage now in machining an internal chamfer on the inner end of the proposed thread in order to provide slightly more tool clearance in its undercut area; therefore, at this stage in the setting-up process, we must now take advantage of the tools current state of readiness by using the tool's *trailing cutting edge* to chamfer the now existing undercut's width to provide a slightly wider clearance area for the tool at the end of its passing cut). To perform this operation we use the actual screw-cutting tool's *trailing cutting edge* to machine a 30-degree chamfer on the inner end of the potential thread material before the thread has been cut. We accomplish this by first turning the *compound slide*'s hand wheel in an anticlockwise direction by half a turn to eliminate any of its existing backlash in its lead-screw, followed by re-zeroing its scale. Doing this now makes it possible to operate its hand-wheel in an anticlockwise direction to allow the tool to move to its right to machine the chamfer, without incurring any lost motion or back-lash in its lead screw.

[It will be found a distinct advantage (to the machinist), when this chamfer has been completed as it now makes the undercut area slightly wider and provides a slightly wider working clearance for the screw-cutting tool at the inner end of its screw-cut thread].

To perform this chamfer operation, we now progress the tool along the bored hole by using the saddle's hand-wheel, until the extreme end of the cutting tool, (which should now possess an off-hand ground shallow point at its extreme end), (similar in shape to the outline drawing shown in fig. 47 number 10), (where it is here described as its end point, and shown on the tool in fig. 43a), we then allow this shallow point to just touch the end of the bored hole. We then securely lock the lathe's saddle to prevent any further longitudinal movement. We select a work-speed of approximately 60 rpm, followed by switching on the lathe's motor in its normal direction of rotation and by switching on the lathe's coolant supply to the workpiece's bore. This is followed by turning the cross-slide's hand wheel very carefully anticlockwise in order to index the tool into the 4 mm wide gap of the under–cut's clearance area, until its zero point is reached on its scale, (the tool's tip will now be positioned at the proposed threads full depth of thread). We then turn the compound slide's hand wheel very carefully (and extremely slowly) in an anticlockwise direction, (the initial movement of the tool will be assisted to some extent by the student listening intently to the noise the tool will now be actually making when it starts cutting.

When the tool is heard to be actually cutting material from the inner end of the proposed thread's material, the tool is then indexed longitudinally outward further by using the compound slide's hand wheel being turned in an anticlockwise direction further until its scale is seen to read approximately one mm of tool movement. This movement of the tool will now have produced the one mm chamfer required at the far end of the 'to be threaded' part of the bore's internal tapping-sized bore. The lathe is then switched off, the tool is then moved out of the 'to be threaded' area of the bore by now turning the cross-slide's hand wheel in a clockwise direction by at least half a revolution, thereby providing the necessary tool point clearance from the under'cut, this is followed by unlocking the saddle, and the tool assembly is moved out of the bored hole by carefully turning the saddle's hand-wheel in a clock-wise direction until the tool's removal from the workpiece is complete.

[The student may be puzzled regarding how the figure of 1 mm was arrived at to perform the correct length of the 30-degree chamfer that has just been machined at the far end of the proposed thread. This is calculated by drawing our familiar triangle and using the fact that we already know its The Angle (30 degrees), this being half of the tool's included angle, and its side c, by using its calculated depth of thread of 1.7 mm, with the need to know the length of its side b, This is done by using side 2 of the probe rotated into the position where its The Angle is at the bottom, and its side c is vertical, followed by using the prompt $\angle$ c, b =, c × TAN $\angle$. We therefore enter into the calculator the series, 30, TAN, (0.577350269) ×, 1.7, =, 0.981495457 mm (now rounded up to 1 mm].

Section 19/9.

Performing a dummy run of the screw-cutting procedure to enable the machinist to become fully aware of the positions of the warning marks previously made on the top of the tool's extended shank indicating the exact point where the tool must be instantly extracted from the thread and the motor and work-piece instantly dead stopped.

Explaining the method of checking the thread's fit by the use of a thread plug gauge

Helpful notes on performing the internal screw-cutting operation, followed by the selection and fitting of the parting-off

**tool into the tool-post, after the screw-cutting procedure is
completed, with instructions on the correct way this should be done**

With the lathe now switched off and with its *lift-up screw-cutting
lever* in its disengaged (down position), the cross-slide's hand wheel
is now turned anticlockwise sufficient to allow the tool to be one
revolution and outboard of and positioned on the machinist's side of
the workpiece in order to give the tool sufficient clearance to perform
a dummy run of the screw-cutting procedure. It is advisable to do
this if the student is still unsure how to safely observe and act upon
the important warning marks made on the chuck's periphery and the
tool's extended shank that indicate the exact point where the tool must
be withdrawn from the thread and the motor instantly stopped. The
student now goes through a simulation of the whole screw-cutting
procedure by first switching on the lathe at a low work-speed of say
60 rpm, with the tool at its starting position of 5 to 6 mm outboard
of and to the right of the workpiece. This is followed by engaging
the *lift-up screw-cutting lever* at the chosen line or number indicated
on the apron's rotating dial, relative to its fixed datum line, to allow
the tool to progress along in a controllable manner by the use of the
lathe's clutch lever (if fitted), until the warning marks on the chuck,
and in particular the first warning mark on the tools shank are
reached, whereupon the motor is then instantly stopped. The warning
mark on the chuck can now be accurately re-positioned (if necessary)
to show the exact point where the instant stop should be made, and
also making the student realize that this must also be the *chosen*
revolution of the chuck and work-piece for the motor to be stopped. A
dummy snatched withdrawal is then made by turning the cross-slide's
hand wheel half a revolution clockwise, followed by the motor being
reversed to return the tool to its start position.

Having completed this practice operation satisfactorily, the cross-
slide's hand wheel is then returned to its original start position; this
is followed by turning its cross-slide's hand wheel in an anticlockwise
direction and down to its previously set scale's reading of 1.70 mm.
A check should now be made to ensure the tool's cutting tip is at its
previously set correct depth setting. (The cross-slide's hand wheel is
now turned by the minimum amount anticlockwise and just past
the 1.70 mm scales mark by 0.05 mm in order to put on its first

cut), thereby making its index scale read 1.65 mm; this should now be marked with pencil.) The lathe is now switched on in its normal direction of rotation, at a work speed of approximately 60 revs per minute, with the coolant switched on to cover both the tool and the workpiece; the tool will now start to move to its left on its first screw-cutting passing cut.

{Note that if the lathe is equipped with a clutch controlled on/off operating lever, this will be found a very useful addition to its controls for screw-cutting, as this allows the work-piece to be slowed down to a very low speed, and even stopped, as the tool nears the critical end of the screw-cut thread, on its approach to the 'safety' of the under-cut area.]

When the tool's withdrawal mark (previously inscribed on the top of the tool's extended shank) is seen to have reached the datum end of the workpiece while also catching a brief glimpse of the chuck's rotating warning marker on its chosen revolution, the motor is instantly stopped by using the 'dead stop' lever or foot control pedal. The cross-slide's hand wheel is then instantly turned half a revolution clock-wise to permit the *snatched method of tool withdrawal* to be used, followed by restarting the motor in its *reverse direction* to take the tool assembly back to its start position under its own power to a position 5 to 6 mm to the right of the work-piece, where the motor is again stopped.

The next cut is then indexed into the workpiece by turning the cross-slide's hand wheel a further 0.05mm (seen indicated on its scale), to the new depth setting of 1.6 mm and again marked with pencil. The motor is then switched back on in its normal direction of rotation to permit the tool to take the next cut. Note that extreme care must be taken when complying with the warning marks made on the tool's shank and those on the chuck's outer periphery, to ensure they are being *very carefully and exactly complied with* at the end of each passing cut.

Further passing cuts are then made, by using the recommended 0.05 mm depth of cut for each passing cut, while also instantly using the *snatched method of tool withdrawal* as the tool reaches the mark indicating that the undercut at the end of the thread has been reached.

Further cuts of 0.05 mm are then used until the full depth of thread has been reached, indicated by the cross-slide's scale now reading zero.

At the end of this last passing cut, the *snatched method of tool withdrawal* is used at exactly the same time and place as before; the motor is then instantly stopped, followed by the motor being reversed to take the tool back to its starting position of 5 to 6 mm to the right of the workpiece, where the motor is then finally stopped.

Having now completed the full series of screw-cutting passing cuts and where the tool has reached its full calculated depth of thread, the lathe's tooling must now be repositioned outboard of the work-piece to permit the component's internal screw-cut thread to be checked by the use of the thread plug gauge for correct fit. We do this by first turning the cross-slide's hand wheel several turns in an anticlockwise direction to allow the tool assembly to fully clear the workpiece, to provide sufficient clearance for the checking thread plug gauge to be carefully *offered into* the newly cut screw-cut thread.

When it is found that the thread plug gauge can be screwed fully into the thread without exhibiting any slackness or end float being present, then the screw-cut thread's fit can be considered capable of being passed by the inspection department.

However, if it is found that the thread plug gauge is still reluctant to screw fully into the thread, this will then indicate that the thread is still too tight and will require a further passing cut (or cuts) to be taken until the thread plug gauge can be carefully screwed fully into the thread without exhibiting any tightness, slackness, or end float being present.

If after taking this extra passing cut or cuts, the plug gauge can now be screwed into the thread without any slackness or end float being found present, then the component will now be ready for the parting-off operation to take place, but before we do this, we must always check that we have completed all of the other drawing's dimensions correctly before we actually part off the component. (The parting-off operation will be fully explained later.).

[Note that the plug gauge used for checking this internal thread will normally be available from either the firm's tool stores, the inspection department, or the firm's standards room, and will be found to possess, engraved on its handle, the name and the pitch of the thread that is about to be checked. This gauge will normally have a knurled central handle with two precision ground and hardened male threaded ends. One threaded end will be marked 'GO', this being the correct one to use for checking the fit of the thread that has just been machined. The thread on the gauge's other end will be marked 'NO GO', and this thread, as its name suggests, should not be able to be screwed into the screw-cut thread by any more than half a revolution. This 'NO GO' thread has been specifically designed and precision ground so that it is oversize, and beyond the recommended limits of fit required for this particular internal thread.

Of course if the 'NO GO' threaded end of the gauge can be screwed into the workpiece, this will (unfortunately), signify to the operator that the thread has been machined too deeply, which makes it too loose, and its fit will therefore be beyond the acceptable limits required by the drawing, and the firm's inspection department.

Therefore if the workpiece is now found to possess a screw-cut thread that is too loose and possesses considerable end float when being checked for fit by the thread plug gauge, then the component will fail the inspection test because it has a *faulty thread*, therefore this workpiece must now be considered to be scrap.

[The component, now having failed the thread plug gauge test, must now be removed from the chuck and replaced with a new billet of similar-grade material for machining into a replacement component. It will be found good practice at this stage of the proceedings, by having scrapped the previous attempt, to allow (this time) an extra 3 mm on its overall length, so that the new workpiece can be machined to include *an extra safety feature* that will allow the machinist to check the depth of thread reached by the tool before it enters the screw-cut portion of the thread. This extra length of material must be bored out to an identical diameter to the bolt's 23.9 mm outside diameter, over the very short 3 mm extra length. This extra operation is of course performed after the component has been bored out to its correct tapping size diameter of 21 mm.

The actual screw-cutting tool can now be maneuvered by hand (and with care) to internally turn this diameter to its required 23.9 mm diameter over its 3 mm length; it will be found that (as a bonus) this turning operation can also provide a 30-degree internal chamfer at the starting end of the new thread].

During the latter stages of the normal internal screw-cutting procedure, the inner surface of this 23.9 mm turned bore should be smeared with micrometer blue and will be seen to show a very faint spiral witness mark when the full depth of thread has been reached by the tool on its final screw-cutting pass or passes; this then indicates to the machinist the full depth of thread has been reached and has not been exceeded.

With the tool's point honed to possess its correct radius of 0.25 mm (or minimally less), this witness mark can then be used with confidence, to provide a very effective final depth of thread indication to the machinist just prior to the thread plug gauge being used for checking the final fit of the thread.

The 3 mm excess material now existing on the outer end of the workpiece can now be faced off by using a right-hand knife facing tool.

*[The student should now be realizing that it is possible to use a similar strategy when screw-cutting the **external** thread (as detailed in fig. 42), where in this case a 3 mm extension **can be** added to the workpiece's overall length which is turned down to leave a 20.3 mm short diameter, over a length of 3 mm, (this being identical to the core diameter of the male thread. This will then allow the external screw-cutting too, (with its minutely honed radius, and with the help of micrometer blue, to leave a helpful witness mark on this short diameter during the last passing cuts, to indicate to the machinist that the tool has now reached its full calculated depth of thread.*

In this case, the added excess material on its end face can now be faced off, and the greased center's drilled support deepened by approximately 3 mm (if required) to give sufficient support during the following parting-off operation.]

We return now to the parting-off operation' to be carried out on the internally threaded workpiece (shown in fig. 44). The workpiece, having now passed the plug gauge test for fit, and the final check made of all the other drawing dimensions now found to be correct, the finished workpiece can now be parted off by using the following procedure:-

Firstly, the lathe's *lift up screw cutting lever* must now be moved into its down or disengaged position, and its *turning mode lever (or its turning mode selector knob)* must now be placed into its neutral position to prepare the lathe for the hand operation of its slides and hand-wheel controls.

[It will be found to the machinist's advantage if the lathe being worked on is equipped with a tool-post that is designed to take quick change tooling, (as shown in use in fig. 46), in this case the existing screw-cutting tool complete with its holder, can now be removed, and to prevent the need to re-chuck the component to chamfer its parted-off end, a chamfer operation can be performed while the workpiece is still in position in the chuck. We do this by now installing a 90-degree included angle pointed V cutting tool previously mounted in a separate holder, secured in the tool-post with the tool's cutting edges set at their correct cutting height. The compound slide's hand wheel is then turned clockwise to eliminate any back-lash in its lead-screw; its slide is checked to ensure that it does not possess any overhang from its support casting. The tool's point is then moved along to its left using the saddle's hand wheel to a longitudinal position along the workpiece of exactly 38 mm from its right-hand end.

With the tool's point now positioned to be just touching the workpiece's diameter, the *cross-slide's* scale is now set to read zero. The saddle is now locked to prevent any further longitudinal movement. The lathe and its coolant are now switched on and the cross-slide's hand wheel turned very carefully clockwise in order to machine a V groove in the workpiece's surface until its scale indicates that it has moved inward by one mm. This operation will have completed the required chamfer on the *to be parted off* end of the workpiece. The saddle's lock is then released and the cross-slide's hand wheel is now turned one revolution anticlockwise, followed by the motor being switched off. The 90-degree V tool and its holder are now removed from the tool-post and replaced with a tool holder that now contains the selected *parting tool* to be used in the following parting-off operation. This tool must be checked for sharpness and adjusted to its correct cutting height by using the *lathe's tool height setting gauge*. (This tool's height adjustment can be made (when using the quick change tooling), by minutely rotating its knurled adjusting nut, followed by locking the assembly to establish the tool's correct cutting height finally. The actual *parting tool* used that contains a cutting edge width of 3 mm, as shown in fig. 47 tool number 5 (or possibly tool number 6, if particularly heavy parting-off cutting is expected, must be allowed to possess only sufficient stick-out for its cutting edge to fully reach and clear the full diameter of the workpiece down to the workpiece's center line. The tool must also possess sufficient side clearance that will allow its front cutting edge to reach the center of the work-piece unhindered. The alignment of the tool's extended shank should also be checked to ensure that its center-line is perfectly in line with the cross-slide's center line. This true alignment of the tool then allows it to possess equal side clearance in its intended parting-off slot, while performing the full parting-off operation.

A revolving pipe center should now be installed in the tail-stock with its large diameter tapered nose entered into the component's chamfered end; the tailstock's hand wheel is then carefully adjusted to provide sufficient support for the workpiece, followed by the tailstock's quill and its tailstock's casting being locked into this position. However, if the pipe center being used is of the fixed solid type, then its support cone should now be greased to provide sufficient lubrication in the area of its contact with the inner surface of the threaded workpiece.

The lathe's compound-slide, including its tool-post, should now be moved into a position where it is being fully supported by its support slide's casting, *(it is vitally important that we provide the absolute maximum rigidity throughout the whole setup when performing a parting-off operation).* The compound slide's hand wheel is then turned anticlockwise by approximately half a turn, immediately followed by it being turned clockwise by a quarter of a turn to remove any inherent backlash existing in its lead-screw; its scale is then set to read zero. This is followed by the whole tool and saddle assembly being moved along to its left by using the saddle's hand-wheel, to a position where the *right-hand side* of the parting tool's front cutting edge is accurately positioned to leave the required 38 mm workpiece length to its right, by placing the tool's right-hand side cutting edge lined up exactly in the center line of the previously machined V in the workpiece, (the required 38 mm length of the component should now be checked for its correct length by using the depth facility of a vernier caliper gauge or a depth micrometer). This is followed by the saddle being locked into this position to prevent any further longitudinal movement.

A spindle speed of (say) 135 rpm should now be selected, the lathe switched on, and a *continuous* supply of coolant supplied to both the tool and the workpiece. The cross-slide's hand wheel is then turned very carefully clockwise to allow the tool's front cutting edge to contact and start cutting into the rotating workpiece's surface, the cross-slide's hand wheel must be turned slowly but positively (without hesitation) clockwise, in order to commence the parting-off operation. The tool's progression into the workpiece should be by approximately 0.025 mm (0.001 inch) per revolution of the workpiece. This amount of infeed given to the tool during the parting-off operation may need to be gradually reduced as the tool nears the workpiece's center line; this is due to the surface speed of the material actually passing the tool's front cutting edge now becoming progressively less as the tool approaches its final parting-off point.

On reaching this point, the workpiece will be seen to drop momentarily very slightly and stop revolving, because it does not now receive any power. The lathe is then immediately stopped and the work-piece is supported by the hand while its tailstock's support is being removed and unlocked followed by the tail-stock assembly being moved to its far right for clearance, to complete the operation.

To obtain further details of 'how to part off a component safely and successfully', the student is advised to refer to section 18 / 27, and section18/36 where a full description is given on virtually every aspect in the use of this critically important and delicate parting-off operation is explained.

HOW TO USE THIS BOOK.

The following list identifies to the reader where the Probe and Prompt system (including the Combination Probe' calculation aid), is used. It also lists examples of the slightly more complicated Wonky Gabled House method of non-right-angle (or unequal angle triangle) calculation, involving the use of a slightly more advanced practical trigonometry system needed to be adopted in order to complete the full series of the example calculations depicted in the fig 37 through to the fig. 37e example drawings, and in the fig. 37f and 37g drawings which complete the full explanation of this calculation process.

Chapter 1.
This chapter contains the introduction into the contents of the book.
It also gives an indication of the author's hopes in the success that can finally be achieved by those students who are prepared to make a thorough study of the contents of this book. It is hoped that the student will eventually realize (after putting this practical method of trigonometry calculation into practice) that he or she will then be able to solve virtually all the practical trigonometry problems encountered during daily work experience.
This section also includes a full description of how to use the newly designed 'Bend Development tables, followed by making useful comments describing how one resolves the inherent lack of information that is often left out of the preproduction and issued production engineering drawings.
Also described is the use of the Probe and Prompt system of triangle calculation, and how it is used to solve the general run-of-the-mill

trigonometry problems encountered on the work-shop floor and in the offices of the engineering industries. This includes details of the author's acquired experiences and engineering skills that have been gained through hands-on working in all aspects of the engineering industry.

Fig. 1

Includes a description of a very simple and 'typical' 3 - 4 - 5 right angle triangle that is used to illustrate the starting point of this book's new and innovative practical right-angle triangle

Probe and Prompt calculation system

Chapter 2.

Fig. 2 describes the design evolution of the probe. This is explored and explained in minute detail to allow its easy interpretation and understanding by the student. Its working functions are fully explained in figures, 2d, e, f, g, and h.

Also included is a *provisional* explanation in the use of the cut-and-fold Combination Probe used for employing the innovative Wonky Gabled House method of calculation used for non-right-angle triangles. The full explanation of this method of calculation is explained in the examples shown in Chapter 14, fig. 37, 37a, 37b, 37c, 37d, 37e, and 37g. Its working formulas are also shown in the drawing of the Combination Probe shown in fig. 37f.

Fig. 2, also illustrates a drawing that shows both of the probe's sides and its prompts (in mirror image). This section also contains a detailed explanation of the design and development of the vital and unique 90-degree corner tag, and its method of establishing a truly positive and recognizable corner in the particular triangle under calculation.

Chapter 3. An explanation of the prompts with examples.

This chapter also contains exceptions to my simplification rule, and gives a full explanation of the prompt's symbols and their use in helping the student to solve right angle triangle calculation problems. This chapter also explains the use of the 90-degree tag and the provisional use of the Combination Probe.

Chapter 4. Design of the Probe and Prompt

This chapter points out the main advantages enjoyed by the student who uses the Probe and Prompt calculation system as an aid to solving those so-called difficult formulas currently used in the engineering industry for the calculation of the right-angle triangle's length of sides and angles.

This chapter also describes and explains the function of the 90-degree tag and its particular use in the identification of the triangle's vital 90-degree corner at all its various orientations. This chapter also has an explanation of design of the probe's off-center prompts.

Fig. 4a, explains the use of the Inside Prompts, shown in the off-center block of The Probe and explains their prime use in obtaining the unknown length of sides in a right-angle triangle, when using either a basic, or a scientific calculator.

Chapter 5.

Describes the method used to solve three dimensional triangle problems, (often found when calculating the dimensions within 3D sketches and drawings), by using the Probe and its Prompts for the necessary guidance during the calculation. See fig. 19 for the calculation examples. These include a method of calculating the overall length of a tubular component, that includes two angles of bend, two tangent lengths, and two arc lengths, contained within this introduced tubular component, that crosses the box diagonally. This is shown in the plane of the paper (IPOP), in the lower drawing of fig. 19.

Chapter 6, using figures 7, 8, and 9, for reference.

This chapter describes and solves a relaxing and practical calculation problem, entitled 'To catch a fly'. (Incidentally, the student will find that this method of catching flies does actually work in practice.). This method of description allows the student to visualize how this particular sequence of calculations can be used with accuracy during the whole calculation.

This working example explains a method that can be used to catch the common house-fly while using the assistance of practical trigonometry, the Probe and Prompt, and the cupped sweeping hand of the 'predator'.

Chapter 7.

The identification of, and the method of calculation used, when dealing with shallow and steep angles; these variations in the angle and length of side of a triangle are often found (by the student) to be extremely difficult to positively identify when attempting to solve a shallow type of right-angle triangle problem. This explanation is aided by studying and using the examples shown in figures, 10 and 11, and the Expandable Probe, (Chapter 8, that gives a detailed description in figures 12, and 13, that provides an explanation of the scope, and mental simulation processes required during its use.

Chapter 8.

This chapter explains the advantages gained by the student, who uses decimal notation throughout the whole calculation (instead of using sexagesimal notation, of degrees, minutes, and seconds), that will later involve the necessity to convert the calculation back into decimal notation, in order to finally calculate the trigonometry problem). Figures 15 and 16 explain the use of The Probe and its Prompts when using practical trigonometry calculations to find the length of side and the unknown angle, in a right-angle triangle.

Chapter 9.

Figs. 17 and 17a, describe in a practical way the stepping stone approach method of triangle calculation, using these as calculation examples.

Chapter 10.

A method used to solve those awkward and difficult-to-resolve triangle problems often found in issued pre-production engineering drawings. This method can be used if one wishes to avoid the need to seek calculation advice or aid from colleagues or from the workshop supervision to get the problem resolved.

An unknown dimension, can be obtained by drawing an accurate 10-times magnification of a drawing, followed by carefully measuring the result.

A further similar explanation is given of how one draws a 100-times magnified drawing of an internal screw-cutting tool's tip (as shown in fig. 43a), to discover the exact dimension needing to be honed onto the tool's tip to produce the correct tool's tip radius.

Fig.18 is the drawing of a metal plate. Fig. 18a and 18b explain the method of calculation required to produce the co-ordinate method of work-piece dimensioning, this includes the use of an edge finder (or Wobbler) as an aid to positioning the machine tool in a similar manner to the method used in the precision engineering and manufacturing industries). In this example, this device is used to position, with precision, the machine's table to accurately locate the holes needing to be drilled in a metal plate with extreme accuracy. This method is normally used in the tool-room to obtain an accurate precision drilling of a metal plate. It is also used as a precision alternative to using just the hand and eye to position the drill relative to the work-piece.

Fig. 18c describes a method of scaling a drawing's dimensions to obtain (when in difficulty), an unknown dimension or angle. However, the scaling of a drawing, is a practice normally frowned upon throughout the whole engineering industry, mainly because of issued drawings not necessarily being true to scale; therefore, one should only carry out the practice of scaling a drawing to obtain a dimension when using one's own known accurately drawn sketch or drawing.

Also included in this text is a method that can be used to obtain an internal screw-cutting tool's tip radius, by scaling an accurately drawn *100*-times magnified drawing. The internal screw-cutting tool referred to is fully featured in drawing fig. 43a, which includes full details of the 24 mm × 3 mm pitch internally screw-cut nut to which this refers. By using this practical method alone, it is possible to accomplish an accurate result without the need to be familiar with or know how to use trigonometry, or for the student to possess the ability to use the Probe and Prompt for the task.

Fig. 18, d, e, and f, these drawings have been produced to demonstrate how this practical method can be used by an inspection department, as an aid to the accurate inspection of the positions and diameters of the drilled or reamed holes in the metal base-plate.

Fig, 18 is a drawing of a metal base-plate requiring three precision drilled holes equally spaced on a pitch circle diameter, (a phrase possibly unfamiliar to the trainee student, but often used by draftsmen in the industry); the script and drawings of 18a and b contain a full explanation of this method of dimensioning, and explain how one can convert a drawing-given pitch circle diameter (PCD), into a co-ordinate dimension for use on the X- and Y-axis slides of a machine tool.

An explanation of the precautions and checks one must make, before marking out this base-plate for its drilling operation.

Examples of issued drawing omissions are also described.

The setting up of a rotary table for drilling purposes.

Marking out the fig. 18 base-plate by using the assistance of the probe and its prompts, to convert the drawing-given pitch circle dimensions, into the required (linear) co-ordinate dimensions to position the drill using just the hand and the eye.

Also shown is the recommended method used for drilling precision holes in the metal plate when an accurate machine tool is available for use, thereby proving that the machine tool will perform the task with greater accuracy.

The regrinding of twist drills to cut accurate diameter holes.

Making one's own drill grinding gauge from two hexagonal nuts is featured on page 65.

Using a radial drill to drill the holes in the metal plate.

Figures.18a, d, e, and fSketched figures and calculations that describe the probe's use as an aid to the precision inspection of the completed metal plate by the Inspection department.

Chapter 11, (see also Fig 19).

View fig. 19, which shows an example of three-dimensional triangle calculations; the script includes instructions on how they are performed by using the aid of the probe and its prompts to obtain its box's correct angles and side lengths. It also describes the 'across the box' method of triangle calculation, and includes a quick method for use in obtaining the diagonal length of the box's corner to-corner dimension, by using the *box calculation method*. This calculation is also aided by using the assistance of the so-called stepping stone methodical approach to solving problems.

This particular box calculation method is also used to explain how, if two bends are now introduced into the calculation at points Y and X, they allow the student to gain considerable 'tube bending' calculation experience. Their method of calculation is shown in the lower fig. 19a drawing, where this tube bending method is being used to calculate small-diameter steel tube. This drawing gives a full explanation of how one calculates a component's overall length, its bend points, its arc lengths, and its tangent lengths, in order to provide the final accurate overall length dimension of the tube from point Z through to point W. The methods used in these final tube calculations can then be adapted for use on any available bending machine.

For the trainee 'tube bending engineer', these calculations and explanations regarding the methods used for practical tube bending, will be found very informative, and will help how one deals with the problems encountered if one is considering taking up tube bending as a career.

Chapter 12.

Fig. 20 shows an (as issued) basic drawing of a sheet-metal component.

A study of this drawing highlights the need for additional dimensions to be ascertained by the student. This can only be done by carrying out all the further calculations required before the manufacture of the component is at all possible. This problem is mainly due to a complete lack of additional or useful information being supplied on the drawing. Fig. 22. This sketch shows a freehand much enlarged representation of the fig. 20 drawing; it also confirms that we need to discover all the unknown dimensions, before an accurate production of the component can be undertaken.

Fig. 23 is a calculated and drawn modification to the fig. 20 drawing. It shows the new relative position of the unknown but urgently required dimension *x*.

Fig. 24 is a very much enlarged sketch of fig. 20 showing how to use practical triangulation calculations, and the Probe and Prompt system to obtain the very elusive dimension *x*.

Fig. 25 shows the full bend calculations required to allow the accurate production of the simple metal bracket shown in figures, 20, 22, 23, and 24. These calculations are needed to obtain all of the necessary dimensions and angles required to allow the precision production of either one, or a complete batch of the required sheet-metal components.

Fig. 26 bend option one. This drawing shows a side view of a Promecam, hydraulic, sheet metal (air) bending machine. This accurately sketched simulation of the setup of the machine shows the bending process needed to bend the sheet-metal bracket with the ultimate precision; this component is also shown in figures. 20, 22, 23, 24, 25, (and fig. 28, the version with stretched, dimensions).

Fig. 27 shows the less-accurate bend option two, describing the folding machine setup required for bending an identical metal bracket (as shown in figures 20, 22, 23, 24, and 25) but with less precision. This option, however, will require the marking out of the material in the flat before bending. (This method of marking out the workpiece is prohibited if the component is to be used in the aircraft industry, the Royal Air Force, or if manufactured for use by the Air Registration Board Authority, (including the Inspection department of the manufacturing company), who cannot accept by law, any scratches or marking out to be present on the surface of the component, for reasons of air safety.

Fig. 28 shows the special 'stretched bend' allowances being used on the fig. 25 metal bracket. These dimensions are used to manufacture this component, and have been designed to take into account, the amount of stretch that occurs in the component during bending.

Fig. 29 shows an example of a sheet metal 30-degree angle of bend, with the calculations required to form its bend accurately. The position of the critical vertex point is also shown to aid in the bend calculations.

Fig. 30 shows an example of a sheet metal 90-degree angle of bend and the calculations required to form its bend accurately. The position of the critical vertex point is also shown to aid in its bend calculations.

Fig. 31 shows an example of a sheet-metal component with an angle of bend of 150 degrees and the calculations required to accurately form its bend. The position of its critical vertex point is also shown to aid in its bend calculations.

Fig. 32, shows a newly computed table of standard metric bend allowances; these are used for calculating the center-line arc' of the general run-of-the-mill sheet-metal components requiring just a normal standard of accuracy, and includes instructions for its use.

Fig.32a, shows a newly computed table of stretched, metric bend allowances. These are used to establish the precise overall length of material required to produce a stretched precision bend in the sheet-metal bracket as detailed in fig. 28, and includes instructions for its use.

The text also gives the author's explanation, for clarity reasons, for pointing out the advantages that could be gained throughout the whole of the sheet metal and tube bending industry if the name *angle of bend* were now used world-wide, in preference to using the much abused confusing name *bend angle* currently used, and the author's

490

wish to clarify the difference between the meaning of the two names, in this age-old engineering anomaly.

Chapter 13.

Fig. 33 shows a very basic as-issued production drawing depicting the dimensions of a coned sheet-metal component to be manufactured. The script includes a description of two of the normally accepted production methods used to produce this cone-shaped component.

Guidance notes and extra sketches have been produced to explain the extra calculations required and the methods used to discover these unknown dimensions and angles to enable the precision manufacture of this cone-shaped sheet-metal component.

Fig. 34 shows a three-dimensional view of the sheet-metal cone; (this has been drawn by the author to enlighten the student to the *hidden* dimensional contents of the issued basic drawing featured in (fig. 33), in order to assist in the clarification of the actual shape and the final dimensions required for its manufacture.

Figs. 35, 35a, 35b, and 35c, show the calculations required to actually make the development blank for the cone shown in figures 33, 34, and 36.

Fig. 36, 36a, and 36b, show the development of the 'in the flat' blank' that is specifically required for the manufacture of the cone shown in figures 33, 34, and 35.

Chapter 14.

Figs 37, 37a, b, c, d, and e, show all the steps needed to calculate the unequal-angle triangle, when using the Wonky Gabled House method of triangle calculation.

Fig. 37g illustrates the gable end of the roof of a domestic terraced house; this shape has been chosen to be used as a visual aid to the whole calculation process.

The aim of this sketch is to familiarize the student with using this particular shape to perform the Wonky Gabled House method of *unequal-angle* or *non-right-angle* triangle calculation. It includes the formulas used for this method of calculation. The explanations also include the basic Probe and 'Prompt formulas that are needed during the latter stages of this unequal-angle triangle calculation to allow the student to quickly obtain its vertical height.

It will be seen how in fig. 37, 37a, b, c, d, e, (in fig. 37f, the cut-out aid), and in 37g, (the drawn Wonky Gabled House example), they describe this new and unique triangle calculation method, used for the accurate calculation of the unknown angles contained within the (normally difficult to calculate) unequal-angle triangle or the non right-angle triangle, particularly when the shape of the calculated example being worked on does not contain a convenient 90-degree corner within its shape.

The use of this unequal-angled triangular shape has been designed and developed purely for the purpose of explaining (in the simplest possible way to the less able), this calculation's method of working. This particular figure can be referred to as being either a non-right-angle triangle, or, alternatively, an unequal-angled triangle. These names have been used purely for convenience purposes and for identification purposes in this book. These easily remembered names have been chosen to avoid the student needing to know the names of the host of conventionally named triangular figures that actually exist in the universe.

Fig. 37f is a cutout version of the Combination Probe. It is intended by the author that this aid (or page) (is duplicated and positioned at the back of the book) so that it can be removed from the book and its Combination Probe section cut out for use in its various folded forms as an aid to solving all triangle calculation problems.

Chapter 15.

Fig. 38 utilizes the Probe and Prompt system as an aid to obtaining sine bar setting angles while using slip gauges in a dimensioned pile. The use of this method allows extremely accurate angular settings to be achieved in the production workpiece, and to the set-up of the machine tool being used in the component's production.

Chapter 16.

The sine vice work-holding device; this chapter details its use in the engineering industry for the accurate setting up of the work-piece, relative to the machine's cutting tool or grinding wheel.

Chapter 17.

Explaining the lathe operation called precision taper turning. (See also fig. 45, which shows the three methods of taper turning on the center lathe.).

The figures, 39, 40, and 41d, e, f, g, h, and j calculated illustrations, give an in-depth explanation of a precision taper-turning operation. This operation is performed on a center-lathe equipped with a taper-turning attachment. (However, some designs of lathe do not have this facility, so in this case, the less accurate 'angling of the compound slide' method, or the offset tail-stock method will have to be used for taper turning operations.) See fig. 45 for the three methods of taper turning.

The sketched calculated triangular figures, included in fig. 41, explain graphically the method used to convert the given conventional 'taper per foot' angle dimensions, (as are being given and used on most issued drawings that are universally produced in the engineering industry), into their metric decimal equivalent, (more suitable) dimensions', for setting-up a lathe or a precision machine, being used for general engineering and inspection purposes.

This chapter provides a general description of how one can use a handheld 25 mm diameter wooden shaft, together with a selected strip of emery cloth held in its slotted end that embodies a transverse longitudinal saw-cut slot, in order to remove a minute quantity of material from a tapered or parallel bore in order to increase its diameter by the minute amount of (say) 0.0127 mm (0.0005 inch), this dimension being less than the lathe is capable of accurately removing while using just its cutting tool to obtain the workpiece's required precise diameter.

Chapter 18, including further sections.
Precision turning, parting off, and screw-cutting on the center lathe. See figures 42, 43, 43a, 44, 45, 46, 47, 48, and 49, (particularly fig. 47 and 47a), that illustrate the author's selection of the most useful lathe tools.

Included in this chapter is an introduction into the intricacies of lathe work. It contains a host of the necessary preparations required by the machinist prior to using the center lathe for precision turning, parting off, boring, facing off, and screw-cutting operations. The text explains the practical way of off-hand grinding being used to sharpen lathe tooling, and describes a practical method used to obtain an accurate tail-stock's parallelism. Also included is a practical method of using in-situ maintenance of a lathe's chuck if it is found to have 'bell mouthing' wear problems, with helpful hints and tips on the precision grinding of twist drills, including descriptions of the traditionally used straight plunge method of external screw-cutting, together with the more popular alternative method of angle approach external screw-cutting.

Section 18/1.
An introduction into the methods used to solve many of the problems found in general lathe work, including the vital importance of the student *getting to know* his lathe.

Section 18/2. Work-holding lathe devices explained.

Section 18/3. Removing and accurately replacing the lathe bed's gap.

Section 18/4. Setting up a work-piece accurately, on the lathe's face-plate.

Section 18/5. The main uses of the lathe's collet chuck.

Section 18/6. The correct way of using the lathe's 'four-jaw' chuck, to hold materials securely for turning.

Section 18/6a. How to make a home-made internal collet chuck on the lathe, for either single or batch production.

Section 18/7. Using a home-made and helpful piece of tooling called a 'sticky pin', used for centralizing the work-piece in the four-jaw chuck, on the face-plate, and other applications on the milling machine.

Section 18/8. The correct way of securely holding irregular shaped components in the lathe's four-jaw chuck. The best method to use when machining 'difficult to turn' flame cut pipe flanges on the center-lathe. The dangers of using a lathe tool with a too-large radius on a poorly supported workpiece.

Section 18/9. The correct way of using the three-jaw chuck, in order to minimize the possibility of causing damage to its jaws.

A method used to remove persistent and unsightly chatter marks from the surface of large-diameter workpieces that are found to possess poorly machined external radii.

Section 18/10, see fig. 48, the practical use of the fixed steady, and notes explaining the pipe center, the greased dead center, and the 'travelling steady'.

Section 18/10a. Helpful notes on lathe drilling and reaming operations when using reamers in a tail-stock supported Jacobs chuck.

Section 18/11. Using the chuck's alternative set of soft jaws, (which can be supplied by the lathe manufacturer on request); these jaws are used in the three-jaw chuck, (as extra equipment).

Section 18/12. The correct way to remove and replace the jaws of the three-jaw lathe chuck.

Section 18/13,Very important safety checks that are often found necessary when using the center-lathe and its equipment to carry out machining operations.

Section 18/14 Useful notes on center lathe turning, with particular regard to repair work that often needs to be undertaken solely by the machinist, including important safety precautions that must be rigorously complied with.

Section 18/15. Checking the accuracy of the lathe's parallelism relative to its tail-stock's alignment, also checking the cleanliness of all the chuck's vital location faces.

Section 18/16. The rectification of a condition known as bell mouthing, previously suffered by the jaws of a poorly used three-jaw chuck.

Section 18/17. Solving the problems encountered when using the center lathe to turn very small diameters. This advice relates to both fully supported and unsupported work-pieces.

Section 18/18. The methods used to establish an off-set tail-stock alignment to enable tapered work to be produced.

Section 18/19. Turning a work-piece held between centers (fig. 45), also the proper use of a test piece to check the lathe for an accurately set tailstock, (called the lathe's parallelism test).

Section 18/20. Rechecking the tailstock's alignment for the machining of long work-pieces held between centers, and the use of a test piece to check its accurate parallelism. See fig. 47 for the author's basic selection of the most useful and suitable lathe tools needed to perform the majority of the turning operations required on the center lathe.

Section 18/21. The inspection, sharpening, and honing of lathe turning tools, including an explanation of how one obtains an accurate point radius on an external screw-cutting tool.

Section 18/22. Information regarding the proper way to use a test piece (as detailed in Section 18/19), to obtain true parallelism of the lathe's tailstock.

Section 18/23. The selection of tools, in order to decide on the most appropriate lathe tool to use for the requirements of the work-piece, (see the fig. 47 selection of lathe tools). Explaining the lathe settings required to screw-cut a 3 mm pitch male metric coarse thread on the center lathe. See also the fig. 42 drawing of external angle approach screw-cutting on the center lathe.

Section 18/24. General notes developed to assist the machinist when he or she is forced to replace a damaged screw-cutting tool midway through a screw-cutting procedure.

Section 18/24a. Method used to accurately replace a damaged screw-cutting tool midway through an angle approach screw-cutting operation, with its lead-screw permanently engaged.

Section 18/24b. Method used to accurately replace a damaged screw-cutting tool while using the straight plunge method of screw-cutting with a disengaged lead-screw

Section 18/24c. Method used to accurately replace a damaged screw-cutting tool while using the straight plunge method of screw-cutting and a permanently engaged lead-screw, where the machinist has failed to note down the depth of thread so far reached by the screw-cutting tool before the tool was removed for inspection.

Section 18/25. Method used to accurately replace a screw-cutting tool partway through an angled approach screw-cutting procedure, while using a lathe with a permanently engaged lead-screw, where the machinist has failed to note down the depth of cut so far reached, before removing the damaged screw-cutting tool for replacement

Section 18/25a. Notes on replacing a damaged screw-cutting tool, midway through an angled approach screw-cutting procedure, while using a lathe with a disengaged lead – screw.

Section 18/25b. Replacing a damaged screw-cutting tool midway through an angled approach screw-cutting procedure when using a permanently engaged lead-screw, where the machinist has failed to note down the depth of thread so far reached by the tool before its removal.

Section 18/26. Important notes relating to the correct selection of work-speed for screw-cutting a thread, particularly when using steel components, including the method used to make a trial cut on a test piece to establish the correct pitch has been obtained.

Section 18/27. Special techniques required when using parting-off tools, and also when turning small diameters. How the problems found when using these particular tools are best dealt with, including general information regarding the safety precautions specifically required when machining long fragile work-pieces, and designed components that must specifically not possess a center-drilled hole in their outermost end.

Section 18/28. Instructions to assist with the setting-up problems encountered while preparing and using a lathe for screw-cutting a 24 mm × 3 mm pitch metric male coarse thread, while referring to figures 42, and 43, drawings that contain the required dimensions and depth of thread information

Section 18/29. Screw-cutting on the center lathe continued. Notes on the off-hand grinding of lathe tools. How one finds the helix angle of the thread being cut, in order to assist in the accurate grinding of the clearance angles required on the tool.

Section 18/30. The advantages of using the angle approach method of screw-cutting as shown in fig. 42. How one obtains the 'Straight Plunge' depth of thread, by using a thread constant, and the aid of the Probe and Prompt system. Also explained is the four main methods generally used to screw-cut a component on the center-lathe.

Section 18/31. The off-hand grinding of lathe tools, including a full and in-depth explanation of the whole tool grinding process. See also Section 18/36 for extensive instructions on the safe use of parting tools). Also a full explanation of the correct way to prepare the grinding machine and its grinding wheels to enable the machine to accurately perform the tool grinding and sharpening process.

Section 18/32. Explaining the possibility (when in an emergency situation), of using an imperial 55-degree included angle screw-cutting tool to cut a metric or an American 60-degree thread in the work-piece, should the machinist find that he is not in possession of the correct tool for the job.

Section 18/33. Describing the disadvantages found when using the straight plunge method of external screw-cutting.

Section 18/34. An alternative method that can be used to obtain a *not shown on the drawing* depth of thread dimension, by the use of calculations, the Probe and Prompt, and a thread constant, in order to discover the very important, useful (and required) *depth of thread*' information.

A selection of other calculated 'depth of thread' constants that can also be used for obtaining other unknown depths of thread, by using their pitch, a sketched right angle triangle, and a series of further triangle calculations.

Section 18/35 A full explanation of the fig. 43 drawing that shows a cross-sectioned and dimensioned view of a typical metric nut that includes its bolt, and it's very important tool 'clearance areas' that must be maintained. This example drawing details a 24 mm × 3 mm coarse pitch metric thread in minute dimensioned detail. Its lower 10 × magnified cross-sectioned drawing explains the calculations and the formulas required to obtain its depth of thread, and its tool's point radius dimension by using these formulas.

Also included is an explanation of the fig. 42 drawing that shows the angle approach setup for external screw-cutting, complete with the advice and calculations required to obtain its important depth of thread by using the Probe and Prompt, two sketched right-angle triangles, and a thread constant. This section also contains advice and notes on how to deal with engineering inspection problems.

Section 18/36. Additional checks that are often required when preparing a center-lathe for a screw-cutting operation; also, the recommended precautions the machinist must take and the methods used before using a *parting-off tool* in order to prevent accidents occurring. (Section 18/27) also contains special techniques required when using parting-off tools.

Section 18/37. General advice and a warning to machinists regarding the possible dangers one can be exposed to when operating a badly worn lathe.

Section 18/38. External straight plunge screw-cutting, using the quick tool withdrawal method, with the workpiece still in rotating mode, and with the thread cutting knock down lever being used to isolate the tool from the lead screw.

Section 18/39. A general description of the external angle approach screw-cutting method that uses the quick tool withdrawal, while the work-piece is still in rotating mode, and the thread cutting knockdown lever being used to isolate the tool from the lead-screw.

Section 18/40. The normally used practice of marking the cross-slide's scale with pencil to indicate the depth of thread the tool has so far reached, particularly when using external straight plunge screw-cutting.

The importance of marking the chuck's outer periphery to indicate to the machinist the exact point where the snatched method of tool withdrawal is to be used when using the disengaged lead screw technique, and to also indicate when the lathe is to be instantly stopped, when using the permanently engaged lead screw method of screw-cutting. This chuck marking technique will be found advantageous by all machinists when using either method of screw-cutting.

Section 18/41. *Option one.* The external straight plunge screw-cutting of a thread, leaving the tool engaged with the lead screw, using the quick tool withdrawal, followed by reversing the motor to take the tool back to its start point to prepare for the next cut to be taken.

Section 18/42. *Option two.* The external straight plunge screw-cutting of a thread, by disengaging the lead screw, using the quick tool withdrawal, returning the tool manually to its start position, using the apron's drop-in rotating screw-cutting dial's indication to choose the correct place on its dial to take the next cut.

Section 18/43. *Option three.* The external angle approach screw-cutting of a thread, leaving the lead screw permanently engaged, using the quick tool withdrawal, followed by reversing the motor to return the tool to its start position ready for the next cut to be taken.

Section 18/44 *Option four.* The external angle approach screw-cutting of a thread, leaving the work-piece rotating, disengaging the lead screw, using the quick tool withdrawal, returning the tool back manually to its start position ready for the next cut to be taken.

Chapter 19. The internal screw-cutting of a metric thread.

Section 19/1.

This chapter explains in detail the straight plunge method of internal screw-cutting, as shown in the fig. 43a and fig. 44 drawings. In this case, the center lathe is being used to machine an internal 24 mm × 3 mm pitch metric coarse thread in the component. This particular component contains a design of nut that is rather more difficult to machine than a straight through nut, and is often referred to (in the engineering industry), as being a 'blind nut', due to it having the absolute minimum of tool clearance available for the tool beyond the extremities of its screw-cut thread. This component is designed to fit the previously screw-cut and similarly pitched male metric screw-cut thread produced from the fig. 42 drawing.

This chapter explains in detail the initial preparation of the work-piece material followed by it being internally screw-cut to possess a 24 × 3 mm pitch metric coarse thread.

The following guidance notes explain the methods used, and the sequences of operations required that will enable an engineering trainee or student to screw-cut an accurate internal metric coarse thread into this component's internal bore.

Section 19/2.

The fig. 44 drawing shows in detail the set-up required to internally screw-cut this component. Its required depth of thread can be calculated by using a thread constant, two sketched right-angle triangles, and the Probe and Prompt method of triangle calculation, as described in the lower part of the drawing. This depth of thread dimension can also be obtained by calculation using the inscribed letters and figures shown in the 10 times enlarged formula, sketched in the lower part of the fig. 43a drawing. These thread formulas can also be used to obtain the vital depth of thread required, by using this alternative method of calculation.

Section 19/3.

Using a test piece, to check that the lathe is cutting the correct pitch of thread required. Setting up the material in the lathe, including a note regarding the importance of taking a light skim, along the outside surface of the material, to allow its now concentric diameter to be used for the workpiece's accurate replacement.

How one drills and bores the correct tapping diameter in the workpiece

Section 19/4.

Selecting a suitable internal under-cut tool for the job in hand, followed by setting up the lathe to allow this tool to produce the correctly sized under-cut at the extreme end of the tapping size bore of the component

Explaining an alternative calculation method, that can be used to obtain the correct tapping diameter for this internally threaded component.

Section 19/5.

Calculating the important depth of thread dimension required prior to screw-cutting the thread in the nut.

How one can ensure that the screw-cut thread will fit the checking thread plug gauge. A warning to the student of the danger of allowing the screw-cutting tool to overrun the thread's under-cut area, and colliding with the end of the bored hole.

A helpful note explaining the tool clearance problems encountered when screw-cutting internal threads.

The marking of the top of the tool's extended shank to indicate to the operator the exact position of the hidden under-cut, in order to assist the machinist in establishing its exact position when the tool is out of sight during the internal screw-cutting operation.

Section 19/6.

A helpful note regarding the clearance problems encountered when internally screw-cutting very small diameter threads

The dangers one can experience during internal screw-cutting, if the tool is allowed to approach too closely to the nut's blind end face.

Section 19/7.

The importance of selecting one specific and chosen revolution of the workpiece, at the end of each screw-cutting passing cut to allow the student to instantly stop the lathe's motor and work-piece within the very short clearance area provided by the narrow 4 mm width of the under-cut.

Section 19/8.

Part two.

Further details explaining the full setup required when internally screw-cutting a thread.

Choosing the correct screw-cutting tool, to screw-cut the internal thread in the component.

Explaining the accuracy required when setting up the tool in the tool-post, also the advantage of machining a chamfer on the inner end of the thread material in the work-piece's bore, to provide a slightly wider under-cut area to allow more tool clearance.

Explaining why it is necessary to eliminate all the back-lash from both of the lathe's slides lead-screws prior to screw-cutting the thread, and a thorough check being made of the tool's hand-honed cutting radius to establish that it is absolutely correct.

Section 19/9.

Performing a dummy run of the whole screw-cutting procedure, to enable the machinist to become fully acquainted with the relative positions of the warning marks on the tool's shank that indicate the exact point where the tool must be extracted from the thread, followed by the motor and work-piece being instantly dead stopped at this point.

An explanation of the method used when checking the thread for correct fit by using the thread plug gauge.

Helpful notes on performing the internal screw-cutting operation, followed by the selection and fitting of the parting-off tool into its tool-post, with instructions on the correct way it should be used.

*   *   *

There follows the author's design of a miniature trigonometry aid, with instructions on how it can be fitted to a wrist-watch strap for unobtrusive use in the work-shop or office.

There follows five full-size trigonometry aids designed to be affixed to either the inside of the lid or the underside of a calculator (by using double-sided adhesive tape), in order to enable the aid be used for basic triangle calculations.

*　　*　　*

A complete list of all the fig. drawings is now shown below.

The following list of drawings explain the development and the method of use of the Probe and Prompt triangle calculation system including the Combination Probe, with its unique system of practical unequal-angle or non-right-angle triangle method of triangle calculation.

Fig. 1 A typical 3 - 4 - 5 right-angled triangle, showing its designated sides *a*, *b*, and *c*, its 90-degree tag, and its The Angle (its sharp end) indicated. Also shown is the newly named Other Angle, placed in its correct position.

Fig. 2 Both sides of the probe are being shown in order to display the clarity of its prompts.

Fig. 2d, 2e, 2f, 2g, and 2h explain the design evolution of the probe, from its original concept through to its final developed shape.

Fig. 3 This author-developed formula is now superseded by the new development of the Probe and Prompt calculation system, which has been designed to discover the component's vertex point, and its so far unknown angle of bend, which in the original drawing's case was only given by *x*, *y*, and R to calculate with. This old calculation method has now been completely superseded by using the new Probe and Prompt triangle calculation system, which now provides all six of the sought-after unknowns, and now requires much less mental dexterity on the part of the student when used. This new triangulation method of calculating the position of point *x* is now being shown in drawings fig. 3a, and fig. 24.

Fig. 3a. This drawing contains the triangulation calculations required to find the triangle's dimensions which are necessary to obtain the component's angle of bend, its tangent point, (x dimension), and to calculate the bend's center line arc, required for the calculation of the sheet-metal bracket detailed in fig. 23. (Note that this drawing is duplicated for further use in the later fig. 24.)

Fig. 4 An explanation of the workings of the probe's prompts, in detail

Fig. 4a shows us that the triangle's The Angle can be found, in this example, by entering the sequence 3 divided by 4 =, giving us (0.75 in the display) followed by using the INV TAN key sequence to give us the answer 36.86989765 degrees.

The length of the triangle's side *a* can be found by using the prompt, $\angle b$, $a = b \div \text{SIN}\angle$; we do this by entering the sequence 36.86989765, SIN,(display shows 0.6) XM, followed by entering 3 ÷ RM =, giving us the answer 5.

Fig. 4b Choosing an inside prompt from the off-center area, to find the length of *a* when The Angle is unknown.

Fig. 5 shows eight rotated right angle triangles together with their 90-degree identifying tags and their designated sides shown as *a*, *b*, and *c*.

Fig. 6 shows eight tumbling right angle triangles together with their identifying 90-degree tags, and their designated sides *a*, *b*, and *c*. These views of the tag's positions still identify their true 90-degree corner although they are being viewed obliquely.

Fig. 6a shows eight tumbling right-angled triangles *without* their identifying 90-degree tags or their designated sides. These are oriented identically to fig. 6, and are also drawn obliquely, which now shows them to be unrecognizable for calculations because of the absence of their tags and their designated sides.

Fig. 6b shows a duplicated version of the fig. 37f Combination Probe to be used for reference purposes while its original fig. 37f is being used in a cut out and folded form for practical calculation purposes.

Fig. 6c An enlarged version of the Combination Probe (shown in fig. 6b) used for general reference purposes that includes its cutting and folding instructions that will be required before using the fig. 37f cut-out and removable Combination Probe aid.

Fig. 7 Using the Probe and Prompt aid, together with using its practical trigonometry calculations, to catch a fly.

Fig. 8 shows a side view of the initial calculated position of the fly before its takeoff, in the tabletop capture scene.

Fig. 9 Calculations showing the ideal angle of palm tilt of the predator's hand.

Fig. 10 showing the probes' extremes of use. In this case, it is showing The Angle of the triangle in its so-called shallow position of 1 degree.

Fig. 11 showing the probes' extremes of use. In this case, it is showing The Angle of the triangle in its so-called steep position, of 89 degrees.

Figures 12 and 13 These figures explain the full scope of the 'to be imagined' shape of the Expandable Probe when it is being used to calculate problems that extend from the extremely shallow to the extremely steep angles, while being used on the paper and rotated at will into both horizontal or vertical positions in order to clarify the actual position required of its Tag, its The Angle, and its designated sides *a*, *b*, and *c*.

Fig. 14 Explaining the method used to convert a triangle's degrees and minutes into decimal degrees, by using the Probe and Prompt system.

Fig. 15 The calculations required to discover an unknown length of side, and an unknown angle in a right-angle triangle.

Fig. 16 Using the The Angle part of the Probe and Prompt aid during a calculation.

Fig. 17 The pyramids basic dimensions prior to using the stepping stone approach method for its full calculation.

Fig. 17a, The pyramids dimensions after using the stepping stone approach method, by using the basic figures shown in fig. 17, thereby proving that the so far unknown dimensions can be discovered by using this particular assistance during the calculation.

Fig. 18 A typical issued drawing of a metal base-plate showing the bare minimum of information supplied for the components production. See figures 18a, b, c, d, e, and f, for the calculations that should be undertaken before the component can be accurately manufactured.

Figures 18a and 18b The extra calculations required to drill 3 – 5.0 mm diameter holes equally-spaced on a 52 mm PCD pitch circle diameter), when a rotary table is not available or the work-piece is much too cumbersome or is too long. In this case, the process is called ordinate drilling, and will require the component to be very accurately marked out by using both its end and edge datum faces.

Fig. 18c A ten-times enlargement of fig. 18b, (this drawing having been rotated anti-clockwise by 90 degrees because of the lack of space on the A4 page), to show how (when in an awkward situation) the student can use an accurate 10 × scaled drawing to obtain a reasonably accurate dimension when the student is not sufficiently experienced in the use of practical trigonometry calculations, to obtain the very accurate dimensions required.

Fig. 18d, also showing 18e, and 18f, drawings that contain triangular dimensioned sketches, to allow the required dimensions to be obtained by using practical trigonometry, thereby obtaining the exact dimensions of the holes in the fig. 18 metal plate, designed for the use of the inspection department's knife edge vernier to check the component's dimensions.

Fig. 19 explains the use of a three-dimensioned box, to aid the Probe and Prompt system in the three-dimensioned trigonometry calculations required to discover the unknown angles and lengths contained within the box, and later when extended beyond the box. These include how to obtain the total center-line length of a bent tubular component that starts at point Z through point Y and point X, and continues on to reach its end at point W. The final total calculated length of the bent tube, after calculation, will include both of the two bends' center-line arcs, minus the two bends' four tangent lengths.

The required dimensions of the tube have been shown in the lower drawing, (this having been rotated clockwise when viewed from the arrow at point Z), to allow the student to experience how it is possible to produce the accurate bending machine settings, required for a precision hydraulic bending machine. Also included is an alternative calculation method that can be used to quickly obtain the exact straight center-line diagonal length across the box from point Y to point X.

Fig. 20 shows a basic sheet metal drawing, identical to that issued to the sheet metal shop for this particular component's manufacture. The student should note the missing but required dimensions are as follows: (1) Angle of bend, (2) Exact length of material required, (3) Folding machine bend point, (4) The hydraulic bending machine's bend point, (5) The marking-out position of the hole in flat material, (6) Bend vertex points dimension. (These dimensions will need to be found by the machinist using the calculations in figures 22, 23, 24, and 25, and if required in the stretched version fig. 28, before the work can proceed). The full sequence of calculations is shown in figures 3a, 22, 23, 24, 25, and in fig. 28, if it is required to manufacture the stretched component.

Fig. 21 Not used.

Fig. 22 A freehand sketch of drawing fig. 20, showing the calculations required to obtain six further important dimensions, which are not shown on the issued drawing

Fig. 23 Sheet-metal bracket as shown in fig. 20, but now indicating the previously missing important dimension (x), while still now requiring the angle of bend, the length of material, the folding machine bend point, and the marking-out dimension for the drilled hole in the flat on the material.

Fig. 24 The series of calculations required to find the angle of bend and the tangent point (x) dimension of the sheet-metal bracket shown in figures 20, 22, 23, 24, 25, and 28, by using the 'Probe and Prompt' system for the calculations.

<<Fig. 25>> The full bend calculations required for the metal bracket shown in fig. 20, 22, 23, 24, and 25, which finally provides all the dimensions required for the component's manufacture.

<<Fig. 26>> The Promecam hydraulic (air) bending machine, being used to bend the metal bracket shown in fig. 25 and if necessary fig. 28.

<<Fig. 27>> The folding machine, set up for bending the metal bracket shown in fig.25. This machine is used when extreme accuracy is *not* the main criteria, and where actually marking out the surface of the material would not be acceptable to the Air Registration Board's safety officer, but for other uses would be acceptable.

<<Fig. 28>> The calculation of the bend using special Stretched Bend allowances required for the original sheet-metal bracket shown in figures 20, 22, 23, 24, and in its normal un-stretched bend form in fig. 25. See also the table of Stretched Bend allowances in fig. 32a.

<<Fig. 29>> An example of a 30-degree angle of bend showing the calculations required for the accurate manufacture of the component.

<<Fig. 30>> An example of a 90-degree angle of bend showing the calculations required for the accurate manufacture of this component.

<<Fig. 31>> An example of a 150-degree angle of bend, showing the calculations required for the accurate manufacture of this component.

<<Fig. 32>> A table of normal metric bend allowances for 1 degree of bend that possesses inside bend radii ranging from 0 to 7 mm.

<<Fig. 32a>> A table of stretched metric bend allowances for 1 degree of bend that possesses an inside bend radius ranging from 0 to 7 mm.

<<Fig. 33>>. The drawing of a sheet-metal cone-shaped component as issued by the drawing office to the workshop for its manufacture. It will be seen that it lacks most of the dimensions required by the sheet metal fitter to manufacture this component.

<<Fig. 34>>. A produced three-dimensional drawing of the cone shown in fig. 33, showing more of the dimensions required to manufacture the component. This view is used as an introduction into explaining the total number of calculations that will be required to allow the machinist to complete the full manufacture of this component, (these are shown in figures. 35 and 36).

<<Fig. 35>> Further calculations required prior to the manufacture the development blank in figures 36, 36a, and 36b, required for manufacturing the coned component in figures, 33, 34, and finally in fig. 36.

<<Fig. 36a, and b>>, The calculation and development, of the 'in the flat' blank, required for the manufacture of the coned component shown in figures. 33, 34, and 35.

<<Fig. 37>>
The fig. 37 drawing shows, in steps 1, 2, 3, 4, and 5 that contain figures 37a, 37b, 37c, 37d, and 37e, explaining in detail the sequence of calculations required to complete the full calculation of the un-equal or non-right-angle triangle when using the Wonky Gabled House method of triangle calculation.

<<Fig. 37f>>, The cut and fold instructions required prior to using the cut-out-and-fold Combination Probe aid, which is used for assisting in the calculation of both right-angle triangles and non-right-angle triangles.

<<Fig. 37g>> Notes that refer to the 'Wonky Gable House' method of triangle calculation. Naming the individual parts of the gable end of the example house roof, and explaining the correct method to use when naming and dimensioning the individual sides of the Wonky Gabled House, unequal-angle triangle method used to solve the problem.

<<Fig. 38>> The sine bar shown positioned to allow the component's accurate angular setting of 36 degrees to be achieved, by using a calculated slip gauge pile to obtain the exact angle.

<<Fig. 39>> Item 1. This component is to be internally taper turned so that it is an exact fit with the 3/8 of an inch taper per foot external taper that exists on item 2, fig. 40. The machining operation must be very accurately performed by the machinist in order to obtain, when the two components are finally assembled together, a precise gap of 15 mm + or − 0.05 mm to exist between the end face of item 1 and the existing flange's datum face of item 2.

<<Fig. 40>> This is a drawing of the supplied pattern component (item 2), that possesses a 3/8 of an inch taper per foot (male taper). This is to be used as a gauge to check the progress of the operation to accurately machine the internal taper in fig. 39 (item 1). When this component is finally assembled into item 1, a gap of 15 mm plus or minus 0.05 mm must be maintained between the datum flange of item 2 and the end face of item 1 as shown in fig. 41.

<<Fig. 41>> showing the trigonometry calculations needed to accurately machine the internal taper of 3/8 of an inch taper per foot in item 1 (fig. 39), to exactly fit item 2, shown in fig. 40.

The following drawings apply specifically to the *external* screw-cutting procedure required, and to the screw-cutting thread's dimensions shown in fig. 43, as explained in chapter 18.
This is followed by the *internal* screw-cutting procedure required for the manufacture of the nut, and complying with its internal thread dimensions that are fully explained in drawing fig. 43a, chapter 19.

<<Fig. 42>> shows a drawing of the set-up that is used for the angled approach screw-cutting of a 24 mm diameter × 3 mm coarse pitch external male thread while using a center-lathe for the screw-cutting operation.

Fig. 42a (This sketch is included at the bottom of the fig. 42 drawing). It shows the calculations required when using a sketched triangle to obtain the correct angle approach depth of thread.

Fig. 42b (This sketch is included at the top of the fig. 42 drawing.). It shows the method of using a sketched triangle, the constant number 31.526618, the thread's pitch, and the Probe and Prompt, to obtain the correct straight plunge depth of thread for the bolt's externally cut thread.

Fig. 43 shows a 9 × enlarged cross-sectional drawing of a 24 mm × 3 mm pitch coarse male metric thread together with its female threaded nut having a similar pitch of 3 mm; this drawing shows the theoretical and the practical dimensions required for accurately completing this screw-cutting operation. It's lower drawing (at 10 times magnification) contains all the necessary formulas used to obtain the male thread's depth of thread, which is achieved by complying with the formulas given, and by the use of further calculations. (The lower drawing's formulas can also be used to obtain the exact depth of thread for all of the coarse series of metric threaded bolts.).

Fig. 43a shows a cross-sectioned drawing (of approximately 9 × magnification) that shows a typical 24 mm × 3 mm pitch metric threaded nut. This drawing contains the very important dimensions of the tool's tip radius that needs to be very carefully hand honed onto the cutting tip of the screw-cutting tool. Shown also is the angle of the tool's approach into the workpiece while it is being used for either the straight plunge or the angle approach method of internal screw-cutting, being indicated by arrows. The lower drawing, (at10 × magnification), contains the necessary formulas required to accurately obtain the nut's very important depth of thread, by complying with the formulas given, coupled with using further calculations to obtain the depth of thread required for all the metric coarse series of threaded nuts.

Fig. 44 shows the center lathe with its screw-cutting tooling setup for internally screw-cutting a metric coarse 24 mm × 3 mm pitch nut, in a component that contains this blind nut in its design; this is shown using the straight plunge method of internal screw-cutting, together with its required tooling setup. Also shown at the drawing's lower left, is the method used to calculate the required depth of thread, when using a given constant, a sketched right-angle triangle, and the Probe and Prompt aid used in the triangle calculation.

Also shown is the method used to calculate the internal angle approach depth of thread and shows the angled inclination of the compound slide if it were being used for the angled approach internal screw-cutting operation outlined by using dashed lines.

Fig. 45 shows three views of the lathe's tooling being used to taper turn a variety of angles, including a vertical view of the taper turning attachment with details of its use; also featured is the rotated compound slide method being used to turn short tapers, and the offset tail-stock method being used for turning shallow-angle long tapers, where the material in this case is being held between centers.

Fig. 46 shows the general arrangement of the center lathe's controls. In this case, the lathe is shown set up for screw-cutting an external metric thread, in a typical shop-floor working environment.

Fig. 47 shows the authors selection of the most useful basic lathe tools, this selection includes the ones most preferred by the majority of machinists working in the machine shop.

Fig. 47a (featured below fig. 47) shows the author's design of a multi-purpose knife lathe tool. With this tool ground to the indicated cutting and relief angles shown, it becomes a dual-purpose knife tool, enabling it to be used for either right-hand or left-hand turning. The tool can also be used for boring short lengths of large-diameter bores, (after it has been rotated into either its left-or right-hand turning modes). The large-diameter boring of short holes is achieved by the tool's shank being suitably fitted and aligned in an *angled* tool-post.

When used in its straight turning mode, this tool can also be used to machine a left-or right-hand chamfer on the ends (or shoulders) of the work-piece @ 45 degrees. Another tooling advantage is that while in storage, this tool will be found to occupy only approximately one fifth of the shelf space that is normally required for the storage of six conventionally designed knife, chamfer, and boring tools used for similar jobs such as right-hand turning, left-hand turning, right-hand chamfering, left-hand chamfering, right-hand boring (of short bores), left-hand boring (of short bores).

If the tool is being used for boring short length *small*-diameter holes, its lower clearance angles will need to be increased substantially by the use of off-hand clearance grinding techniques to provide the necessary clearances required. In this case, the clearance grinding of the heel of the tool must be sufficient to form a typical 6 mm clearance radius in a position where it allows the tool's underside to clear the turned bore's diameter fully.

When changing the tool from right-to left-hand use, the tool's point radius and its clearances will need to be very minutely re-honed (or very lightly re-ground) in order to provide the new cutting tip with its required small radius and its necessary clearances to suit its new mode of operation. The author has found this design of lathe tool to be very useful when no other proper tool is available for the job in hand and the machinist is called upon to finish machine a particular job very quickly because of the very short time available.

Fig. 48 shows the author's tool-room-made fixed steady being used in its working environment, to support a work-piece that is being internally bored. This tool was designed and made with the aim of producing a steady, with a perfectly clean and clear front working face, that would allow the turning or boring tool to be used in very close proximity to its front face; this design contributes to its ease of use by enabling the machinist to achieve full work-piece rigidity, during the whole machining operation, (this advantage is considered to be its main feature). This particular design of steady is not seen in any other manufacturer made or supplied fixed steadies, due in this case to its ability to be attached to the lathe's bed merely by being locked into position, solely from the top, without the need for the machinist to grovel under the oily and swarf-strewn lathe bed, to tighten the normally difficult-to-locate securing clamp and nut featured on other manufacturer's supplied fixed steadies.

Fig. 49 shows the author's precision ground (home-made in the tool-room) sine bar. Also shown is a center finder (or wobbler), used to set up work-pieces on the milling or drilling machine's table where ordinate dimensioned work-pieces can be extremely accurately positioned; also featured is a square shaped screw-cutting tool setting gauge; and a thread pitch measuring and setting gauge.

Fig. 50 shows the author's design of a miniature trigonometry aid that contains all six of the vitally important trigonometry formulas required for the calculation of the right-angle triangle. It has been specifically designed for those who have by now become accustomed to using the Probe and Prompt system of triangle calculation with its unique method of orienting the problem and the probe (in one's mind's eye) to match the problem's exact orientation. The student should now realize (after prolonged use of the Probe and Prompt system), that they no longer need its familiar triangular shape to aid them in the orientation of the problem triangle on the paper. This aid has been designed so that it can be affixed to the strap of a wrist-watch, (the author has also supplied an adequate supply of additional aids in the lower part of the drawing to use for renewal, should the original become worn out or degraded).

The author has also included informative notes giving simple advice regarding the actual fitting of the aid to the strap of the watch, to enable it to become very useful as a memory jogger and for it to remain unobtrusive in use by students and the other workers in the industry.

It is considered by the author that it is inadvisable to use this particular calculation aid in the school examination room or in similar restricted areas, as this type of assistance, if used during an examination, is not generally allowed. Students are therefore warned that its use in an examination room could cause a possible disqualification from the test room. This problem can be overcome to a certain extent by the student committing to memory (over a period of time) the six formulas shown, (this should not be as difficult as it first appears).

Committing these six formulas to memory, will enable the student to use this triangle calculation system legally, in any environment, without fear of any penalty or repercussions.

The original full-size image shown at the end of this paragraph was reduced by photo-copying to miniaturize its six written formulas to a sufficient size for it now to occupy the very small format of approximately 11 mm wide by 16mm long. Each miniature aid will need to be cut out individually from the page to establish an outer size of approximately17 mm wide × 21mm long; the excess added to the original dimensions is allowed to provide sufficient attachment surface to match the full 18 mm width of the strap. Its actual fitting to the watch-strap is best accomplished with the strap buckled up into its normally closed position, as used by the wearer. A small strip of double-sided adhesive tape is then stuck to the back of the aid, and, with its protective paper removed, the aid is then attached to the strap in a position where the center-line of its written formulas is positioned approximately 15 mm from the edge of the fastened buckle. A strip of black (or one's preferred color to suit the strap), adhesive duct tape, is then cut to an overall size of approximately 36 × 36 mm. This is followed by the piece being re-stuck onto the roll; it is then marked out by using (say) a blue ball-point pen to provide a centrally placed oblong window of approximately 11 mm × 17 mm. A sharp craft knife or scalpel is then used to cut through the marked-out rectangle, taking care to only cut through just one thickness of the tape, the resulting small oblong shape is then removed and discarded. The now windowed piece of tape can then be carefully peeled off the roll and offered over the previously fitted aid while allowing its excess material to be carefully wrapped around the inside of the strap, positioned with its right-side edge being butted up as far as possible under the buckle, (while ensuring that its six formulas are still visible and in full view in the cutout window).

The whole sheet of paper triangle calculation aids provided in the lower part of the fig. 50 drawing, will be found more durable and longer-lasting if the removed sheet is first heat-sealed in transparent plastic on its outer printed surface, to prolong its usable life. For convenience, this heat sealing should be carried out while the sheet of aids is still in its complete sheet form. Following this, each section can then be cut out individually, as required, prior to its attachment to the strap.

The author uses this watch aid daily as a trigonometry memory jogger, and still finds it very useful.

*   *   *

The four full-size trigonometry aids featured are best reproduced on a separate sheet of A4 paper. These aids can be cut out and stuck (using double-sided adhesive tape) to the underside, or to the inside lid of your calculator, enabling this aid to be used as a quick reference guide during right-angle triangle trigonometry calculations. They contain the author's condensed version of the probe's' six trigonometry formulas that are needed to fully calculate the right-angle triangle.

How one converts metric dimensions into imperial (inch) dimensions

The reader should note that it will be found extremely useful if the machinist can convert decimal (mm) dimensions into decimal imperial (inch) dimensions. This can be achieved by multiplying the mm figure by 0.03937. Alternatively, if one needs to convert a decimal (imperial inch) dimension into a decimal (mm) dimension, we divide the imperial inch figure by 0.03937. It will also be found an advantage if the student writes the following helpful note on the face of his or her calculator, for use as an easy reference. Note that these formulas have been reduced to their bare minimum for the abbreviated version shown below.

TO CONVERT MM DIMENSIONS into IMPERIAL DIMENSIONS, × by 0.03937

TO CONVERT IMPERIAL DIMENSIONS into MM DIMENSIONS, ÷ by 0.03937

There follows an explanation of how one can roughly convert imperial or American inch dimensions, into their metric equivalent dimensions in order to make an approximate measurement of a component if a metric tape measure is not available when making an estimated measurement of a large item while using an imperial (inch) tape measure. A quick way of establishing the item's length in meters is to consider that the meter (which actually measures 39.37 inches) is rounded up (in one's mind), to actually measure 40 inches in length, the difference being 0.63 inches (16.002 mm); this rough estimation then enables us to deduce that half a meter must be roughly 20 inches, and a quarter of a meter must be roughly 10 inches. We can now also establish that a tenth of a meter (0.1 of a meter) must be equivalent to roughly 4 inches, and 0.2 of a meter (two tenths of a meter) must be equivalent to roughly 8 inches and so on up to 0.9 of a meter, which must be equivalent to roughly 36 inches (the imperial yard).

The imperial inch, when divided into a thousand parts, becomes what is called (in engineering speak, a thou (0.025 mm). Therefore, when measuring small items using this same technique, (while also using decimal notation), it will be found that a tenth of an inch (100 thou, 0.1 of an inch) actually contains 2.540 mm. (roughly two and a half mm). A 1 mm. length of material therefore contains exactly 0.03937 inches (39.37 thou), (this being *approximately* 0.040 inches).

If we now divide 0.03937 inches by 10, we get 0.003937 which is *almost* equivalent to 0.004 inches (4 thou in engineering speak), which makes this dimension roughly equivalent to 0.1 mm. The length difference between these two dimensions being 0.016002032 mm (0.000629999 inches), (six tenths of a thou in engineering speak).

Knowing these *approximations of length*, now makes it possible for a machinist operating a machine that is equipped with imperial or American inch graduations on its scale to move the machine's traverse slide along by, say, 0.004 inches in order to establish an approximate movement of 0.1 mm. of its slide. Knowledge of this close approximation in length now makes it possible to machine a relatively small metric-dimensioned component while using an imperial/American scaled machine for the operation. One must of course bear in mind that there will be an accumulated error occurring during this particular exercise that amounts to plus 0.016002032 mm (0.000629999 inches) occurring for every 0.004 inch indicated by the machine slide's scale; therefore when indexing the slide along by 0.004 inches, the distance the slide will actually be travelling along will include an additional 0.016002032 mm (0.000629999 inches) (six tenths of an inch) in excess of the 0.004 inches measured on its scale. This is a useful exercise if the component being machined is not required to be extremely accurate.

FOUR FULL SIZE TRIGONOMETRY AIDS.

∠ a, b =, a × SIN ∠.
∠ a, c =, a × COS ∠.
∠ b, a =, b ÷ SIN ∠.
∠ b, c =, 90 - ∠, TAN × b.
∠ c, a =, c ÷ COS ∠.
∠ c, b =, c × TAN ∠.

∠ a, b =, a × SIN ∠.
∠ a, c =, a × COS ∠.
∠ b, a =, b ÷ SIN ∠.
∠ b, c =, 90 - ∠, TAN × b.
∠ c, a =, c ÷ COS ∠.
∠ c, b =, c × TAN ∠.

∠ a, b =, a × SIN ∠.
∠ a, c =, a × COS ∠.
∠ b, a =, b ÷ SIN ∠.
∠ b, c =, 90 - ∠, TAN × b.
∠ c, a =, c ÷ COS ∠.
∠ c, b =, c × TAN ∠.

∠ a, b =, a × SIN ∠.
∠ a, c =, a × COS ∠.
∠ b, a =, b ÷ SIN ∠.
∠ b, c =, 90 - ∠, TAN × b.
∠ c, a =, c ÷ COS ∠.
∠ c, b =, c × TAN ∠.

The triangle calculation aids shown above can be photo copied and affixed to the underside of a scientific calculator for easy triangle calculation purposes. They contain all six of the formulas required to calculate the right-angle triangle when using the 'Probe and Prompt' triangle calculation system.

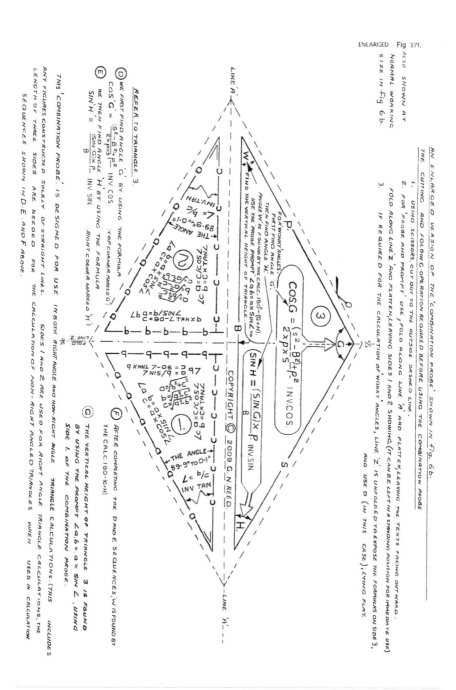

ALSO SHOWN AT NORMAL WORKING SIZE IN fig 6 b.

AN ENLARGED VERSION OF THE 'COMBINATION PROBE' SHOWN IN fig. 6b.

THE CUTTING AND FOLDING OPERATION REQUIRED BEFORE USING THE COMBINATION PROBE

1. USING SCISSORS, CUT OUT TO THE OUTSIDE DASHED LINE.

2. FOR 'PROBE AND PROMPT' USE, 'FOLD ALONG LINE 'A' AND FLATTEN, LEAVING THE TEXTS FACING OUTWARD. FOLD ALONG LINE 'Z' AND FLATTEN, LEAVING SIDES 1 AND 2 SHOWING. (IT CAN BE LEFT IN A STANDING POSITION FOR IMMEDIATE USE)

3. IF REQUIRED FOR THE CALCULATION OF WONKY ANGLES, LINE 'Z' IS UNFOLDED TO EXPOSE THE FORMULAS ON SIDE 3, AND USE D (IN THIS CASE), LYING FLAT.

REFER TO TRIANGLE 3.

(D) WE FIRST FIND ANGLE 'G' BY USING THE FORMULA
$$COS\ G = \frac{(S^2-B^2)+P^2}{2\times P\times S}\ INV.COS.$$
(TOP CORNER MARKED 'G')

(E) WE THEN FIND ANGLE 'H' BY USING THE FORMULA
$$SIN\ 'H' = \frac{(SIN\ G)\times P}{B}\ INV\ SIN$$
(RIGHT CORNER MARKED 'H')

THIS 'COMBINATION PROBE', IS DESIGNED FOR USE IN BOTH RIGHT ANGLE AND NON-RIGHT ANGLE TRIANGLE CALCULATIONS. (THIS INCLUDES ANY FIGURES CONSTRUCTED SOLELY OF STRAIGHT LINES. SIDES 1 AND 2 ARE USED FOR RIGHT ANGLE TRIANGLE CALCULATIONS, THE LENGTH OF THREE SIDES ARE NEEDED FOR THE CALCULATION OF NON-RIGHT ANGLED TRIANGLES WHEN USED IN CALCULATION SEQUENCES SHOWN IN D.E. AND F. ABOVE.

(F) AFTER COMPLETING THE D AND E SEQUENCES, W IS FOUND BY THE CALC. 180-(G+H)

(G) THE VERTICAL HEIGHT OF TRIANGLE 3 IS FOUND BY USING THE PROMPT ∠a.b = a × SIN L, USING SIDE 1, OF THE COMBINATION PROBE.

LINE 'A'

③ $$COS\ G = \frac{(S^2-B^2)+P^2}{2\times P\times S}\ INV.COS.$$
$$SIN\ H = \frac{(SIN\ G)\times P}{B}\ INV\ SIN$$

FOR WONKY ANGLES, FIRST FIND ANGLE 'G', THEN FIND ANGLE 'H'. ANGLE W IS FOUND BY THE CALC. 180°-(G+H). USE THE PROBE PROMPT ∠a.b = a× SIN ∠ TO FIND THE VERTICAL HEIGHT OF TRIANGLE.

THE ANGLE 89.9°-0-0.1°
$$∠ = b/c\ INV\ TAN$$

②
$$∠a\ \frac{c/cos\angle}{b/sin\angle}$$
$$b = \sqrt{a^2+c^2}$$
$$c = \sqrt{a^2-b^2}$$
$$a = c×cos\angle$$
$$b = c×tan\angle$$
$$∠b = b/sin\angle$$
$$C = 90 - ∠\ tan×b$$

①
$$∠a\ \frac{c/cos\angle}{b/sin\angle}$$
$$b = \sqrt{a^2+c^2}$$
$$c = \sqrt{a^2-b^2}$$
$$a = c×cos\angle$$
$$b = c×tan\angle$$
$$∠b = b/sin\angle$$
$$C = 90 - ∠\ tan×b$$

THE ANGLE 89.9° TO 0.1°
$$∠ = b/c\ INV\ TAN$$

B

COPYRIGHT © 2009 G.N. REED.

FOLD LINE Z

LINE 'A'

INDEX

C

calculating a sheet metal component that contains one bend (see figs. 20, 22, 23, 24, 25, and 28)   134

calculating a triangle problem by using decimal degrees throughout   62

calculating the depth of thread for imperial Whitworth threads   394

calculating the important depth of thread required for the internally screw-cut nut (see figs. 43a and 44)   461

calculating three-dimensional triangles in a box, also dimensioning a 3D tubular component allowing it to be accurately bent (see fig. 19)   122

calculation sequence   xxviii, 16, 23, 30, 32, 54, 204, 206, 208, 210-11, 508

checks and adjustments often needed on the center lathe when setting up (see figs. 45, 46, and 48)   239, 257

choosing the correct tool for use in internal screw-cutting, and the importance of using extreme accuracy when setting up the internal screw-cutting tool in its tool-post   468

choosing the correct work speed for screw-cutting steel components   351

chuck:
center line of   113, 298
collet   270-3, 304, 494
four-jaw   238, 267, 271, 273-4, 276, 278, 401, 404, 455, 494
jaws of   273, 281-2, 288-91, 295, 298-301, 305-6, 310, 355, 361, 404, 406, 455

three-jaw   262, 273, 281, 286, 288, 291, 296, 299, 302, 304, 306, 309-10, 404, 494-5

chuck key   289-92, 300, 307

Combination Probe   xvii, xxv, 12-17, 19, 38, 200, 203, 205, 207, 210, 483-4, 492, 503-4

coordinate method of marking out the work-piece   84

D

dangers one can experience if operating a badly worn lathe   409

development calculations required to manufacture the sheet metal cone (shown in fig. 34)   187

dimensioning a tubular component using the box method (see fig. 19)   122

disadvantages of using the straight-plunge method of screw-cutting   386

drilling   78, 84-5, 91, 94, 99, 101-3, 107-8, 110-11, 113-16, 120, 176, 285, 382, 449

drilling and reaming on the lathe   285

drilling operation   78, 101, 103, 108, 110, 114-15, 155, 457, 487

drills   xxx, xxxiii, 71, 78, 93-4, 96, 99-111, 114-17, 149, 154-5, 248, 261, 380-3, 457-8, 487-8
twist   99-100, 103, 108-9, 364, 379, 470, 488, 494

E

evolution of the Probe and Prompt system (see figs. 1, 2, 2a, b, c, d, e, f, g, and h)   10

Expandable Probe   7, 52, 55-6, 486, 505

Expandable Probe explained (see figs. 12 and 13)   52, 56